# Water Management: A Futuristic Approach

# Water Management:
# A Futuristic Approach

Edited by **Keith Wheatley**

R**C**ALLISTO **R**EFERENCE

New York

Published by Callisto Reference,
106 Park Avenue, Suite 200,
New York, NY 10016, USA
www.callistoreference.com

**Water Management: A Futuristic Approach**
Edited by Keith Wheatley

International Standard Book Number: 978-1-63239-607-5 (Hardback)

Printed in the United States of America.

# Contents

# Preface

A futuristic approach in water management has been discussed in this descriptive book. It is estimated that there is nearly 1.4 billion km$^3$ of water in this world but only about 3% (39 million km$^3$) of it is available as fresh water. Furthermore, most of this fresh water is found as ice in the arctic regions, atmospheric water or deep groundwater. Since water is a source of life and is vital for all forms of life across the planet, its sustainable use is an extremely crucial issue. Water Management is the fundamental term employed for elucidating all the activities involved in managing the optimum use of the world's water resources. However, only a small percentage of fresh water available can be subjected to water management. It is still a huge amount, but the distinct characteristic of water is that unlike other resources, it is irreplaceable. This book describes an elementary analysis on several topics related to water management from across the globe. The topics elucidated in this book cover politics, recent models for water resource management of rivers and reservoirs as well as issues associated with agriculture. Growing water demands, water quality problems and water pricing have also been considered. Reputed scientists and experts from across the world have contributed descriptive information in this book. It covers a broad range of current issues, describing the current problems and illustrating the complexity of water management.

Various studies have approached the subject by analyzing it with a single perspective, but the present book provides diverse methodologies and techniques to address this field. This book contains theories and applications needed for understanding the subject from different perspectives. The aim is to keep the readers informed about the progress in the field; therefore, the contributions were carefully examined to compile novel researches by specialists from across the globe.

Indeed, the job of the editor is the most crucial and challenging in compiling all chapters into a single book. In the end, I would extend my sincere thanks to the chapter authors for their profound work. I am also thankful for the support provided by my family and colleagues during the compilation of this book.

Editor

# Part 1

# Water Resource Management for Rivers and Reservoirs

# Generalized Models of River System Development and Management

Ralph A. Wurbs
*Texas A&M University*
*United States*

## 1. Introduction

This comparative review of capabilities for computer simulation of the control, allocation, and management of the water resources of river basins focuses on user-oriented generalized modeling systems developed in the United States that are applicable anywhere in the world. The objectives of this chapter are to assist practitioners in selecting and applying models in various types of river/reservoir system management situations and to support research in continuing to improve and expand modeling capabilities. The chapter begins with a broad general review of the massive literature and then focuses on comparing several generalized modeling systems that have been extensively applied by water management agencies in a broad spectrum of decision-support situations. Modeling capabilities are explored from the perspectives of computational methods, model development environments, applications, auxiliary analyses, and institutional support. The chapter highlights advances in modeling complex issues in managing rivers and reservoirs that are significantly contributing to actual practical improvements in water management.

Reservoir/river system modeling encompasses various hydrologic, physical infrastructure, environmental, and institutional aspects of river basin management. Dams and appurtenant structures are required to control highly fluctuating river flows to reduce flooding and develop reliable water supplies. Institutional mechanisms for allocating and managing water resources are integrally connected to constructed facilities. Management of the water and related land and environmental resources of a river basin integrates natural and man-made systems.

This review of computer modeling of river system development and management focuses on user-oriented generalized modeling systems developed in the United States. *Generalized* means that a model is designed for application to a range of concerns dealing with river systems of various configurations and locations, rather than being site-specific customized to a particular system. Model-users develop input datasets for the particular river basin of interest. *User-oriented* implies that a model is designed for use by professional practitioners other than the model developers and is thoroughly tested and well documented. User-oriented generalized modeling systems should be convenient to obtain, understand, and use and should work correctly, completely, and efficiently.

This state-of-the-art assessment begins with a brief overview of the extensive literature and then focuses on the four modeling systems listed in Table 1. ResSim, MODSIM, WRAP, and RiverWare were developed and are extensively applied in the United States, are also applied in other countries, provide a broad range of analysis capabilities, and are representative of the state-of-the-art from the perspective of practical applications dealing with complex river systems. The four alternative modeling systems reflect a broad spectrum of computational methods, modeling environments, and analysis capabilities.

| Short Name | Descriptive Name | Model Development Organization |
|---|---|---|
| ResSim | Reservoir System Simulation | U.S. Army Corps of Engineers (USACE) Hydrologic Engineering Center (HEC) http://www.hec.usace.army.mil/ |
| MODSIM | River Basin Management Decision Support System | Colorado State University (CSU) and U.S. Bureau of Reclamation (USBR) http://modsim.engr.colostate.edu/ |
| WRAP | Water Rights Analysis Package | Texas Water Resources Institute (TWRI) and Texas Commission on Environmental Quality http://ceprofs.tamu.edu/rwurbs/wrap.htm |
| RiverWare | River and Reservoir Operations | University of Colorado CADSWES and USBR http://riverware.org/ |

Table 1. Selected representative generalized modeling systems

## 2. General characteristics of modeling systems

The generalized river/reservoir system management models explored in this chapter are based on volume-balance accounting procedures for tracking the movement of water through a system of reservoirs and river reaches. The model computes reservoir storage contents, water supply withdrawals, hydroelectric energy generation, and river flows for specified water demands, system operating rules, and input sequences of stream inflows and net reservoir surface evaporation rates.

From the perspective of the water management modeling systems addressed in this chapter, the spatial configuration of a river/reservoir system is represented by a set of model control points connecting river reaches as illustrated in Figure 1. Control points represent the sites of reservoirs, hydroelectric power plants, water supply diversions and return flows, environmental instream flow requirements, conveyance canals and pipelines, stream confluences, river basin outlets, and other system components. Stream inflows at control points are provided as input. Reservoir storage and stream flows are allocated between water users based on rules specified in the model. The models described in this chapter have been applied to river systems ranging in complexity from a single reservoir or run-of-river water supply diversion to river basins containing many hundreds of reservoirs and water supply diversion sites with operations governed by complex multiple-purpose reservoir system operating rules and institutional water allocation mechanisms.

The models of this chapter combine a specified scenario of water resources development, control, allocation, management, and use with a specified condition of river basin hydrology which is most often historical hydrology representing natural unregulated conditions. River basin hydrology is represented by stream flow inflows and net reservoir surface evaporation-precipitation rates for each time step of a hydrologic period-of-analysis.

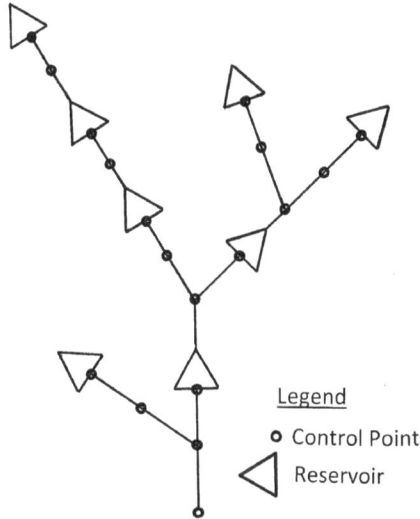

Legend
o Control Point
◁ Reservoir

Fig. 1. Illustrative schematic of a river system as viewed from a modeling perspective

The hydrologic simulation period and computational time step and may vary greatly depending on the application. Storage and flow hydrograph ordinates for a flood event occurring over a few days may be determined at intervals of an hour or less. Water supply capabilities may be modeled with a monthly time step and many-year hydrologic period-of-analysis reflecting a full range of fluctuating wet and dry periods including extended multiple-year drought.

A river/reservoir system model simulates a physical and institutional water management system with specified conditions of water demand for each sequential time step of a hydrologic period-of-analysis. Post-simulation stream flow and reservoir storage frequency analysis and supply reliability analysis capabilities are typically included in the modeling systems addressed by this chapter. Reservoir storage and stream flow frequency statistics and water supply reliability metrics are developed for alternative river/reservoir system management strategies and practices.

Other auxiliary modeling features are also, in some cases, incorporated in the river/ reservoir management models. Some models include features for economic evaluation of system performance based on cost and benefit functions expressed as a function of flow and storage. Stream inflows are usually generated outside of the reservoir/river system management model and provided as input to the model. However, reservoir/river system models may also include capabilities for simulating precipitation-runoff processes to generate inflows. Though hydraulics issues may be pertinent to reservoir operations, separate models of river hydraulics are applied to determine flow depths and velocities.

Some reservoir/river system management models simulate water quality constituents along with water quantities. However, generalized water quality models, not covered in this chapter, are designed specifically for particular types of river and/or reservoir system water quality analyses. The typically relatively simple water quality features of the models explored in this chapter are secondary to their primary function of detailed modeling of water development, regulation, allocation, and management.

Modeling applications often involve a system of several models, utility software products, and databases used in combination. A reservoir/river system management model is itself a modeling system, which often serves as a component of a larger modeling system that may include watershed hydrology and river hydraulics models, water quality models, economic evaluation tools, statistical analysis methods, databases and various software tools for managing time series, spatial, and other types of data.

The models discussed here are used for various purposes in a variety of settings. Planning studies may involve proposed construction projects or reallocations of storage capacity or other operational modifications at existing projects. Reservoir operating policies may be reevaluated periodically to assure responsiveness to current conditions and objectives. Studies may be motivated by drought conditions, major floods, water quality problems, or environmental losses. Operating plans for the next year or next season may be updated routinely based on a modeling system. Models support the administration of treaties, agreements, water right systems, and other water allocation mechanisms. Real-time modeling applications may involve decision-support for water management and use curtailment actions during droughts. Likewise, real-time flood control operations represent another type of application.

## 3. Models for analyzing development and operation of reservoir systems

Pioneering efforts in computer simulation of reservoir systems include U.S. Army Corps of Engineers studies of six reservoirs on the Missouri River initiated in 1953, International Boundary and Water Commission simulations of the Rio Grande in 1954, and a simulation study of the Nile River Basin in 1955 (Maass et al., 1966). Several books on modeling and analysis of reservoir operations are available (Votruba and Broza, 1989; Wurbs, 1996; ReVelle, 1999; Nagy et al., 2002). Labadie (2004) summarizes the extensive and complex research literature on reservoir system optimization models. Wurbs (1993, 2005a) presents state-of-the-art reviews of reservoir system analysis from a practical applications perspective.

### 3.1 Optimization and simulation

Reservoir system analysis models have traditionally been categorized as simulation, optimization, and hybrid combinations of both. Development and application of decision-support tools within the water resources development agencies in the United States have focused on simulation models. The published literature on modeling reservoir systems is dominated by optimization techniques.

The term *optimization* is used synonymously with *mathematical programming* to refer to a mathematical algorithm that computes a set of decision variable values which minimize or maximize an objective function subject to constraints. Optimization is covered by water resources systems books (Karamouz et al., 2003; Jain & Singh, 2003; Simonovic, 2009) as well as numerous operations research and mathematics books. Thousands of journal and conference papers have been published since the 1960's on applying variations of linear programming, dynamic programming, gradient search algorithms, evolutionary search methods such as genetic algorithms, and other optimization techniques to reservoir system analysis problems. Various probabilistic methods for incorporating the stochastic nature of stream flows and other variables in the optimization models have been proposed (Labadie2004).

This chapter focuses on generalized simulation models. A simulation model is a representation of a system used to predict its behavior under a given set of conditions. Alternative executions of a simulation model are made to analyze the performance of the system under varying conditions, such as for alternative operating plans. Although optimization and simulation are two alternative modeling approaches with different characteristics, the distinction is obscured by the fact that models often contain elements of both. An optimization procedure may involve automated iterative executions of a simulation model. Optimization algorithms may be embedded within simulation models either to perform certain periphery computations or to provide the fundamental computational framework for the simulation model.

## 3.2 Network flow linear programming

Of the many mathematical programming methods available, linear programming (LP), particularly network flow LP, has been the method most often adopted in practical modeling applications in support of actual water management activities. The general LP formulation described in many mathematics and systems engineering textbooks is as follows.

$$\text{Minimize or Maximize } Z = \sum_{j=1}^{n} c_j x_j \tag{1}$$

$$\text{subject to } \sum a_{ij} x_j \le b_i \text{ for } i = 1,\dots,m \text{ and } j=1,\dots,n \tag{2}$$

$$x_j \ge 0 \text{ for } j = 1,\dots,n \tag{3}$$

A LP solution algorithm finds values for the n decision variables $x_j$ that optimize an objective function subject to m constraints. The $c_j$ in the objective function equation and $a_{ij}$ and $b_i$ in the constraint inequalities are constants.

A number of generalized reservoir system simulation models including several discussed later in this chapter are based on network flow programming, which is a computationally efficient form of LP. Network flow programming is applied to problems that can be formulated in a specified format representing a system as a network of nodes and arcs having certain characteristics. The general form of the formulation is as follows.

$$\text{Minimize or Maximize } \sum \sum c_{ij} q_{ij} \text{ for all arcs} \tag{4}$$

$$\text{subject to } \sum q_{ij} - \sum q_{ji} = 0 \text{ for all nodes} \qquad (5)$$

$$l_{ij} \le q_{ij} \le u_{ij} \text{ for all arcs} \qquad (6)$$

where    $q_{ij}$ is the flow rate in the arc connecting node i to node j
            $c_{ij}$ is a penalty or weighting factor for $q_{ij}$
            $l_{ij}$ is a lower bound on $q_{ij}$
            $u_{ij,}$ is a upper bound on $q_{ij}$

The system is represented as a collection of nodes and arcs. For a reservoir/river system, the nodes are sites of reservoirs, diversions, stream tributary confluences, and other pertinent system features as illustrated by the control points of Figure 1. Nodes are connected by arcs or links representing the way flow is conveyed. Flow may represent a discharge rate, such as instream flows and diversions, or a change in storage per unit of time.

A solution algorithm determines the values of the flows $q_{ij}$ in each arc which optimize an objective function subject to constraints including maintaining a mass balance at each node and not violating user-specified upper and lower bounds on the flows. The weighting factors $c_{ij}$ in the objective function are defined in various ways such as unit costs in dollars or penalty or utility terms that provide mechanisms for expressing relative priorities. Each arc has three parameters: a weighting, penalty, or unit cost factor $c_{ij}$ associated with $q_{ij}$; lower bound $l_{ij}$ on $q_{ij}$; and an upper bound $u_{ij}$ on $q_{ij}$. Network flow programming problems can be solved using conventional LP algorithms. However, the network flow format facilitates the use of much more computationally efficient algorithms that allow analysis of large problems with thousands of variables and constraints.

### 3.3 Caution in applying simplified representations of the real world

Models are necessarily simplified representations of real world systems. Many references discuss shortcomings of the mathematical representations used to model systems of rivers and reservoirs. Rogers and Fiering (1986) outlined reasons that water management practitioners were reluctant to apply mathematical optimization algorithms proposed by researchers that included deficiencies in databases, modeling inadequacies, institutional resistance to change, and the fundamental insensitivity of many actual systems to wide variations in design choices. Iich (2009) explores limitations of network flow programming. McMahon (2009) highlights the various complexities of applying computer models and concludes that models can be quite useful despite their imperfections when considered in the context of data uncertainties, real-world operator experience, social priorities for water management, and externally imposed constraints on actual operational practice.

Powerful generalized software packages are playing increasingly important roles in water management. Computer models greatly contribute to effective water management. However, models must be applied carefully with professional judgment and good common sense. Model-users must have a thorough understanding of the computations performed by the model and the capabilities and limitations of the model in representing the real-world.

### 4. Generalized user-oriented river/reservoir system models

Many hundreds of reservoir/river system models are described in the published literature. However, only a small number of these models fit the definitions of *generalized* and *user-*

*oriented* presented at the beginning of this chapter. Many models are developed for a specific reservoir system rather than being generalized. Most of the numerous reservoir system optimization models reported in the literature were developed in university research studies and have not been applied by model-users other than the original model developers.

Under the sponsorship of the U.S. Army Corps of Engineers (USACE) Institute for Water Resources, Wurbs (1994, 1995) inventoried generalized water management models in the categories of demand forecasting, water distribution systems, ground-water, watershed runoff, stream hydraulics, river and reservoir water quality, and reservoir/river system operations. Wurbs (2005a) later reviewed generalized reservoir/river system operations models in greater detail for the USACE. Most of the models cited in these inventories were developed by government agencies in the United States and are in the public domain, meaning they are available to interested model-users without charge.

Public domain generalized modeling systems play important roles in many aspects of water management in the United States (Wurbs, 1998). Of the many water-related models used in the U.S., the Hydrologic Modeling System (HMS) and River Analysis System (RAS) are probably applied most extensively. These and other models developed by the Hydrologic Engineering Center (HEC) of the USACE are available at the website shown in Table 1. HEC-HMS watershed precipitation-runoff and HEC-RAS river hydraulics modeling systems are combined with HEC-ResSim in the integrated Corps Water Management System for modeling reservoir system operations described later. However, most applications of HEC-HMS and HEC-RAS by government agencies and consulting firms are for urban floodplain delineation or design of urban stormwater management facilities. The number of agencies and individuals that model operations of major multiple-purpose reservoir systems is much smaller than the number of users of HEC-HMS, HEC-RAS, and various other generalized models used for other purposes. However, generalized reservoir system models are significantly contributing to effective river basin management.

A Hydrologic Modeling Inventory (HMI) is maintained at Texas A&M University at the web site http://hydrologicmodels.tamu.edu/ in collaboration with the U.S. Bureau of Reclamation. The HMI is updated periodically, including an update during 2010. Models are organized in various categories with summary descriptions provided for each model. The HMI includes the MIKE BASIN, CALSIM, MODSIM, RiverWare, and WRAP models cited later in this chapter. In addition to developing and maintaining the HMI, Singh and Frevert (2006) edited a book inventorying models focused primarily on watershed hydrology but also including several river/reservoir system management models including RiverWare (Zagona et al., 2006), MODSIM (Labadie, 2006), and WRAP (Wurbs, 2006).

The following review focuses on several of the generalized reservoir/river management modeling systems that have been extensively applied by water management agencies and/or their consultants to support actual planning and/or operations decisions. The models cited below along with other similar models are discussed in more detail by Wurbs (2005a).

This presentation focuses on modeling systems developed in the United States largely because the author's professional experience has been limited primarily to the United States. The U. S. is somewhat unique compared to most other countries in that generalized models are available in the public domain free-of-charge. Most, though not all, water management software products developed with government funding in the U. S. are made accessible to the professional water management community without charging a fee for the software.

## 4.1 Models developed by international research and consulting organizations

However, reservoir/river system models are developed throughout the world. Three examples of the many non-U.S.-based modeling systems are cited as follows. The proprietary MIKE BASIN, WEAP, and OASIS software products were developed and are marketed by organizations that provide consulting services in applying the models. The developers and others have applied the models to reservoir/river systems throughout the world.

The Danish Hydraulic Institute (http://www.dhi.dk/) has developed a suite of models dealing with various aspects of hydraulics, hydrology, and water resources management. MIKE BASIN, the reservoir/river system component of the DHI family of software, integrates geographic information system capabilities with modeling river basin management. MIKE BASIN simulates multiple-purpose, multiple-reservoir systems based on a network formulation of nodes and branches. Time series of monthly inflows to the stream system are provided as input. Various options are provided for specifying reservoir operating rules and allocating water between water users.

The Water Evaluation and Planning (WEAP) System developed by the Stockholm Environmental Institute (http://www.weap21.org/) is a reservoir/river/use system water balance accounting model that allocates water from surface and groundwater sources to different types of demands. The modeling system is designed as a tool for maintaining water balance databases, generating water management scenarios, and performing policy analyses.

The Operational Analysis and Simulation of Integrated Systems (OASIS) model developed by HydroLogics, Inc. (http://www.hydrologics.net/) is based on linear programming. Reservoir operating rules are expressed as goals and constraints defined by the model-user using a patented scripting language that is similar to the Water Resources Engineering Simulation Language (WRESL) in the WRIMS-CALSIM model discussed next.

## 4.2 Models developed by state water agencies in the United States

CALSIM consists of the generalized Water Resources Integrated Modeling System (WRIMS) combined with input datasets for the interconnected California State Water Project and federal Central Valley Project. The California Department of Water Resources in partnership with the U.S. Bureau of Reclamation developed the WRIMS and CALSIM modeling system (Draper et al., 2004) to replace an earlier California Department of Water Resources model.

The generalized WRIMS and California CALSIM are designed for evaluating operational alternatives for large, complex river systems. The modeling system integrates a simulation language for defining operating criteria, a linear programming (LP) solver, and graphics capabilities. The monthly time step simulation model is based on a LP formulation that minimizes a priority-based penalty function of delivery and storage targets. The LP model is solved for each month. Adjustment computations are performed after the LP solution to deal with nonlinear aspects of modeling complex system operations. A feature called the

Water Resources Engineering Simulation Language (WRESL) was developed for the model to allow the user to express reservoir/river system operating requirements and constraints. The user-supplied statements written in the WRESL language are used by the model to define the LP formulation. Time series data are stored, manipulated, and plotted using the Hydrologic Engineering Center (1995, 2009) Data Storage System (HEC-DSS), which is also used with WRAP, discussed later, as well as with HEC-ResSim and other HEC simulation models.

The Texas Water Development Board (TWDB) Surface Water Resources Allocation Model and Multiple-Reservoir Simulation and Optimization Model simulate and optimize the operation of an interconnected system of reservoirs, hydroelectric power plants, pump canals, pipelines, and river reaches using a monthly computational time step. The daily time step MONITOR also simulates complex surface water storage and conveyance systems operated for hydroelectric power, water supply, and low flow augmentation (Martin, 1983, 1987). The TWDB has adopted the WRAP modeling system, described later, for statewide and regional planning studies conducted in recent years, replacing these early TWDB models.

The early TWDB models, original California Department of Water Resources model, and the original versions of HEC-PRM and MODSIM discussed later are all based on the same network flow programming solution algorithm. An early version of WRAP was also developed using the same algorithm, but another simulation approach was actually adopted for WRAP. The original solution algorithms in HEC-PRM and MODSIM were later replaced with much more computationally efficient network flow programming algorithms.

## 4.3 Models developed by federal agencies in the United States

Most of the large federal reservoirs in the U.S. were constructed and are operated by the U.S. Army Corps of Engineers (USACE) or U.S. Bureau of Reclamation (USBR). The USACE has over 500 reservoirs in operation across the nation as well as many navigation locks, hydropower plants, and flood control structures. The USACE operates essentially all of the reservoir projects that it has constructed. The USBR has transferred operation of many of its projects to non-federal sponsors upon completion of construction but continues to operate about 130 reservoirs and appurtenant structures in the 17 western states. The USACE plays a dominant role in the U.S. in operating large reservoir systems for navigation and flood control. The USBR water resources development program was originally founded upon constructing irrigation projects to support development of the western U.S. The responsibilities of the two agencies evolved over time to emphasize comprehensive multiple-purpose water resources management.

The USACE and USBR developed many models for specific reservoir systems during the 1950's-1970's (Wurbs, 1996, 2005a). Many of these system-specific models have since been replaced with generalized models. The USBR currently uses RiverWare and MODSIM, which are described later in this chapter, and several remaining system-specific models. The USACE Hydrologic Engineering Center (HEC) maintains a suite of generalized simulation models that are widely applied by water agencies, consulting firms, and universities throughout the U.S. and the world. This chapter later focuses on HEC-ResSim but several other HEC products are also noted below.

The Corps Water Management System (CWMS) is the automated information system used by the USACE nationwide to support real-time operations of flood control, navigation, and multiple-purpose reservoir systems (Fritz et al., 2002). The CWMS is an integrated system of hardware and software that compiles and processes hydrometeorology, watershed, and project status data in real-time. A map-based user-friendly interface facilitates modeling and evaluation of river/reservoir system operations. CorpsView, a spatial visualization tool developed by the HEC based on commercially available geographic information system (GIS) software, provides a direct interface to GIS products and associated attribute information. The CWMS combines data acquisition/management tools with simulation models which include HEC-HMS (Hydrologic Modeling System), HEC-ResSim (Reservoir Simulation), HEC-RAS (River Analysis System), and HEC-FIA (Flood Impact Analysis).

The HEC-5 Simulation of Flood Control and Conservation Systems model (Hydrologic Engineering Center 1998) has been used since the 1970's in many USACE and non-USACE studies, including investigations of storage reallocations and other operational modifications at existing reservoirs, feasibility studies for proposed new projects, and support of real-time operations. The HEC plans to eventually replace HEC-5 with HEC-ResSim. However, HEC-5 is still available at the HEC website and continues to be applied by various model-users.

HEC-5 simulates multiple-purpose reservoir system operations for inputted unregulated stream flows and reservoir evaporation rates using a variable time interval. A monthly or weekly time step may be used during periods of normal or low flows in combination with a daily or hourly time step during flood events. HEC-5 makes release decisions to empty flood control pools and to meet user-specified diversion and instream flow targets based on reservoir storage levels and stream flows at downstream locations. Flood routing options include modified Puls, Muskingum, working R&D, and average lag. Optional analysis capabilities include computation of expected annual flood damages and water supply firm yields.

The HEC Prescriptive Reservoir Model (HEC-PRM) was developed in conjunction with studies of reservoir systems in the Missouri and Columbia River Basins. Later applications include studies of systems in California, Florida, and Panama (Draper et al., 2003; Watkins et al., 2004). HEC-PRM is a network flow programming model designed for prescriptive applications involving minimization of a cost based objective function. Reservoir release decisions are made based on minimizing costs associated with convex piecewise linear penalty functions associated with various purposes including hydroelectric power, recreation, water supply, navigation, and flood control. Schemes have also been devised to also include non-economic components in the objective function. HEC-PRM applications to date have used a monthly time interval.

## 5. Selected state-of-the-art generalized modeling systems

The four user-oriented generalized models in Table 1 provide comprehensive capabilities for a broad spectrum of river/reservoir system modeling applications. ResSim, MODSIM, WRAP, and RiverWare are distinctly different from each other. However, as a group, the four alternative modeling systems are representative of the current state-of-the-art of professional practice in the United States in analyzing complex problems and issues in managing rivers and reservoir systems.

The four modeling systems were developed by water agencies and university research entities and have been extensively applied in both the U.S. and other countries. The software was developed for application by model-users other than the original developers and is accessible by water management professionals throughout the world. The ResSim, MODSIM, and WRAP software and documentation can be downloaded free-of-charge at the websites listed in Table 1. RiverWare is a proprietary software product which is available for a licensing fee as described at the website shown in Table 1. The four software packages all run on personal computers operating under Microsoft Windows and all have also been executed with other computer systems as well. RiverWare was developed primarily for Unix workstations though it also is used on personal computers with Microsoft Windows.

The four alternative modeling systems and their predecessors have evolved through many versions over more than twenty years of research and development, with new versions being released periodically. The modeling capabilities provided by each of the models have changed significantly over time in the past and continue to be improved and expanded.

### 5.1 Hydrologic Engineering Center (HEC) Reservoir Simulation (ResSim) model

The USACE HEC initiated development of ResSim in 1996. ResSim was first released to the public in 2003 with the intention of eventually replacing HEC-5, which has been extensively applied for over 30 years. Documentation currently consists of a Users Manual (Hydrologic Engineering Center 2007) and other information found at the website in Table 1. ResSim is designed for application either independently of the previously discussed Corps Water Management System or as a component thereof. Applications have included the Sacramento and San Joaquin River Basins in California and Tigris and Euphrates River Basins in Iraq.

ResSim is comprised of a graphical user interface, computational program to simulate reservoir operation, data management capabilities, and graphics and reporting features. Multiple-purpose, multiple-reservoir systems are simulated using algorithms developed specifically for the model rather than formal mathematical programming methods. Meeting the needs of USACE reservoir control personnel for real-time decision support has been a governing objective in developing ResSim. The model is also applicable in planning studies. The full spectrum of multiple-purpose reservoir system operations is modeled. Particularly detailed capabilities are provided for modeling flood control operations.

The user-selected computational time-step may vary from 15 minutes to one day. Stream flow routing options include Muskingum, Muskingum-Cunge, modified Puls, and other methods. Stream flow hydrographs provided as input to ResSim can come from any source, including being generated with the HEC-HMS Hydrologic Modeling System. Multiple-reservoir systems, with each reservoir having multiple outlet structures, are modeled. Release decisions are based on specified storage zones that divide the pool by elevation and a set of rules that specify the goals and constraints governing releases when the storage level falls within each zone.

### 5.2 MODSIM river basin management decision support system

MODSIM is a general-purpose reservoir/river system simulation model based on network flow programming developed at Colorado State University (Labadie 2006; Labadie & Larson 2007). The model has evolved through many versions, with initial development dating back to the 1970's. The USBR has been a primary sponsor of continued model improvements at Colorado State University. MODSIM has been applied in studies of a number of river systems in the western U. S. and throughout the world by university researchers in collaboration with various local, regional, and international water management agencies. The software, users manual, tutorials, and papers describing various applications are provided at the website in Table 1.

MODSIM provides a graphical user interface and a general framework for modeling. A river/reservoir system is defined as a network of nodes and links. The objective function (Equation 1) consists of the summation over all links in the network of the flow in each link multiplied by a priority or cost coefficient. The objective function coefficients are factors entered by the model-user to specify relatively priorities that govern operating decisions. The coefficients could be unit monetary costs or more typically numbers without physical significance other than simply reflecting relative operational priorities. An iterative algorithm deals with nonlinearities such as evaporation and hydropower computations. The network flow programming problem is solved for each individual time interval. Thus, decisions are not affected by future inflows and future decisions.

Monthly, weekly, or daily time steps may be adopted for long-term planning, medium-term management, and short-term operations. A lag flow routing methodology is used with a daily time step. The user assigns relative priorities for meeting diversion, instream flow, hydroelectric power, and storage targets, as well as lower and upper bounds on the flows and storages computed by the model. Optional capabilities are also provided for simulating salinity and conjunctive use of surface and ground water.

## 5.3 Water rights analysis package (WRAP) modeling system

Development of WRAP at Texas A&M University began in the late 1980's sponsored by a cooperative research program of the U.S. Department of the Interior and Texas Water Resources Institute (TWRI). WRAP has been greatly expanded since 1997 under the auspices of the Texas Commission on Environmental Quality (TCEQ) in conjunction with implementing a statewide Water Availability Modeling (WAM) System (Wurbs, 2005b). The Texas Water Development Board, USACE, and other agencies have also sponsored improvements to WRAP. The software and documentation (Wurbs 2009, 2010, 2011a, 2011b; Wurbs and Hoffpauir 2011) are available at the website in Table 1.

WRAP is generalized for application to river/reservoir systems located anywhere in the world, with model-users developing input datasets for the particular river basins of concern. For studies in Texas, publicly available TCEQ WAM System datasets are altered as appropriate to reflect proposed water management plans of interest, which could involve changes in water use or reservoir/river system operating practices, construction of new facilities, or other water management strategies. The WAM System consists of the generalized WRAP along with input datasets for the 23 river basins of Texas that include naturalized stream flows at about 500 gauged sites, watershed parameters for distributing these flows to over 12,000 ungauged locations, 3,450 reservoirs, water use requirements associated with about 8,000 water right permits reflecting two different water right systems, two international treaties, and five interstate compacts. WRAP is applied routinely with the WAM System input datasets for the individual river basins by water management agencies and consulting engineering firms in regional and statewide planning studies, administration of the water right permit system, and other water management activities.

WRAP simulates water resources development, management, regulation, and use in a river basin or multiple-basin region under a priority-based water allocation system. In WRAP terminology, a water right is a set of water use requirements, reservoir storage and conveyance facilities, operating rules, and institutional arrangements for managing water resources. Stream flow and reservoir storage is allocated among users based on specified priorities, which can be defined in various ways. Simulation results are organized in optional formats including entire time sequences, summaries, water budgets, frequency relationships, and various types of reliability indices. Simulation results may be stored as DSS files accessed with HEC-DSSVue for plotting and other analyses (Hydrologic Engineering Center, 1995, 2009).

WRAP modeling capabilities that have been routinely applied in the Texas WAM System consist of using a hydrologic period-of-analysis of about 60 years and monthly time step to perform water availability and reliability analyses for municipal, industrial, and agricultural water supply, environmental instream flow, hydroelectric power generation, and reservoir storage requirements. Recently developed additional WRAP modeling capabilities include: short-term conditional reliability modeling; daily time step modeling capabilities that include flow forecasting and routing and disaggregation of monthly flows to daily; simulation of flood control reservoir system operations; and salinity simulation.

## 5.4 RiverWare reservoir and river operation modeling system

The U.S. Bureau of Reclamation (USBR) and Tennessee Valley Authority (TVA) jointly sponsored development of RiverWare at the Center for Advanced Decision Support for Water and Environmental Systems (CADSWES) of the University of Colorado (Zagona et al., 2001; Zagona et al., 2006). RiverWare development efforts date back to the mid-1990's, building on earlier software developed at CADSWES that extends back to the mid-1980's.

RiverWare provides the model-user with a software tools for constructing a model for a particular river/reservoir system and then running the model that include a library of modeling algorithms, solvers, and a language for coding operating policies. The tools are applied within a point-and-click graphical user interface. RiverWare routs inflows, provided as input, through a system of reservoirs and river reaches. The primary processes modeled are volume balances at reservoirs, hydrologic routing in river reaches, evaporation and other losses, diversions, and return flows. Optional features are also provided for modeling groundwater interactions, water quality, and electric power economics.

Computational algorithms for modeling reservoir/river system operations are based on three alternative approaches: (1) pure simulation, (2) rule-based simulation, and (3) optimization combining linear programming with preemptive goal programming. Pure simulation solves a uniquely and completely specified problem. In rule-based simulation, certain information is generated by prioritized policy rules specified by the model-user. Preemptive goal programming considers multiple prioritized objectives based on multiple LP solutions (Eschenbach et al., 2001). As additional goals are considered, the optimal solution of a higher priority goal is not sacrificed in order to optimize a lower priority goal.

The TVA applies RiverWare in optimizing the daily and hourly operation of the TVA system of multiple-purpose reservoirs and hydroelectric power plants. The USBR has used RiverWare as a long-term planning model and mid-term operations model of the Colorado River as well as a daily operations model for both the Upper and Lower Colorado Regions. The USBR has also applied the model in the Rio Grande, Yakima, and Truckee River Basins. The USACE has recently sponsored addition of features to RiverWare for simulating flood control reservoir operations. Other entities have also applied the model in various river basins for various purposes.

## 6. Comparative summary of modeling capabilities

ResSim, MODSIM, WRAP, RiverWare, and other similar models provide flexible capabilities for analyzing multiple-purpose river/reservoir system operations. The models are water accounting systems that compute reservoir storages and releases and stream flows for each time step of a specified hydrologic period-of-analysis for a particular scenario of water resources development, management, allocation, and use. Though fundamentally similar, ResSim, MODSIM, WRAP, and RiverWare differ significantly in their organizational structure, computational algorithms, user interfaces, and data management mechanisms. The alternative modeling systems provide general frameworks for constructing and applying models for specific systems of reservoirs and river reaches. Each of the generalized modeling systems is based upon its own set of modeling strategies and methods and has its own terminology or modeling language.

### 6.1 Types of applications

Water development purposes are a key consideration in formulating a modeling approach. The distinction between flood control and conservation purposes such as hydroelectric power and water supply is particularly important. Hydrologic analyses of floods focus on storm events, and analyses of droughts are long-term time series oriented. Modeling flow attenuation is important for flood control. Evaporation is important for conservation operations. Flood control operations are typically modeled using a daily or smaller time step. Conservation operations are sometimes modeled with a daily interval, but monthly or weekly time steps are more common.

All four of the alternative modeling systems are designed to simulate flood control, hydropower, water supply, environmental flows, and other reservoir management purposes. However, whereas development of the other three models was motivated primarily by conservation purposes, ResSim is flood control oriented. ResSim is limited to daily or shorter time steps and provides greater flexibility for flood routing and simulating flood control operations. RiverWare and WRAP have been recently expanded to increase their flexibility for modeling flood control.

In addition to the basic water accounting computations, the modeling systems include various optional features for reliability and frequency analyses, economic evaluations, water quality, and surface/groundwater interactions. These features may involve either computations performed during the simulation or additional post-simulation computations performed using simulation results. WRAP has particularly comprehensive options for reliability and frequency analyses. The relative priorities represented by the objective function coefficients in MODSIM and the RiverWare LP option may optionally be economic costs or benefits. MODSIM and WRAP simulate salinity. RiverWare options include various water quality constituents. Groundwater sources and channel losses are included in the models. Surface/ground water interactions have been approximated in various ways. MODSIM has a groundwater routine, and has been linked with the U.S. Geological Survey MODFLOW groundwater model.

System analysis models are often categorized as being prescriptive or descriptive. With the exception of the optimization option in RiverWare, the four models are essentially descriptive simulation models that demonstrate what will happen if a specified plan is adopted. Prescriptive optimization models automatically determine the plan that will best satisfy the decision criteria. Although it may be desirable for models to be as prescriptive as possible, real-world complexities of reservoir system operations typically necessitate model orientation toward the more descriptive end of the descriptive/prescriptive spectrum.

## 6.2 Computational structure

The term *ad hoc* in Table 2 refers to computational strategies developed specifically for a particular model, as contrasted with linear programming (LP) which is a generic algorithm incorporated in numerous models. ResSim and WRAP are organized based upon ad hoc model-specific computational frameworks. MODSIM is based on network flow LP. RiverWare has two alternative solution options based on ad hoc algorithms and a third option that uses LP. The LP-based models have additional ad hoc algorithms used along with their LP solver, but the LP solver accounts for a major portion of the computations.

Repetitive loops and iterative solution procedures are incorporated in all of the models. Iterative algorithms are required for evaporation and hydropower computations. Evaporation depends upon end-of-period storage, but end-of-period storage depends upon evaporation. Reservoir storage volume versus surface area and elevation relationships are nonlinear. In the LP models, the entire LP solution of the whole system is repeated iteratively. With the ad hoc simulation procedures, the computations for an individual reservoir are repeated iteratively.

| Modeling System | Programming Language | Computational Approach | Computational Time Step |
|---|---|---|---|
| ResSim | Java | ad hoc | 15 minutes to day |
| MODSIM | C++.NET, Basic.NET | network LP | month, week, day |
| WRAP | Fortran | ad hoc | month, day, other |
| RiverWare | C++ | ad hoc and LP | hour to year |

Table 2. Alternative development frameworks

ResSim and RiverWare generally follow an upstream-to-downstream progression in considering requirements for reservoir storage and releases, diversions, and hydropower generation. WRAP and MODSIM simulation computations are governed by user-specified priorities in considering water management requirements. The WRAP and MODSIM priority-based frameworks are beneficial in modeling complex water allocation systems.

RiverWare includes an optional prescriptive optimization feature that combines LP and goal programming. Computations are performed simultaneously for all the time intervals. Thus, model results show a set of reservoir storages and releases which minimize or maximize a defined objective function assuming all future stream flows are known as release decisions are made simultaneously during each period. With the exception of options for short-term flow forecasting, ResSim, MODSIM, WRAP, and the simulation options in RiverWare step through time performing computations at each individual time step. Thus, operating decisions are not affected by future inflows and future operating decisions.

Many other prescriptively oriented optimization models reported in the research literature, including the HEC-PRM Prescriptive Reservoir Model described earlier in this chapter, adopt the approach of optimizing an objective function while simultaneously considering all time steps of the entire hydrologic period-of-analysis. Thus, these models reflect perfect knowledge of future hydrology. Since the future is not known in the real-world, these models reflecting knowledge of the future provide an upper-limit scenario on what can be achieved. Descriptive simulation models are more realistic in that current operating decisions in the model are not affected by future hydrology and future operating decisions.

## 6.3 Modeling environment and interface features

A model for a particular reservoir/river system consists of a generalized modeling system and an input dataset describing the reservoir/river system. The generalized modeling system provides an environment or framework for assembling input data, executing the simulation computations, and organizing, analyzing, and displaying results.

Each of the four modeling systems has its own unique framework within which the user constructs and implements a model for a particular reservoir/river system. With ResSim, various elements provided by watershed setup, reservoir network, and simulation modules are used to construct and execute a model. MODSIM is based on network flow programming with a reservoir/river system represented by a network of nodes and links with information compiled through an object-oriented interface. WRAP is about managing programs, files, input records, and results tables, with water management and use practices being described in the terminology of water rights. RiverWare has an object/slot-based environment for building models within the context of object oriented programming and provides three optional solution options.

The user interfaces of the models reflect both similarities and significant differences. ResSim, RiverWare, and MODSIM provide sophisticated graphical user interfaces with menu-driven editors for entering and revising input data and displaying simulation results in tables and graphs and features allowing a river/reservoir system schematic to be created by selecting and connecting icons. WRAP has a simple user interface for managing programs and files, which relies upon standard Microsoft Office programs for entering, editing, and displaying data. WRAP as well as ResSim connect with and rely upon graphics capabilities of the Hydrologic Engineering Center (HEC) Data Storage System (DSS). Geographic information system (GIS) tools are included in all four of the modeling systems.

The compiled executable software products were developed in the programming languages shown in Table 2. ResSim, MODSIM, and RiverWare also have their own simulation rule language to allow users to express reservoir/river system operating requirements as a series of statements with if-then-else and similar constructs.

Data management efficiency, effective communication of results, documentation, and ease-of-use are important factors in applying a modeling system. Documentation includes both instructions for using the software and detailed technical documentation for understanding modeling methods. The software should be as near error-free as possible assuming absolutely error-free software may be an idealistic goal yet to be achieved. Dealing with errors introduced by users in model input data is important. The modeling systems contain various mechanisms for detecting and correcting blunders and inconsistencies in input data.

The organizations and individuals that originally developed the four modeling systems continue to improve the models and support their application. ResSim, MODSIM, and WRAP software and manuals are available free-of-charge at the websites listed in Table 1. Licensing fees and training required to implement RiverWare are described at its website. The HEC periodically provides training courses in the application of HEC-ResSim. The TWRI periodically provides training courses in the application of WRAP.

RiverWare is designed for Unix workstations but is also used on personal computers with Microsoft Windows. The other three modeling systems are usually executed on personal computers with Microsoft Windows but can also be applied with other computer systems.

## 7. Conclusions

The evolution of computer modeling of systems of rivers and reservoirs that began in the 1950's is still underway and is expected to continue. Modeling systems continue to grow in response to advances in computer technology and intensifying water management and associated decision-support needs. The published literature on modeling reservoir systems is massive and complex. ResSim, MODSIM, WRAP, RiverWare, and other similar models, though continuing to be improved and expanded, are well established and significantly contributing to water management in the United States and throughout the world. These generalized modeling systems are readily available for application by water management professionals to river systems located anywhere in the world.

Generalized modeling systems reflect the types of applications that motivated their development. ResSim serves as the reservoir system operations component of the Corps Water Management System implemented in the USACE district offices nationwide to support real-time operations of multiple-purposes reservoirs and flood control and navigation projects. ResSim is also used in USACE planning studies. RiverWare was developed as a partnership between CADSWES and the USBR and TVA. The TVA uses ResSim to support real-time hydroelectric power system operations within the setting of multiple-purpose reservoir system operations. The USBR applies RiverWare for both long-term planning and short-term operational planning for its multiple-purpose reservoir systems. The network flow programming based MODSIM was developed at Colorado State University in collaboration with the USBR and has been applied primarily by university researchers in studies both in the United States and abroad. WRAP supports statewide and regional planning and water allocation regulatory activities in Texas that require detailed modeling of diverse and complex institutional water allocation arrangements and reservoir/river system management practices.

ResSim, RiverWare, MODSIM, and WRAP provide general frameworks for constructing and applying models for specific systems of reservoirs and river reaches. Each of these four generalized modeling systems is based upon its own set of data management and computational techniques and has its own terminology, but they all provide flexible broad-based generic capabilities for modeling and analysis of river system development and management.

## 8. References

Draper, A.J., Munevar, A., Arora, S.K., Reyes, E., Parker, N.L., Chung, F.I. & Peterson, L.E. (2004). CalSim: Generalized Model for Reservoir System Analysis. *Journal of Water Resources Planning and Management*. ASCE, 130(6), 480-489, ISSN 0733-9496.

Draper, A.J., Jenkins, M.W., Kirby, K.W., Lund, J.R. & Howitt R.E. (2003). Economic-Engineering Optimization for California Water Management. *Journal of Water Resources Planning and Management*. ASCE, 129(3), 155-164, ISSN 0733-9496.

Eschenbach, E.A., Magee, T., Zagona, E., Goranflo, M. & Shane, R. (2001). Goal Programming Decision Support System for Multiobjective Operation of Reservoir Systems. *Journal of Water Resources Planning and Management*. ASCE, 127(2), 108-120, ISSN 0733-9496.

Fritz, J.A., Charley, W.J., Davis, D.W., and Haimes, J.W.( 2002). New Water Management System Begins Operation at U.S. Projects. *International Journal of Hydropower & Dams*, Aqua Media International, Issue 3, 49-53.

Hydrologic Engineering Center (1995). HECDSS User's Guide and Utility Program Manuals, U.S. Army Corps of Engineers, Davis, CA, USA.

Hydrologic Engineering Center (1998). HEC-5 Simulation of Flood Control and Conservation Systems, User's Manual, Version 8, U.S. Army Corps of Engineers, Davis, CA, USA.

Hydrologic Engineering Center (2007). HEC-ResSim Reservoir System Simulation, User's Manual, Version 3, U.S. Army Corps of Engineers, Davis, CA, USA.

Hydrologic Engineering Center (2009). HEC-DSSVue HEC Data Storage System Visual Utility Engine, User's Manual, U.S. Army Corps of Engineers, Davis, CA, USA.

Iich, N. (2009). Limitations of Network Flow Programming in River Basin Modeling. *Journal of Water Resources Planning and Management*. ASCE, 135(1), 48-55, ISSN 0733-9496.

Jain, S.K. & Singh, V.P. (2003). *Water Resources Systems Planning and Management*, Developments in Water Science 51, Elsevier, Philadelphia, PA, USA, ISBN 0444514295.

Karamouz, M., Szidarovszky, F. & Zahraie, B. (2003). *Water Resources Systems Analysis*. Lewis Publishers. Boca Raton, FL, USA, ISBN 1566706424.

Labadie, J.W. (2004). Optimal Operation of Multireservoir Systems: State-of-the-Art Review. *Journal of Water Resources Planning and Management*. ASCE, 130(2), 93-111, 0733-9496.

Labadie, J.W. (2006). Chapter 23 MODSIM: River Basin Decision Support System in: *Watershed Models*, V.P. Singh and D.K. Frevert (Eds.) CRC Taylor & Francis, London, 569-592, ISBN 0-8493-3609-0.

Labadie, J.W. & Larson, R. (2007). MODSIM 8.1: River Basin Management Decision Support System, User Manual and Documentation. Colorado State University, Fort Collins, CO, USA.

Maass, A., Hufschmidt, M.M., Dorfman, R., Thomas, H.A., Marglin, S.A. & Fair, G.M. (1966). Design of Water Resource Systems. Harvard University Press, Boston, MA, USA.

Martin, Q.W. (1983). Optimal Operation of Multiple Reservoir Systems. *Journal of Water Resources Planning and Management.* ASCE, 109(1), 58-74, ISSN 0733-9496.

Martin, Q.W. (1987). Optimal Daily Operations of Surface-Water Systems. *Journal of Water Resources Planning and Management.* ASCE, 113(4), 453-470, ISSN 0733-9496.

McMahon, G.F. (2009). Models and Realities of Reservoir Operation. *Journal of Water Resources Planning and Management.* ASCE, Vol. 135, No. 2, 57-59, ISSN 0733-9496.

Nagy, I.V., Asante-Duah, K. & Zsuffa, I. (2002). *Hydrologic Dimensioning and Operation of Reservoirs: Practical Design Concepts and Principles.* Water Science and Technology Library Vol. 39, Kluwer Publishers, London, UK, ISBN 1-4020-0438-9.

ReVelle, C.S. (1999). *Optimizing Reservoir Resources.* John Wiley & Sons, New York, NY, ISBN 0-471-18877-8.

Rogers, P.P., & Fiering, M.B. (1986). Use of Systems Analysis in Water Management. *Water Resources Research.* AGU, 22(9), 146-158.

Singh, V.P. & Frevert, D.K., Editors (2006). *Watershed Models.* CRC Taylor & Francis, London, UK, ISBN 0-8493-3609-0.

Simonovic, S.P. (2009). *Managing Water Resources: Methods and Tools for a Systems Approach.* United Nations Educational, Scientific and Cultural Organization, Paris, France, ISBN 9781844075539.

Votruba, L. & Broza, V. (1989). *Water Management in Reservoirs,* Developments in Water Science, Vol. 33, Elsevier Science, Philadelphia, PA, USA, ISBN 0444989331.

Watkins, D.W., Kirby, K.W. & Punnett, R.E. (2004). Water for the Everglades: Application of the South Florida Systems Analysis Model. *Journal of Water Resources Planning and Management.* ASCE, 130(5), 359-366, ISSN 0733-9496.

Wurbs, R.A. (1993). Reservoir System Simulation and Optimization Models. *Journal of Water Resources Planning and Management.* ASCE, 119(4), 455-472, ISSN 0733-9496.

Wurbs, R.A. (1994). Computer Models in Water Resources Planning and Management. Report 94-NDS-7, U.S. Army Corps of Engineers Institute for Water Resources, Alexandria, VA, USA.

Wurbs, R.A. (1995). *Water Management Models.* Prentice Hall, Upper Saddle River, NJ, USA, ISBN 0-13-161621-8.

Wurbs, R.A. (1996). *Modeling and Analysis of Reservoir System Operations.* Prentice-Hall, Upper Saddle River, NJ, USA, ISBN 0-13-605924-4.

Wurbs, R.A. (1998). Dissemination of Generalized Water Resources Models in the United States. *Water International.* IWRA, 12(3), 190-198, ISSN 0250-8060.

Wurbs, R.A. (2005a). Comparative Evaluation of Generalized River/Reservoir System Models. TR-282, Texas Water Resources Institute, College Station, TX, USA.

Wurbs, R.A. (2005b). Texas Water Availability Modeling System," *Journal of Water Resources Planning and Management,* ASCE, 131(4), 270-279, ISSN 0733-9496.

Wurbs, R.A. (2006). Water Rights Analysis Package (WRAP) Modeling System in: V.P. Singh & D.K. Frevert (Eds.) *Watershed Models.* CRC Taylor & Francis, London, 593-612 ISBN 0-8493-3609-0.

Wurbs, R.A. 2011 Water Rights Analysis Package (WRAP) Modeling System Reference and Users Manuals. Technical Reports 255 and 256, 8th Ed., Texas Water Resources Institute, College Station, TX, USA.

Wurbs, R.A. (2009). Salinity Simulation with WRAP. Technical Report 317, Texas Water Resources Institute, College Station, TX, USA.

Wurbs, R.A., and Hoffpauir, R.J. (2011). Water Rights Analysis Package (WRAP) Daily Simulation, Flood Control, and Short-Term Reliability Modeling. Technical Report 402, Texas Water Resources Institute, College Station, TX, USA.

Zagona, E.A. Fulp, T.J., Shane, R., Magee, T. & Goranflo, H. M. (2001). Riverware: A Generalized Tool for Complex Reservoir System Modeling. *Journal of the American Water Resources Association.* 37(4), 913-929, ISSN 1752-1688.

Zagona, E.A., Magee, T., Goranflo, H.M., Fulp, T., Frevert, D.K. & Cotter, J.L. (2006). Chapter 21 RiverWare in: V.P. Singh & D.K. Frevert (Eds.), *Watershed Models.* CRC Taylor & Francis, London, 527-548, ISBN 0-8493-3609-0.

# 2

# Web-Based Decision Support Framework for Water Resources Management at River Basin Scale

José Pinho, José Vieira,
Rui Pinho and José Araújo
*Department of Civil Engineering, University of Minho*
*Portugal*

## 1. Introduction

In recent years a major effort has been done to make water quality modelling tools available for water resources management at a river basin scale. The European water framework directive clear states that these tools must be used in making the diagnostic of surface water bodies water quality status and to anticipate the impact of measures to be implemented in order to achieve a good ecological status by 2015 in European waters (European Commission 2000; Rekolainen 2003; Horn et al. 2004; Ravesteijn & Kroesen 2007; Volk et al. 2008).

In order to make modelling tools usable by all the actors enrolled in water resources management process at river basins, these tools must be simple, user-friendly and robust. Water managers have shown strong interest in the integration of model tools throughout the use of graphical interfaces in their activities (Borowski & Hare 2007). These tools must be able to establish water management scenarios but it is well known that water managers have lack of time for training in the use of complex systems. Also software and hardware costs appear to be an important constrain in implementing these tools.

Several solutions are being implemented throughout European countries, resulting from the development and integration of different software packages using different integration technologies (Dudley et al. 2005, Berlekamp et al. 2007, De Kok et al. 2009).

This chapter presents a hydroinformatic environment (Price, 2000) specifically designed for a Portuguese north-western river basin (river Cávado) in order to define, simulate and analyse hydrodynamics and water quality management scenarios – the ODeCav System. The software solution was designed to be operated in a web environment, taking advantage of the integration capabilities of this software environment and the user friendliness of web interfaces. It is composed of the following main components: water monitoring data-bases, hydrodynamics and water quality river models, and reports facilities for the presentation of output results. The applicability of this hydroinformatic environment is exposed in the study of waste water treatment plants (WWTP) discharges impacts on the river water quality for different river flow regimes.

## 2. Methods

### 2.1 Water quality models

The flow is modeled using one-dimensional formulations of free surface flows based on the conservation of mass and momentum equations:

$$\frac{\partial A_f}{\partial t} + \frac{\partial Q}{\partial x} = q_{lat} \tag{1}$$

$$\frac{\partial Q}{\partial t} + \frac{\partial}{\partial x}\left(\frac{Q^2}{A_f}\right) + gA_f\frac{\partial h}{\partial x} + \frac{gQ|Q|}{C^2 RA_f} - W_f\frac{\tau_{wi}}{\rho_w} = 0 \tag{2}$$

where,

| | |
|---|---|
| Q | Water flow discharge in m³/s; |
| t | Time in s; |
| x | Distance in m; |
| $A_f$ | Wetted area in m²; |
| $q_{lat}$ | Lateral flow discharge in m²/s; |
| g | Gravity acceleration in m/s²; |
| h | Water depth in m; |
| C | Chezy coefficient in m^{1/2}/s; |
| R | Hydraulic radius in m; |
| $W_f$ | Flow width in m; |
| $\tau_{wi}$ | Wind shear stress in N/m²; |
| $\rho_w$ | Water density in kg/m³. |

In addition to these equations the flow discharges at hydraulic structures included in the model segmentation are computed using specific expressions for each type of structure: bridges, culverts, siphons, orifices, pumps, and weirs. In these structures the flow depends on the upstream and downstream levels, on its dimensions and on a set of specific parameters.

The water quality model is based on the one-dimensional transport equation:

$$\frac{\partial\left(A_f C\right)}{\partial t} = -\frac{\partial(QC)}{\partial x} + \frac{\partial}{\partial x}\left(DA_f\frac{\partial C}{\partial x}\right) + SA_f \tag{3}$$

where,

| | |
|---|---|
| C | Substance concentration in kg/m³; |
| D | Diffusion coefficient in m²/s; |
| S | Source, sink and reaction term in kg/m³/s. |

The last term of Eq. 3 refers to the sources and sinks and the dependence on the processes occurring in the water column related with the modeled substance. For each water quality problem a set of equations are considered, one for each substance involved in the water quality problem to simulate. The different biogeochemical processes relevant to the study of surface water quality problems have a great diversity. In this work it was chosen a processes framework, as much inclusive as possible that cover simple water quality processes, such as the modeling of accidental releases of conservative pollutants, or more complex processes, such as degradation of organic matter.

For problems involving conservative substances it is only considered the transport of the substance in the water through advection and diffusion. The evaluation of the extensions and durations of accidental discharge can be carried out recurring to a model in which the accidental discharge is modeled by a conservative substance. In addition to the cases of accidental discharge, these simple models also have practical interest to quantify the residence times and to analyze the effect of different hydrodynamic conditions in the water masses mixing conditions.

The majority of elements and substances in aquatic environments have reactions with other elements and/or substances, resulting in their transformation (decrease or increase in concentration). Bacterial contamination arising from discharges of domestic wastewater or diffuse sources, for example, can be modeled by taking up a 1st-order decay law. The behavior of many other substances (or species) can be approximated by considering the decay or growth of a second order, such as biochemical oxygen demand (BOD) or algae. The reaction coefficients should be established mainly through the available field data or laboratory tests. Dissolved oxygen (DO) is a common environmental element used to characterize the water quality in water systems. The analysis of the impact caused by discharges with a high concentration of organic matter may be made to quantify the effects in terms of variations in concentrations of dissolved oxygen in the water column, due to the decomposition of organic matter contained in wastewater discharges. The water quality processes library used in this work is one of the most complete for surface water quality modeling and includes all the relevant processes allowing the establishment of either complex water quality processes or simple ones depending on data availability.

## 2.2 Web-based hydroinformatic environment

The developed technological platform is mainly based on a database system and a set of hydrological, hydrodynamic and water quality models, operated using web interfaces. It comprises functionalities for query and analysis of the river network (information system), hydrodynamic and water quality models operation (modelling system), and results analysis (analysis system). Besides this, the platform provides the following additional services: user's management, document management and monitoring data store and display. The user's management service allows restricting or not the access to various content, as well as the permissions to run/execute or to view/consult/query the simulations results. The document management service allows the users to perform simple tasks, such as publication of papers and upload/download of work documents. The monitoring data functionalities provides a framework for gauge and water quality stations installed in the river basin, allowing its analysis, validation, and integration using a specifically designed database for this purpose.

The website responsible for the system interaction with the end users is divided into four main sections that allow access to the features mentioned above: Project, Information, Modelling and Analysis. These sub-menus are the main options from the navigation menu (Figure 1).

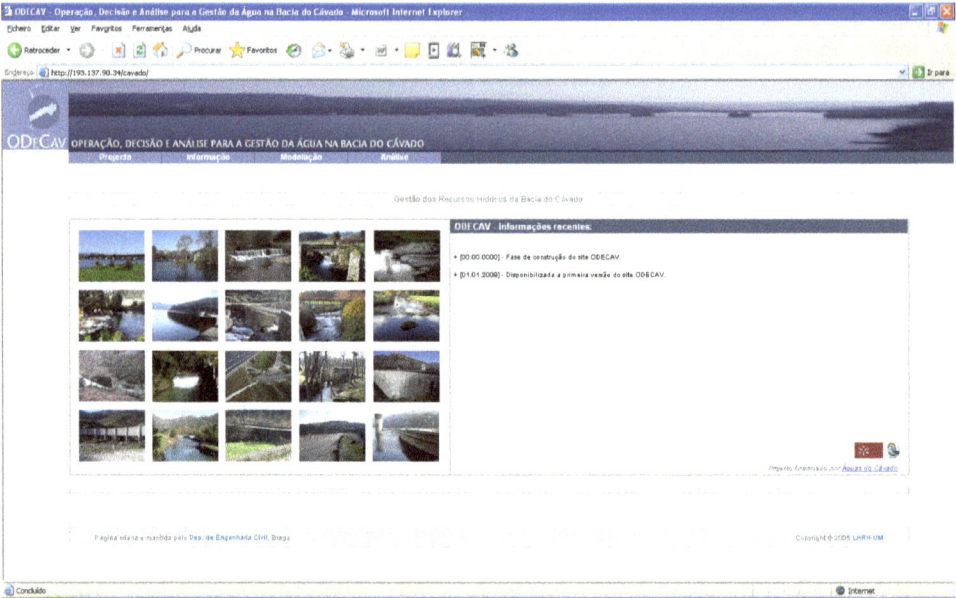

Fig. 1. Main window of the ODeCav modelling system

Implemented data-bases store geometric features of the river network, hydraulic structures, monitoring stations data, and point and non-point pollutant discharges into the river network. This was done using MySQL server (MYSQL 2009). The one-dimensional hydrodynamics and water quality models were implemented applying SOBEK software that numerically solves Eq. 1 to 3. This is commercial software developed and maintained by DELTARES (2009). Specific web-based model operation interfaces were developed in order to directly establish the simulation boundary conditions, run the model and visualize model results. All the developed applications are freely available since it will be used for research purposes.

The Project section presents information about the aim of the developed platform and its main characteristics. It also gives access to the document management tool. In the Information section it is possible to query the geographical database of the river basin. It includes information about details of the river network, data from monitoring stations and the river basin Geographic Information System. The Modelling section allows the users to define new simulations and remotely start the execution of the hydrodynamic and water quality models. This application not only allows the remote operation of the models, but also displays the simulation results from different users. Finally, the Analysis section provides access to several simulation results for a posterior detailed analysis and generation of reports. Each one of these sections includes graphical user interfaces, specifically designed and developed for that purpose, considering imposed requisites of its potential users.

Most of the content of the website has conditioned/restricted access. This means that it is available only through user authentication. A user without a valid login can only view/access the information provided under the Project menu (Figure 1). Any user who was granted with permissions (which has a user registration and password) will have access to the full content of the interface. However, this does not mean the user has permission to view, use or access all the features, functionalities and tools available in the interface. There are interfaces that differentiate the privileges of users. For example, while some users may start the execution of the models, others may only see theirs results. These permissions are managed by the system administrator. Figure 2 depicts used technologies in the implementation of the modelling platform and their interdependencies.

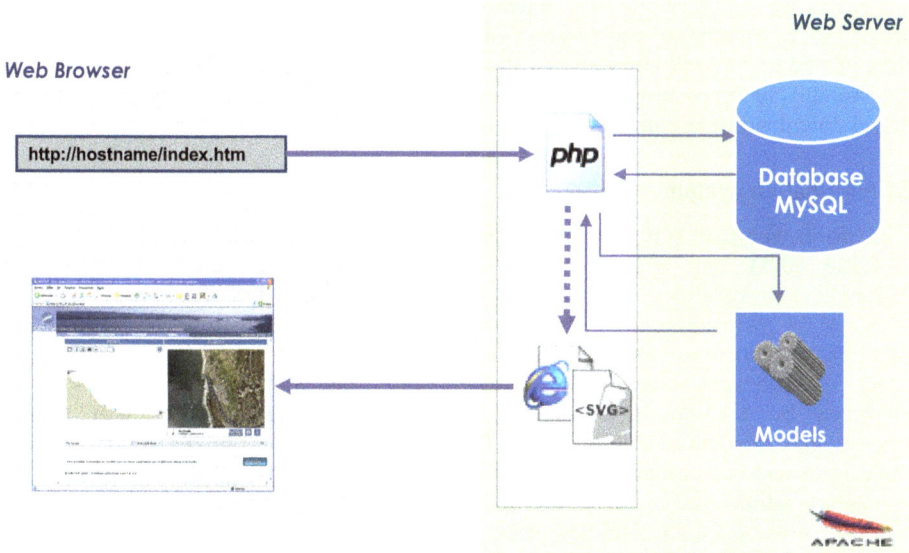

Fig. 2. Adopted technologies for the development of the modelling platform

This system is supported by a web server that provides all the web pages that integrate the interfaces. The basic task of this web server is to receive HTTP requests and to produce HTTP responses, which are mostly HTML documents or other types, such as images, documents, PDF, SVG vector images and simple text. Apache was the selected web server (due to its portability, wide range of services available, and because is a free web server).

All the information provided by the web server is stored in a relational database supported by MySQL, the world's most popular open source database server. Besides, MySQL provides many features not available in other systems and is completely free for both commercial and private use. It is a fast and robust system that can handle with unlimited number of users and records. Among the main features of this system are: portability (support for practically any platform); compatibility (there are drivers and interface modules for various programming languages); the excellent performance and stability; user friendly; and requires few resources in terms of hardware.

PHP was adopted as the main programming language (used to generate the HTML pages sent to the clients machines), while the JavaScript language was selected to develop code responsible for the interaction between the user and the interface. The Sobek software is applicable for hydrodynamic and water quality modelling in rivers, and it´s composed by seven modules: hydrology, hydrodynamics in channels, hydrodynamics in rivers, sewers, real-time control, water quality and floods. Its integrated approach allows the simultaneously simulation of real problems involving different modules. It is based on a robust and reliable numerical method that allows achieving solutions even for highly complex situations.

The consideration of graphical data within the developed interfaces was considered fundamental for a fast and intuitive understanding of the simulated features. In this case, the choice was the image technology SVG (Scalable Vector Graphics). SVG is an XML specification for graphics and has outstanding features as: ability to make enlargements without loss of resolution, the creation of motion graphics that facilitate presentation of results related to dynamic simulations, allows interactivity with the objects represented and the possibility of associating alphanumeric information to graphics. Beyond that, has a reduced download time, compared to other more conventional types of images.

### 2.2.1 Information system

The platform web map server was developed based on the GeoClient application (Rogers and Rosie, 2001). This application has been improved, and some new features and functionalities were added. Noteworthy, was the translation into Portuguese of all the menus and the inclusion of the possibility of automatic export of features to a Google Earth compatible data format. The graphical information themes are organized into databases. Updates of geographical data, adding and deleting data are operations that should be undertaken by the administrator of the database, in accordance with standard methodologies for databases operations. The map visualized in the client side is defined using a form and is automatically generated by an application developed for this purpose. The database administrator can set pre-defined maps. It is also possible for the user to view the results of water quality simulations available on the modelling system. The user must select the rivers and the instant of simulation to be represented in the map interface. The requested map information is communicated to the server and is transferred to the client computer at once. All subsequent operations are dependent on the processing capacity of the client´s machine.

The Geoclient application further developed at University of Minho, and integrated in this platform comprises the following major functionalities (Figure 3): full navigation (zoom in, zoom out and pan); map display of alphanumeric data associated to geographic entities; queries with results in chart or table; set labels display properties, export the map to a file, compliant with the Google Earth application; configurable legend presentation; presentation of a navigation map; selection of display options.

Navigation on the map is accomplished through some buttons on the toolbar which allow, for example, zoom in, zoom out and restore the original view. The toolbar shows the cursor coordinates and the labels associated to the graphical entities that are under the cursor. The panning can also be accomplished using the navigation map. The data associated with each graphic object are available, once the information button is pressed. The data are shown on a table and it is possible to generate a chart with the numeric values of all occurrences of the displayed map. The search tool allows the identification of objects throughout a table or a graph/chart, which satisfy the given search criteria. This criterion can be based on a comprehensive set of comparison operators. Search results can be quickly identified (by zoom in to the object or by its placement in the central area of the map) (Figure 4).

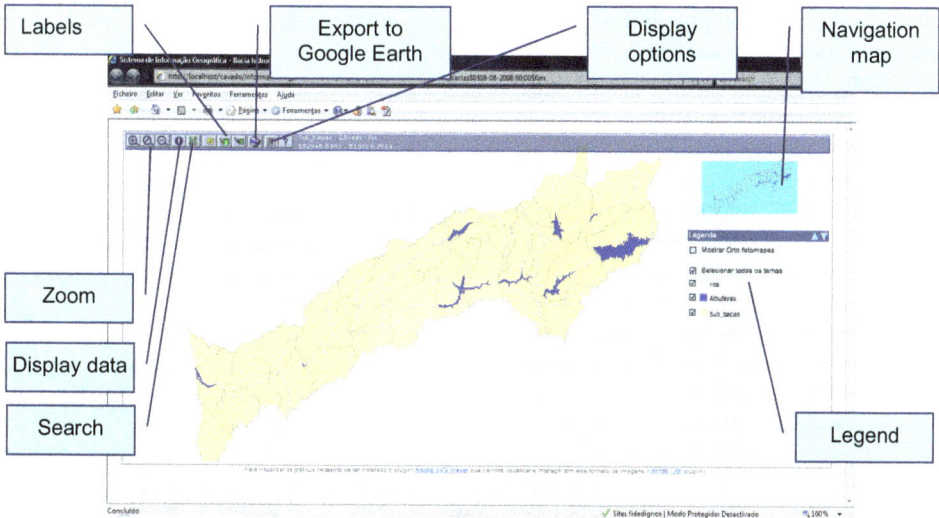

Fig. 3. Web-GIS service for the developed platform

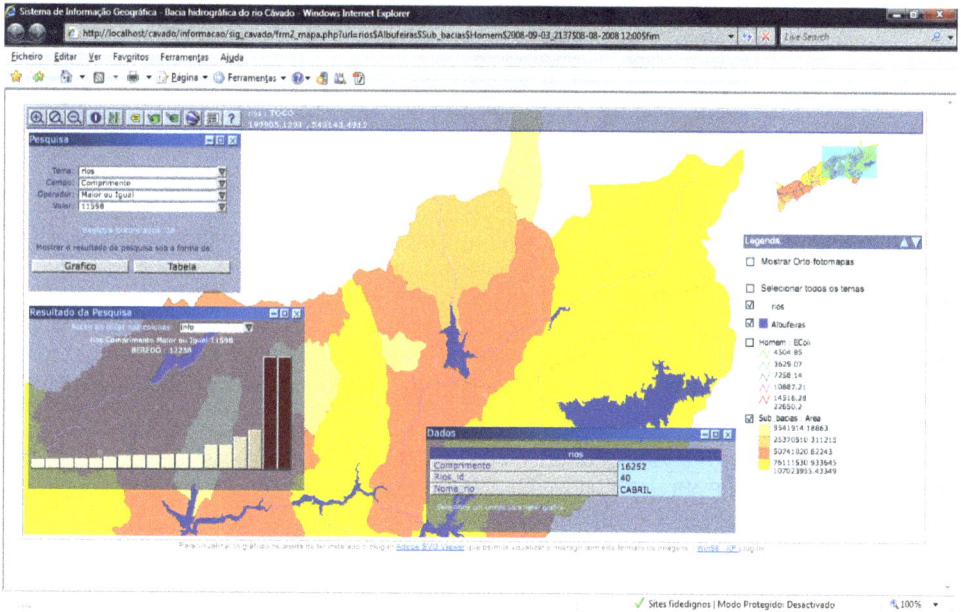

Fig. 4. Search Tool of GeoClient – Web-GIS service application of ODeCav modelling system

Map legend can be customized by the user; it's possible to change the colour of the displayed themes, set their classification according the values of numeric fields and show graphics (bars or circles) for numeric fields. For each object represented on the map there is a field associated on the database that is used for representing their labels on the main view. This feature can be assigned automatically for all entities of the map or placed entity by entity. The user has the possibility to define graphical properties of these labels. Finally, the application allows the automatic export of the map for later viewing on the Google Earth application. The export module is designed to convert the different coordinate data used by the two applications (Figure 5).

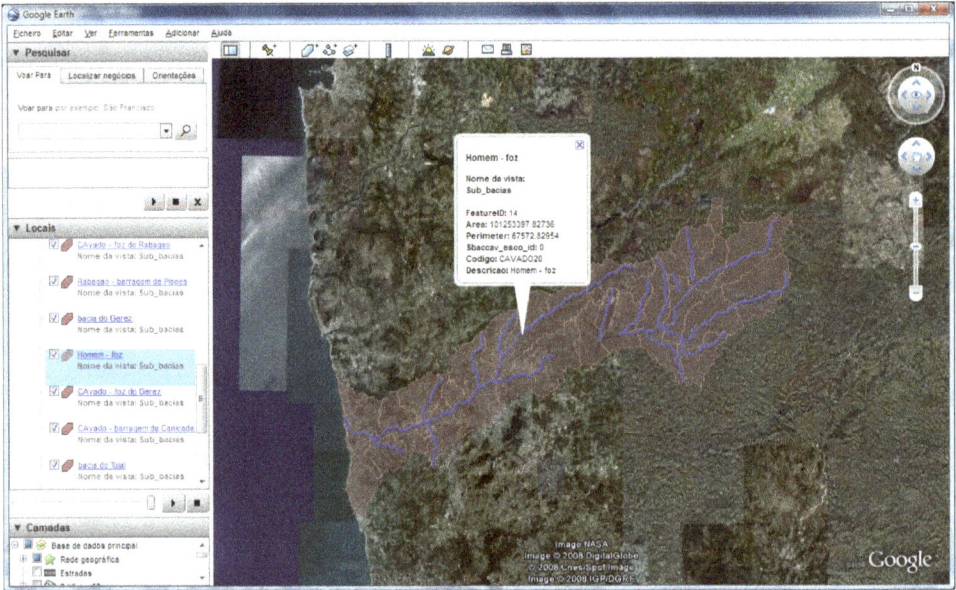

Fig. 5. Export of geographical features to Google Earth compatible data format

In addition it was developed an application to present detailed information about the river network. The application allows making a virtual visit to the rivers, enabling to locate relevant features that are relevant for the rivers hydrodynamics (eg. dams, weirs, controlled gates) and water quality (eg. waste water treatment plants, industries). Navigation is driven by selection of river segments. The user selects the river to visualize from an initial map. Then the navigation tools and the interactive legend allow the user to activate/deactivate the selected themes.

This application allows the visualization of all the alphanumeric data associated with all the entities and to display photos if available. The information management operations are performed on the database. Operations like insert, update, edition, addition or deleting are performed with proper tools for databases management. Aerial photos are stored outside the database in individual and separated files. Only their basic properties are stored in the database.

Figure 6 presents, for illustrative purposes, some views of river detailed information at river Cávado basin.

Meteorological, hydrometric and water quality information are available on the platform. This database contains historical records of the variables monitored for active monitoring stations. These data are the basis for water quality diagnostics assessment, and to the definition of modelling scenarios to predict the water quality dynamics.

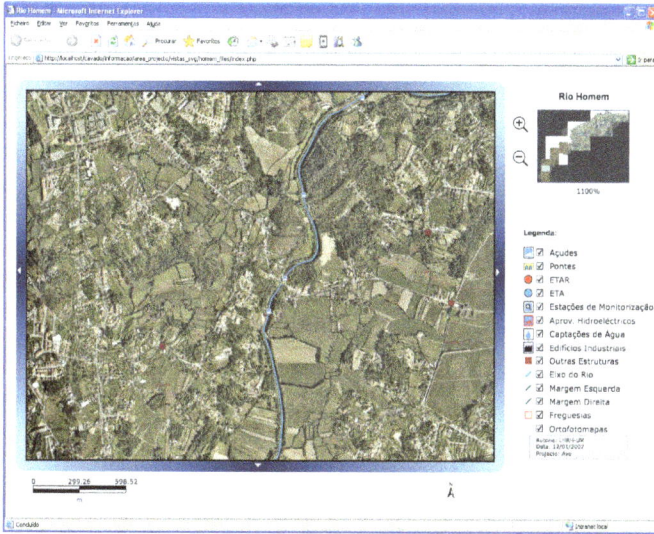

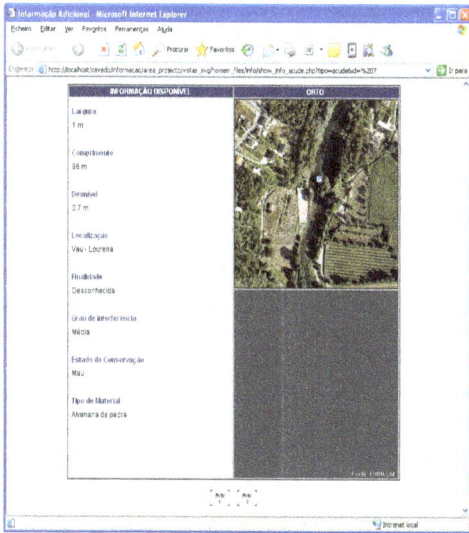

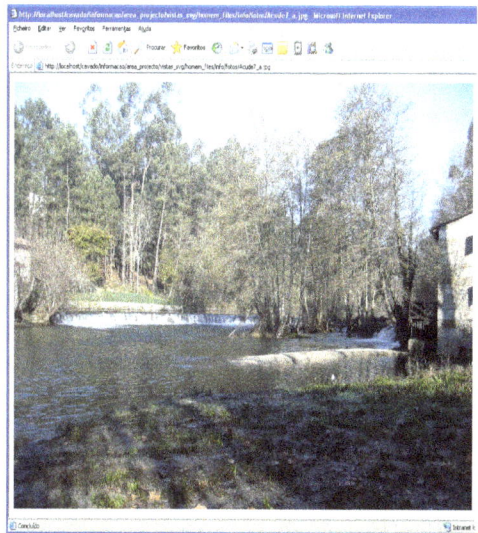

Fig. 6. River detailed information

A web application to access monitoring data was developed. The service provided by this application includes the following functionalities: graphical selection of the monitoring station from the map or from a list of available monitoring stations; graphic visualization of the data series for the selected parameter at a given station, data presentation in tabular layout and the possibility of exporting to a file compatible with MS Excel. It also allows the calculation of statistical parameters of the active data series and the automatic generation of a report. After selection of a particular parameter the associated data is displayed in a graphical layout in the client side (Figure 7).

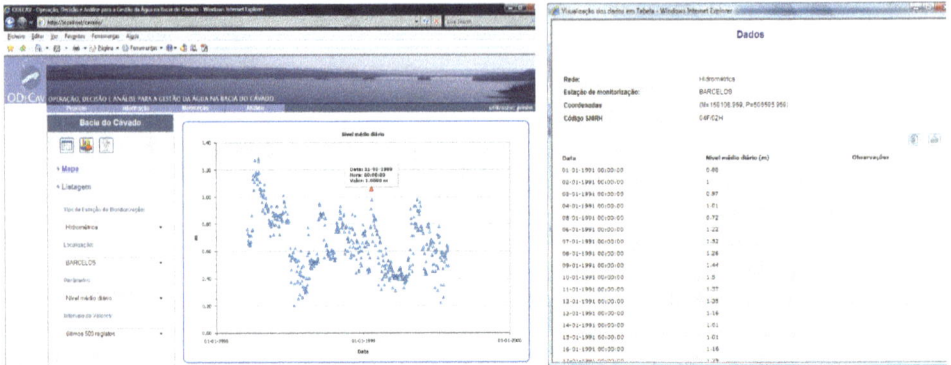

Fig. 7. Monitoring data interface

## 2.2.2 Modeling system

The modelling system consists in mathematical models of the rivers network to simulate river water discharge flows and levels (hydrodynamics) and the transport of substances or properties that are used as water quality indicators.

River Cávado basin is located in the north-western region of Portugal. This basin occupies an area of 1589 km² and the river network considered in the model is about 360 km length corresponding to 16 rivers.

Figure 8 depicts the river basin and the modelled river network. The model includes the rivers Cávado and Homem and the following main tributaries: Beredo, Borralha, Cabreira, Cabril, Cavadas, Caveiro, Covo, Febras, Gerês, Milhazes, Pontes, Rabagão, Toco, and Tojal.

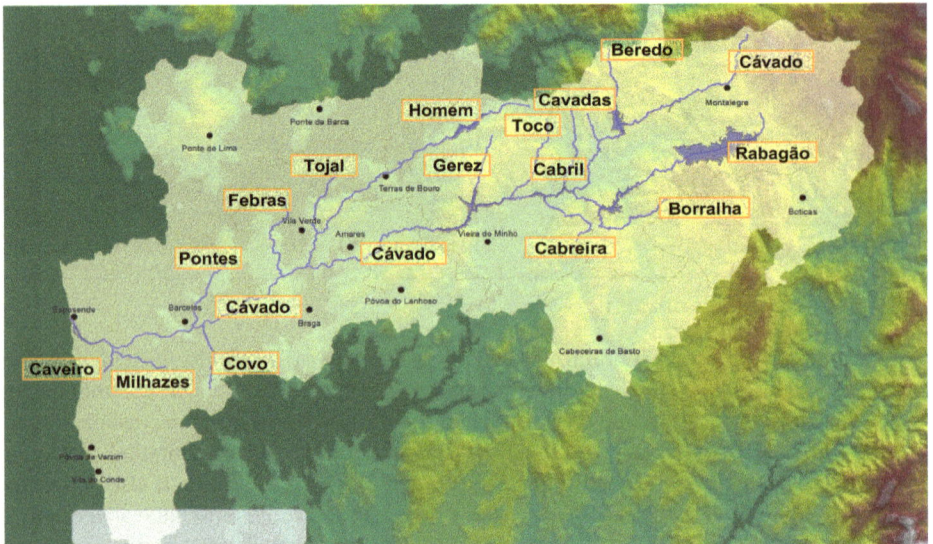

Fig. 8. River Cávado basin and modelled river network

Cross sections of the river channels considered in the developed model were established using bathymetric and topographic data available for this river basin. The one-dimensional grid comprises 1722 computational nodes, 22 open boundaries, 51 controlled discharges at hydraulic structures and 105 non controlled hydraulic structures. The rivers channels geometry was introduced considering 1854 cross sections. Pollutant sources are simulated considering 84 different locations in the river basin. All hydraulic structures with a significant influence in the rivers flows regime were considered with emphasis for dams and hydropower generation plants.

A web interface to present model results and operate the model execution was developed. This tool interacts with the model installed on a remote server, and allows setting up and activating new simulations. The user has access to applications that permit change the hydraulic structures operational settings during simulations, define hydropower generation time scheduling, establish tributaries discharge flows under different scenarios and define pollutant concentrations and discharges.

Figures 9 and 10 shows the main view of the designed interface for hydrodynamic and water quality models, where several functionalities are available (graphic display of the river profile, plan view, tables and graphics results and animations). This interface allows either setting up the data needed to define new simulations and viewing the results of a selected simulation. The main window of the interface corresponds to the graphical display of the profile (Figure 10).

Each one of the included icons has one or more PHP modules associated responsible to produce the desirable output. As already mentioned, the developed interface allows both setting up the input data for new model simulations and presenting final results. To define

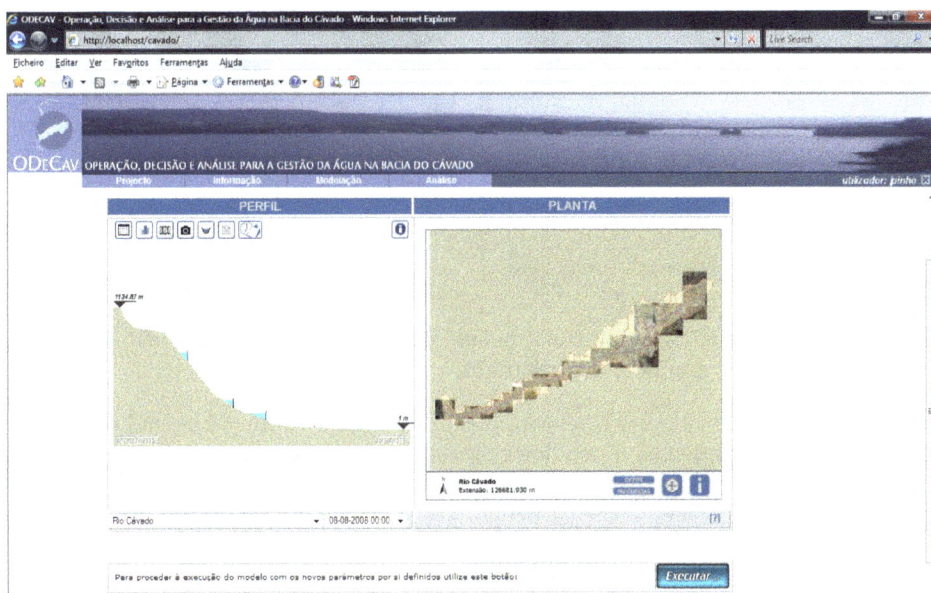

Fig. 9. Web interface for hydrodynamic models

new simulations an intuitive application was developed as depicted in Figure 11. The user has to follow up the successive forms to define all the data involved in a simulation. New simulations are defined using default data associated with the selected one.

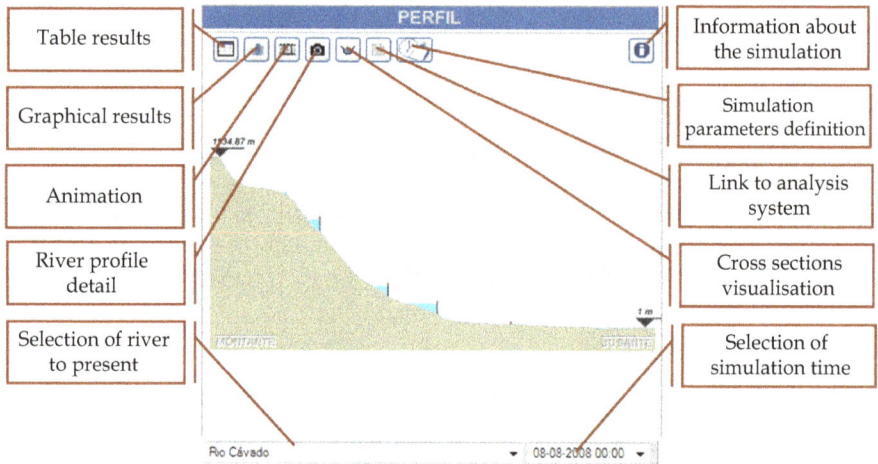

Fig. 10. Main functionalities of the hydrodynamic models web interface: profile view

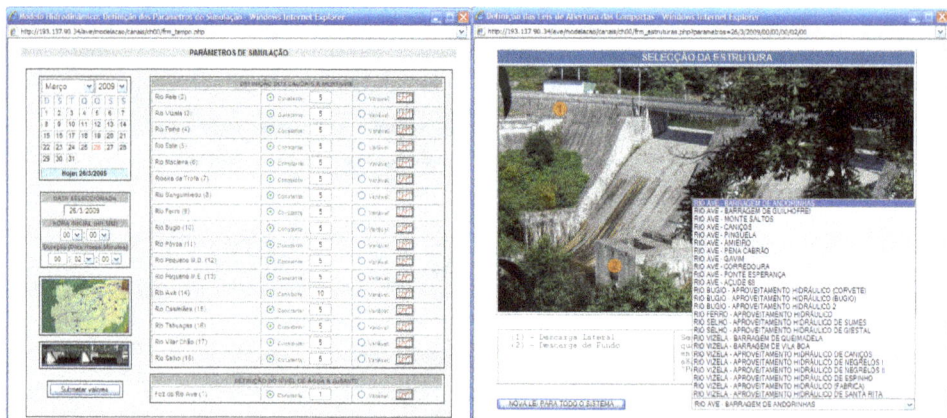

Fig. 11. Main form for definition of new simulations data (left) and interface to define data time series related to hydraulic structures (right)

The script used by this interface is responsible for setting the starting date and duration of the simulations. The presented form allows the definition of the values of river flows at the most upstream open boundaries. It is possible to define a constant value for the entire simulation or a variable law over that period. The image of a dam (as well as images for other hydraulic structures) on the form gives access to a specific new form for setting the opening laws of the gates and orifices (Figure 11 right). Completed the previous steps, the user can execute the model with these new parameters, pressing the "Run" button, located at the bottom of the main window (Figure 9).

## 2.2.3 Analysis system

The main purpose of this system is to simplify the analysis of complex water systems through presentation of synthesized model results and make available tools for evaluation of different management alternatives. The complexity of environmental systems is related to the uncertainty of the system behaviour caused by lack of essential information to describe natural phenomena and due to the simplifications adopted by mathematical models. This complexity is often intensified by the fact that, there are several actors with shared responsibility in solving the existing problems but usually these actors do not work together.

In the design of the analysis system it was considered the integration of different scales involved in the water resources management problems in a simple and intuitive manner, allowing moving from basin scale to the scale of hydraulic structures within the same application. The analysis of management measures under various environmental scenarios, results in a set of simulations that can be stored by the manager of the developed platform. These stored simulations can be analysed by different users of the modelling system.

Through a specific interface (Figure 12) it is possible to manage (store, consult and remove) different simulations generated in the modelling system. This information is organized by rivers and the analysis is carried out through reports available online. Two separated applications were considered for system analysis: one dedicated to the exploration and analysis of results from hydrodynamic simulations and the second committed to the analysis and exploration of results of water quality. The design and features included in the two interfaces are identical, differing only in the type of results available. This division is justified by the expected use of two distinct groups of users (one group focused on quantitative analysis and the other more interested in qualitative analysis).

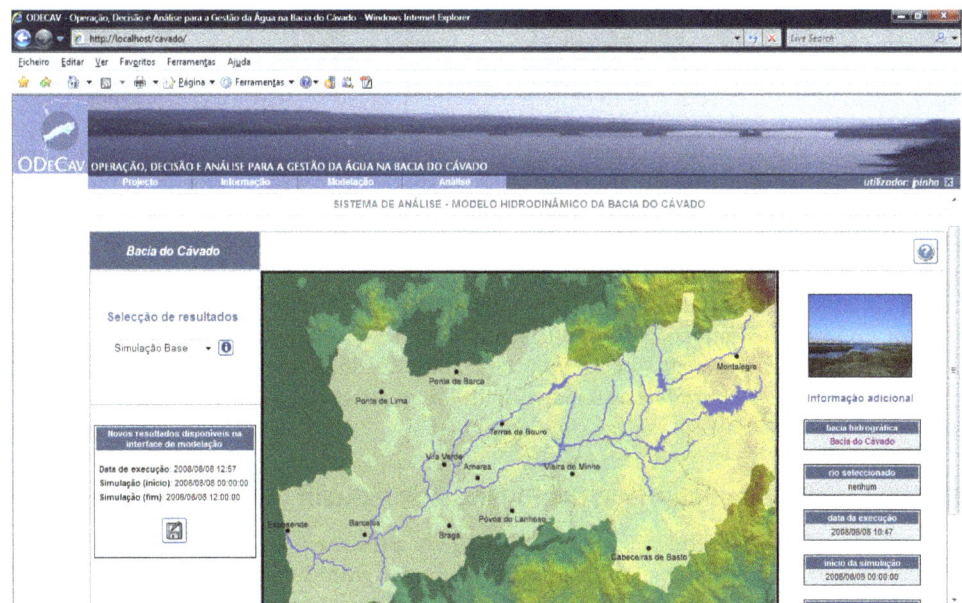

Fig. 12. Main interface of the analysis system

The main window displays a list box with the available simulations results. It also provides an area with indication of possible simulations available for storage (if the user has permission to store simulations) and an area with additional information about the selected river and simulation data. After selecting the river to be analysed, a plan view of the river is displayed with graphical features that represents river reaches, nodes and hydraulic structures (Figure 13).

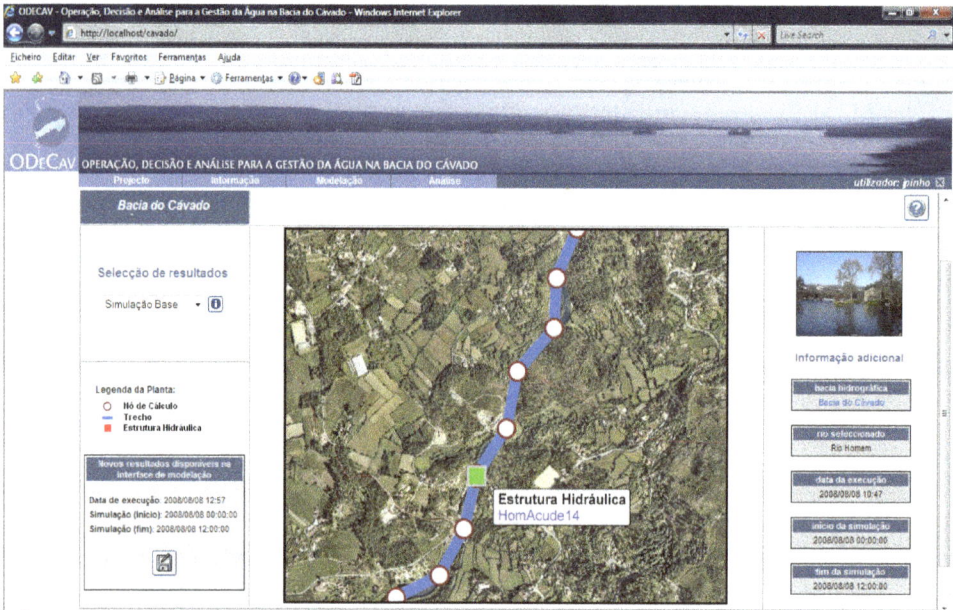

Fig. 13. Plan view for selection of river features in the analysis system

The plan view, also allows visualization of aerial photos, location names and the river name that are used as auxiliary information for easily locate the conceptual elements of the model (nodes, reaches and hydraulic structures). Based on these elements it is possible to automatically generate a report with results of the selected simulation. The report consists of: (i) general information about the simulation and information about the number of model nodes; (ii) reaches or hydraulic structures; (iii) plant and profile location of the selected element; (iv) representation of the profile view at the initial time; (v) intermediate instant and at the final instant of the simulation; (vi) table of results for the active element; (vii) graph of results for all variables associated with the selected element; (viii) and a statistical summary of these variables over the simulation period.

## 3. Water quality simulations at the River Cávado Basin

General procedure for setting up models is briefly presented. In order to illustrate the potentialities of the developed technological platform hydrodynamics and water quality simulations results are presented. Some particular problems of the presented river basin are discussed.

## 3.1 Models implementation

Models implementation followed a comprehensive procedure that includes four main phases: (i) monitoring data analysis; (ii) identification of key parameters including internal processes parameters or open boundary conditions variables; (iii) model calibration base on monitored data; and (iv) scenarios definition and simulations execution.

All available monitored data was included in the system databases witch facilitate the identification of the relevant hydraulic structures to be included in the model as well as the relevant waste water discharges. The segmentation of the model was defined considering the important influence of the upstream reservoirs in the river flows and the intense occupation (industrial, agriculture and urban areas) of the basin in the downstream areas.

In the calibration procedure a hybrid approach was followed: several parameters were established according to proposed values in the literature (Thomann and Mueller, 1987, Chapra, 1997), from previously developed works about river Cávado water quality (Vieira et al, 1998) and/or based on available field data.

## 3.2 Hydrodynamics

The hydrodynamic behaviour of the rivers at river Cávado basin is influenced primarily by rainfall regimes in the region and the exploitation of its hydropower facilities. A conceptual framework (Figure 15), where are identified the main structures influencing the hydrodynamic regimes (numbered from 1 to 28) was adopted in order to characterize the data to define the simulation scenarios. For each scenario the rainfall hydrographs and the pumps and turbines operation must be defined.

To illustrate the application of the developed platform thirteen hydrodynamic scenarios were defined in order to estimate average monthly flows and the average flow considering the data included in the information system for the entire available monitoring period.

Turbines were considered fully operational during simulation periods and the water levels at reservoirs are considered near its average level for the month associated with each simulation. It was also considered that other outflow gates at dams are closed. The exceptions were the Venda Nova and Vilarinho das Furnas (Figure 15).

Finally, on the oceanic boundary of the model, no tidal variations were assumed. It was considered that the water level remains at the mean sea level during the simulation period.

All simulations were defined to have duration of 7 days and a computational time step of 15 minutes.

Figure 14 presents a comparison between the estimated (estimation based on monitoring data) and the simulated values at different rivers locations. The presented simulated results refer exclusively to the last instant of the simulation.

From the observation of Figure 14 it is possible to find a close proximity between the estimated and simulated results, with the exception of locations 5, 6 and 12. In the first two locations the difference of 5.6 m$^3$/s is due to the introduction of the flow contribution of the sub-basin limited by the Caniçada reservoir and the Febras river at this last location. A similar situation occurs in location 12. In general, all the other simulations results present a notable approximation to the monitored values.

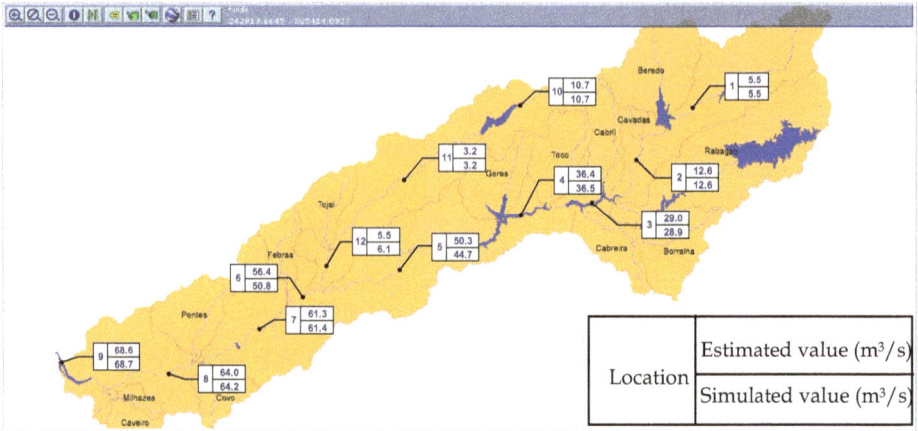

| Location | Estimated value (m³/s) |
|---|---|
| | Simulated value (m³/s) |

Fig. 14. Estimated average rivers flows and simulated values for a scenario corresponding to the annual mean values of the available monitoring period

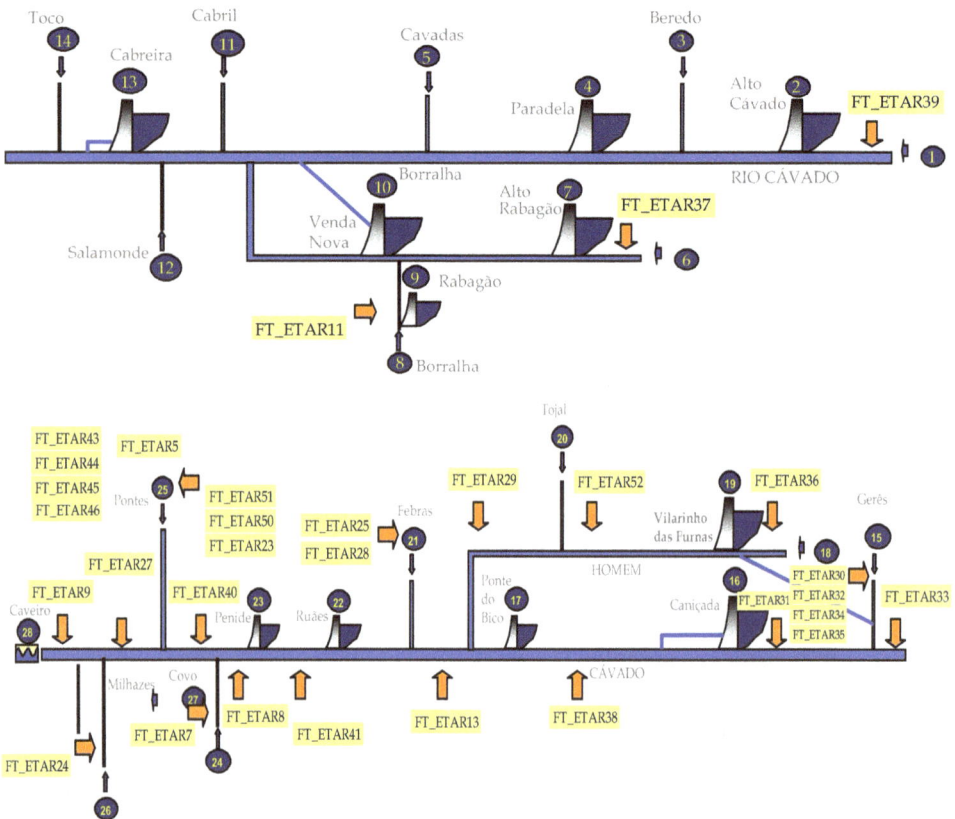

Fig. 15. River Cávado model scheme: WWTP discharges

## 3.3 Water quality

Simulations of water quality were based on hydrodynamic simulations previously presented and on the consideration of different characteristics for the discharges of waste water treatment plants (WWTP), industries, livestock units, and other contributions associated with different tributaries. The establishment of different values for these pollutant discharges results in different simulation scenarios.

In order to present the potential of the developed modelling system different scenarios were defined for average annual and monthly rivers flows discharges and considering the WWTP efficiency in compliance with their treatment schemes (scenarios 1 to 13). Additionally, for the wettest month (January) and the driest month (August) the effects of discharges from WWTP considering extreme values of its efficiencies were worked out (scenarios 14 to 17). Finally, it was considered three scenarios (scenarios 18 to 20) associated with the failure of each one of the three main WWTP: Frossos, Esposende and Vila Frescaínha (Table 1).

| Scenario | Hydrodynamics conditions | WWTP efficiencies |
|:---:|:---:|:---:|
| 1 | Annual | Average |
| 2 | January | Average |
| 3 | February | Average |
| 4 | March | Average |
| 5 | April | Average |
| 6 | May | Average |
| 7 | June | Average |
| 8 | July | Average |
| 9 | August | Average |
| 10 | September | Average |
| 11 | October | Average |
| 12 | November | Average |
| 13 | December | Average |
| 14 | January | Maximum |
| 15 | August | Maximum |
| 16 | January | Minimum |
| 17 | August | Minimum |
| 18 | Annual | Frossos WWTP failure |
| 19 | Annual | Vila Frescainha WWTP failure |
| 20 | Annual | Esposende WWTP failure |

Table 1. Water quality modelling scenarios

The qualitative characteristics of the pollutant discharges from WWTP were estimated for the following parameters: biochemical oxygen demand (BOD$_5$), dissolved oxygen (DO), and total coliform bacteria (TCB), faecal coliform bacteria (FCB) and streptococci bacteria (SB).

Figure 15 shows the approximate location of all wastewater treatment plant specified in the water quality model.

The loads associated with industrial pollutant discharges were estimated. It was used a simple method based on the industrial activities to compute its waste water discharges. However, most of the industries effluents are treated in WWTP.

Another pollutant sources estimated for each simulation was the livestock farms discharges. The effluent loads were estimated considering the number of heads of cattle per farm.

In the upstream open boundary of the modelled rivers it was considered that the water presents characteristics of unpolluted water, with zero values of concentrations of BOD$_5$ and bacteriological variables. For dissolved oxygen, it was considered a value close to the saturation concentration (10 mg/L) and a water temperature of 10 °C.

The initial conditions of the model were established by considering the values of average concentrations at the monitoring stations for each water quality variable.

Results were obtained for the twenty scenarios, using a simulation time of seven days for each scenario.

It should be stressed that the results prove to be more sensitive to the values adopted for the concentrations of wastewater discharges than to the calibration coefficients. Once that data series related to those discharges are available it is possible to improve the performance of the water quality model through the fine tune of the calibration coefficients.

Table 2 presents the obtained simulated results for scenario 1 at the last simulation time and the average values monitored at monitoring stations. It appears that for the rivers Homem and Cávado (where monitoring data is available) the water quality profile related with organic matter discharges, inferred from the results of concentrations of BOD$_5$, presents nearly uniform values (around 1 mg/L). It was possible to achieve good approximation between the monitored values and model results for this scenario.

The installed treatment capacity of waste water (mainly from domestic sources) in the river basin, based on the removal of organic matter is reflected in the monitored values, resulting in low concentrations of BOD$_5$.

Regarding the results of bacteriological variables the differences between simulated and observed values in scenario 1, are more significant in the regions located in the most urbanized areas.

The differences between measured and simulated values arise primarily from the uncertainty associated with quantification of bacterial loads (a sensitivity analysis was performed concerning adopted calibration parameters, and it was concluded that these parameters cannot justify the differences found). The values used in the estimation of these loads (mean values of bacterial loads) are highly variable and may justify the simulated behaviour (underestimation of bacterial loads).

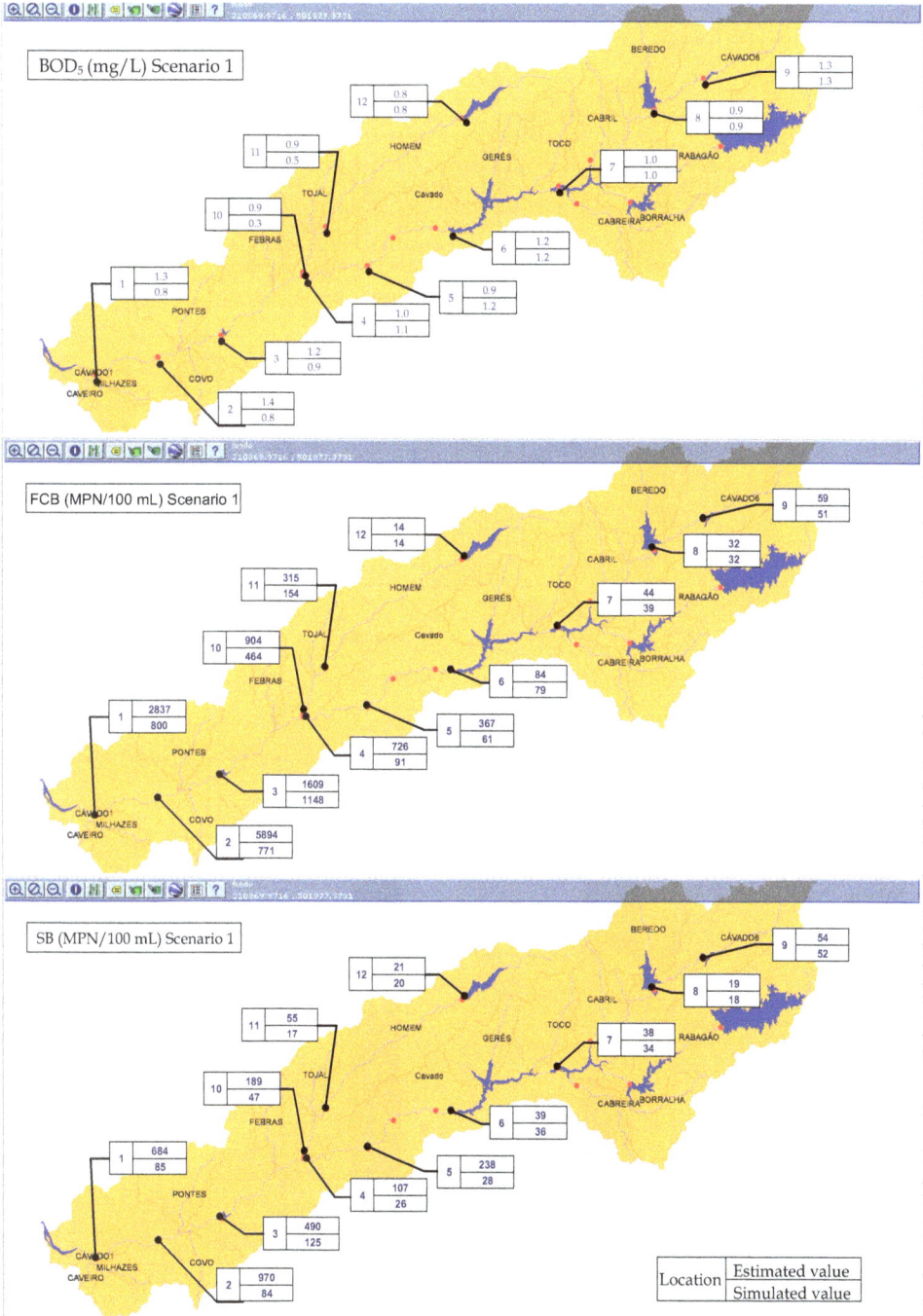

Fig. 16. Water quality model results in the vicinity of monitoring stations for scenario 1

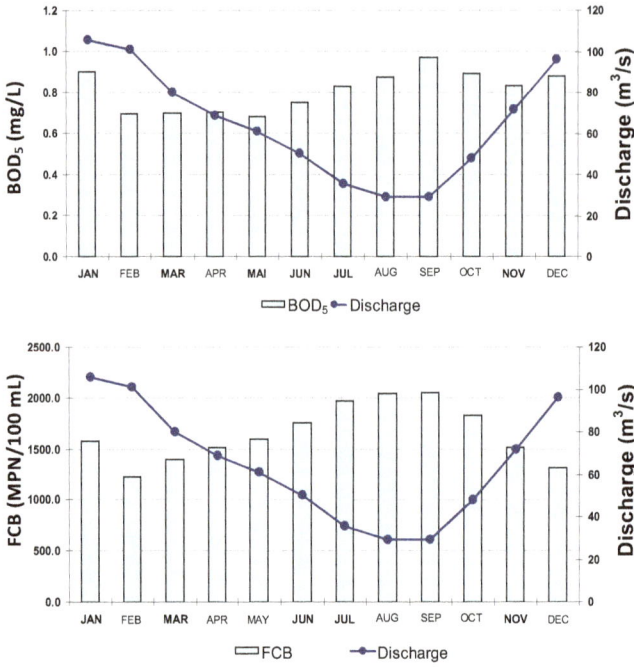

Fig. 17. Concentrations results for BOD$_5$ and FCB for scenarios 2 to 13 at Ponte Nova de Barcelos in river Cávado

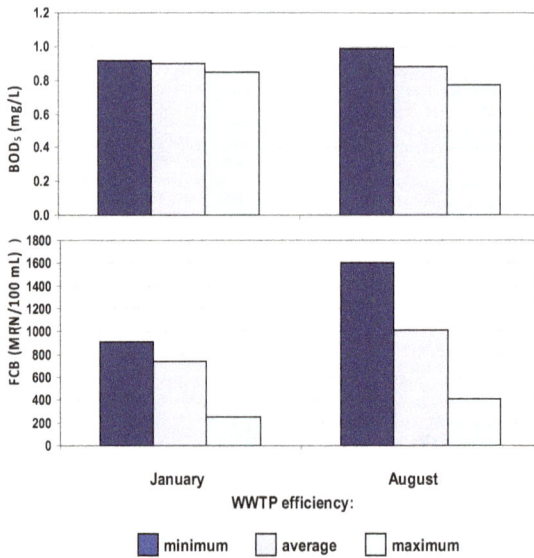

Fig. 18. Concentrations results for BOD$_5$ and FCB under scenarios 14 to 17 at Ponte Nova Barcelos in river Cávado

Results for this scenario showed that the bacteriological pollution in the river Cávado stretch downstream the confluence of the river Homem it is a reality that the wastewater treatment solutions have failed to solve in this basin. The bacteriological pollution also occurs during the dry season which decreases the likelihood of contamination come from diffuse sources.

Adopting the same conditions of pollutant discharges of scenario 1 but now considering the average monthly rivers flows discharges, with the simulated scenarios 2 to 13 it is possible to assess the influence of seasonal flow variation on water quality behaviour. Figure 17 shows results in these scenarios in a specific location of the river Cávado: Ponte Nova de Barcelos.

The results reveal a strong influence of the seasonal river flow regime variation on water quality concentrations. The variations reach values of around 70% of the concentration values in the wettest months in the case of bacteriological variables and about 40% for $BOD_5$.

Identical results can be achieved with simulations involving different time scales with the ODeCav System. It is particularly interesting to evaluate the influence of hydropower generation plants operational rules on water quality for the rivers Cávado and Homem, since for these two rivers the flow regime is strongly influenced by the operations of those hydraulic structures. Also the influence of hourly variations in flow regimes resulting from energy production on water quality of rivers Cávado and Homem, can be easily assessed once the pollutant discharges in the rivers are known with a similar temporal resolution.

With the scenarios 14 to 17 it is possible to evaluate the impact of different performances of the WWTP assuming different river flow regimes (January and August) in the resultant water quality. Figure 18 depict the obtained results in these scenarios in a location of the river Cávado: Ponte Nova Barcelos.

The results show a greater sensitivity of the river receiving waters to wastewater discharges in the dry season, for all simulated variables.

Although the efficiencies of the WWTP considered in each of the scenarios have been defined from literature values, there is a great variability in the resulting receiving waters concentrations, especially for bacteriological variables. For the presented location, the concentrations of bacterial variables when treatment plants operate with a minimum efficiency it is approximately four times the concentration when treatment plants operate at maximum efficiency. The developed modelling system reveal to be very important for the evaluation of alternative wastewater treatment investments since it allows, in a simple way, to anticipate the effects of these structures to improve water quality in river receiving waters.

Finally, it is presented in Figure 19 results for the Frossos WWTP rupture scenario (scenario 18). The rupture is simulated for a situation of average runoff and therefore not reflect the worst situation in terms of impact on water quality.

The presented results refer to two distinct locations: one in the vicinity of the discharge (immediately downstream) of the WWTP and the other on a stretch away. Based on this kind of results protective measures can be planned and the affected water uses can be anticipated, minimizing the impact of this potential accidental situation.

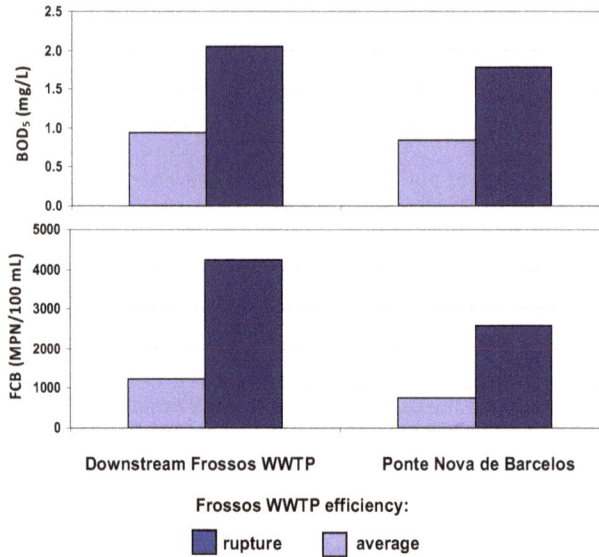

Fig. 19. Concentrations results of BOD$_5$ and FCB for the scenario 18 (rupture of Frossos WWTP) at two different locations

It should be noted that the results presented in this chapter are not exhaustive. All simulations are made available with the installation of ODeCav System and can be found on the web based platform.

The presented examples illustrate the potential of the developed system for management of water quality at river Cávado basin.

## 4. Conclusions

The developed operational modelling platform allows the simulation of an unlimited number of scenarios. Several scenarios were implemented in order to evaluate its performance after a judicious effort for model calibration and validation.

Main obtained results for the rivers within the basin reveal an almost uniform longitudinal profile for BOD concentration with a good agreement between observed and simulated results. The installed wastewater treatment capacity in the watershed (domestic and industrial wastewater) is reflected in the observed values leading to low concentrations of BOD. However, a quite different situation is observed for bacteriological indicators. In the lower part of the basin, concentrations results for all simulated scenarios are always lower than the observed ones even in the dry season. The reason for this fact can be the existence of untreated discharges or the lack of disinfection in WWTP.

The presented decision support tool for the river Cávado basin constitutes a robust and efficient technological platform to support water management at a river basin scale. Obtained results allow predicting that this new tool will be extremely effective and important to achieve the objectives of water management at river basin scale. In the next years, the use of all the potentialities of this kind of platforms in practical situations under different water management problems constitutes a major challenge for its evaluation. Moreover, water authorities once decide to use this kind of management tools will certainly see improved their analysis capabilities, strengthening their technological skills for the adoption of more sustainable water management policies. Especially it is adequate for developing big projects since it facilitates collaborative studies in one common platform and the modelling results became much more transparent for all project partners.

## 5. Acknowledgment

The authors thank to Comissão de Coordenação e Desenvolvimento Regional do Norte, Águas do Ave, SA and Águas do Cávado, SA, for the financial support.

## 6. References

Berlekamp, J., Lautenbach, S., Graf, N., Reimer, S. & Matthies, M. 2007 Integration of MONERIS and GREAT-ER in the decision-support system for the Elbe river basin. Environ Model Softw22(2):239–247.

Borowski, I. & Hare, M. 2007 Exploring the gap between water managers and researchers. Difficulties of model-based tools to support practical water management. Water Resources Management 21 (7), 1049-1074.

Chapra, S. 1997. Surface water-quality modeling. The McGraw-Hill Companies, Inc.

De Kok, J. L., Kofalk, S., Berlekamp, J., Hahn, B. M. & Wind H. 2008 From Design to Application of a Decision-support System for Integrated River-basin Management, Water Resources Management, (23), 1781-1811.

DELTARES 2009 SOBEK software. Available from:< http://delftsoftware.wldelft.nl/>.

Dudley, J., Daniels, W., Gijsbers, P. J. A., Fortune, D., Westen, S. & Gregersen, J. B. 2005 Applying the Open Modelling Interface (OpenMI), Proceedings of the MODSIM 2005 conference, Melbourne, Australia.

European Commission 2000 Directive of the European Parliament and of the Council 2000/60/EC Establishing a Framework for Community Action in the Field of Water Policy, Official Journal 2000 L 327/1, European Commission, Brussels.

Horn, A., Rueda, F.J., Hörmann, G. & Fohrer, N. 2004 Implementing river water quality modelling issues in mesoscale watershed models for water policy demands – an overview on current concepts, deficits, and future tasks. Physics and Chemistry of the Earth 29, 725–737.

MYSQL 2009 MySQL database server. Available from:<http://www.mysql.com/>.

Price, R. K. 2000 Hydroinformatics and urban drainage: an agenda for the beginning of the 21st century, Journal of Hydroinformatics, vol. 2, No. 2, 133-147.

Ravesteijn, W. & Kroesen O. 2007 Tensions in water management; Dutch tradition and European policy. Water Science and Technology 56(4): 105–111.

Rekolainen, S., Kämäri J. & Hiltunen M. 2004 A conceptual framework for identifying the need and role of models in the implementation of the Water Framework Directive. International Journal of River Basin Management 1(4):1–6.

Thomann R. V., and J. A. Mueller. 1987. Principles of Surface Water Quality Modeling and Control. Harper and Row, Inc., New York.

Vieira, J. M. P., Pinho, J. L. S., and Duarte, A. 1998. Eutrophication vulnerability analysis: a case study. Water Science and Technology, 37(3), 121-128.

Volk, M., Hirschfeld, J., Dehnhardt, A., Schmidt, G., Bohn, C., Liersch, S. & Gassman, P.W. 2008 Integrated ecological-economic modelling ofwater pollution abatement management options in the Upper Ems river basin. Ecological Economics Special Issue Integrated Hydro-Economic Modelling 66, 66-76.

# 3

# Integrated Water Resources Management as a Basis for Sustainable Development – The Case of the Sava River Basin

Dejan Komatina
*International Sava River Basin Commission*
*Croatia*

## 1. Introduction

According to a widely used definition, the integrated water resources management (IWRM) „is a process which promotes the coordinated development and management of water, land and related resources in order to maximise the resultant economic and social welfare, paving the way towards sustainable development, in an equitable manner without compromising the sustainability of vital ecosystems" (Global Water Partnership, 2000). The IWRM approach helps to manage and develop water resources in a sustainable and balanced way, taking account of social, economic and environmental interests. Although the IWRM concept has been formulated as early as in mid twentieth century (Biswas, 2004), the approach has been granted a due attention in early 1990-ies (*The Dublin Statement on Water and Sustainable Development*, 1992; UNECE, 1992) and, since then, a remarkable work has been done to examine different concepts of IWRM (for review, see Global Water Partnership & International Network of Basin Organizations, 2009).

Particular challenges of IWRM are associated with transboundary basins, especially due to decreasing resources and growing demands. A great number of international basin organizations have been established to manage water resources in transbouDary basins. A general distinction can be made between implementation-oriented basin organizations, responsible for development, implementation and maintenance of joint projects, often having a development focus and going beyond pure water resources management, and coordination-oriented basin organizations, in charge of coordinating water resources management tasks that are developed and implemented on national level, but coordinated and harmonized on transboundary level (Schmeier, 2010).

Given the nature of the conventions dealing with transboundary basins in Europe, such as the Danube, Rhine or Elbe basins, the respective basin organizations are obviously focused, either on sustainability issues (i.e. protection of the rivers), or on development activities (i.e. development of navigation, or tourism). However, recent processes, led by European Union, namely the *EU 2020 Strategy* (EC, 2010a) and the *EU Strategy for the Danube Region* (EC, 2010b) yielded new frameworks tending to integrate sustainability and development.

In comparison with other European river basins, however, the situation in the Sava river basin was rather peculiar. The political changes in the region of the former Yugoslavia in the 1990-ies, which turned the Sava river from the largest national river into an international river, challenged the water management in the Sava river basin substantially, by seriously affecting its basic elements (hydro-meteorological data exchange system, monitoring and early warning systems, etc.) and confining it to a national level, unlike the integrated approach, emerging in Europe at the same time. The changes have also caused a sharp decrease of economic activities in the region, such as navigation, unlike the other parts of Europe, where the inland waterway transport has proven to be a competitive transport mode, being environmentally friendly and capable of reducing congestion on densely used roads (EC, 2006). Since then, the Sava river has been hardly used for transport, for a number of reasons, including a lack of maintenance and investments, resulting in a poor quality of infrastructure, poor intermodal road and railway connections, as well as damaged ports and river infrastructure and presence of unexploded ordnances, endangering safe navigation.

For these reasons, a new international framework became necessary in order to ensure a sustainable use, protection and management of water resources in the Sava river basin, and thus enable better life conditions and raising the standard of population in the region. After a process of negotiations, the new framework has finally been provided by the development of the *Framework Agreement on the Sava River Basin* (*FASRB*, 2002), and subsequent establishment of the International Sava River Basin Commission (ISRBC), as an international organization with responsibility to coordinate the implementation of the *FASRB*.

The overall objective of the *FASRB* is to establish and maintain the transboundary cooperation in the water sector, in order to provide conditions for sustainable development of the region within the Sava river basin. The main purpose of this Chapter is to present the approach to water resources management, based on the *FASRB*, which appears to be a good basis for a progress toward sustainable development of the region within the basin.

## 2. Natural basis for cooperation in the Sava river basin

This part provides a review of basic facts on the basin, including the information on its biological and landscape diversity, as well as main uses of water resources in the basin, illustrating also the relevance of the Sava river as a Danube tributary (ISRBC, 2009d).

### 2.1 General characteristics of the basin

The Sava river basin is a major drainage basin of the South-Eastern Europe covering the **total area** of approximately 97,713 km², and represents one of the most significant sub-basins of the Danube river basin, with the share of 12% (Fig. 1). The basin is **shared among five countries**, a negligible part of the basin area also extending to Albania (Table 1), and hosts the population of roughly 8.5 million.

The landscape within the Sava river basin is diverse, the **elevation** varying between approx. 71 m above sea level (m a.s.l.) at the mouth of the Sava river in Belgrade (Serbia) and 2,864 m a.s.l. (Triglav, Slovenian Alps). Mean elevation of the basin is approximately 545 m a.s.l. In terms of **land cover/land use**, most of the basin is covered by the forest and semi-natural areas (54.7%) and agricultural surfaces (42.4%), while the share of artificial surfaces is 2.2%.

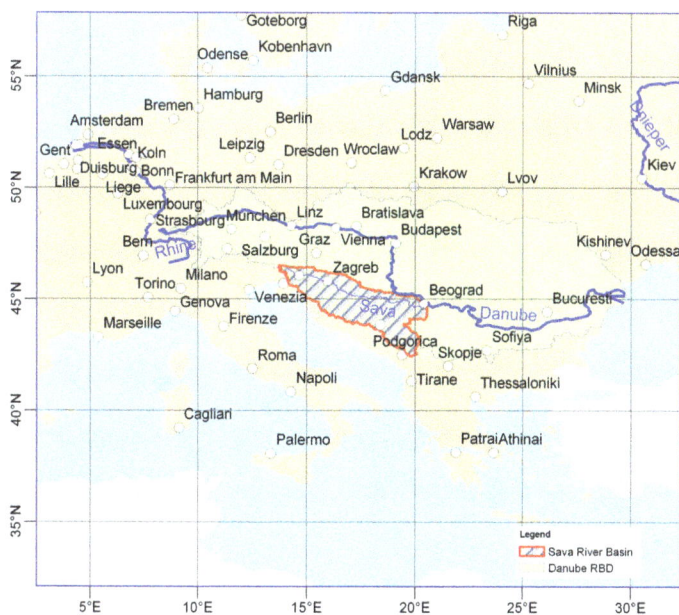

Fig. 1. Location of the Sava river basin (ISRBC, 2009e)

|  | SI | HR | BA | RS | ME | AL |
|---|---|---|---|---|---|---|
| Total country area [km²] | 20,273 | 56,542 | 51,129 | 88,361 | 13,812 | 27,398 |
| Share of national territory in the basin [%] | 52.8 | 45.2 | 75.8 | 17.4 | 49.6 | 0.6 |
| Area of the country in the basin [km²] | 11,734.8 | 25,373.5 | 38,349.1 | 15,147.0 | 6,929.8 | 179.0 |
| Share of the basin [%] | 12.0 | 26.0 | 39.2 | 15.5 | 7.1 | 0.2 |

Notation: SI – Slovenia; HR – Croatia; BA – Bosnia and Herzegovina; RS – Serbia; ME – Montenegro; AL – Albania.

Table 1. Share of the Sava countries belonging to the Sava river basin (ISRBC, 2009d)

The Sava river basin is situated within a wide region where the moderate **climate** of the northern hemisphere prevails. The average annual **air temperature** for the whole basin is 9.5°C. Mean monthly temperature in January falls to about -1.5°C, while in July it can reach almost 20°C.

**Precipitation** amount and its annual distribution are very variable within the basin (Fig. 2), while the basin average is about 1,100 mm/year. Spatial distribution of **unit-area-runoff** largely follows the pattern of precipitation spatial distribution. It varies from 150 mm/year (below 5 l/s/km²) up to 1,200 mm/year (almost 40 l/s/km²), as shown in Fig. 3. Spatial distribution of **evapotranspiration** is heterogeneous, too (Fig. 4), with the basin average of about 530 mm/year. The long-term average discharge of the Sava river at the mouth is about 1,700 m³/s, which is equivalent to effective rainfall of about 570 mm/year, and to the unit-area-runoff for the whole basin of about 18 l/s/km².

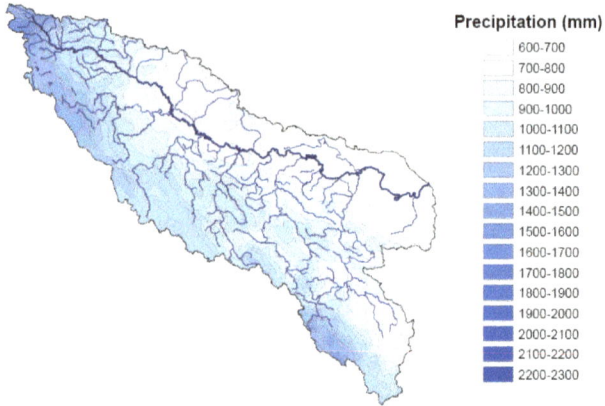

Fig. 2. Mean annual precipitation in the Sava river basin (UNESCO, 2006)

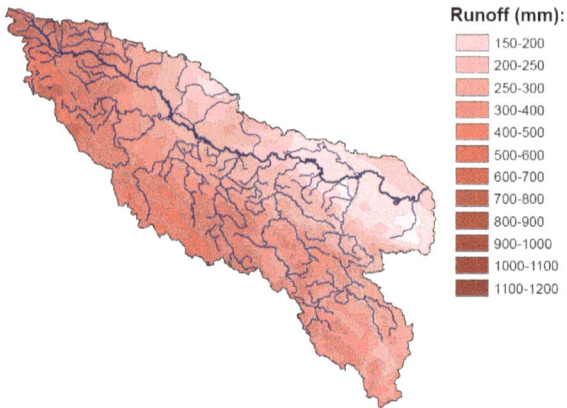

Fig. 3. Mean annual runoff in the Sava river basin (UNESCO, 2006)

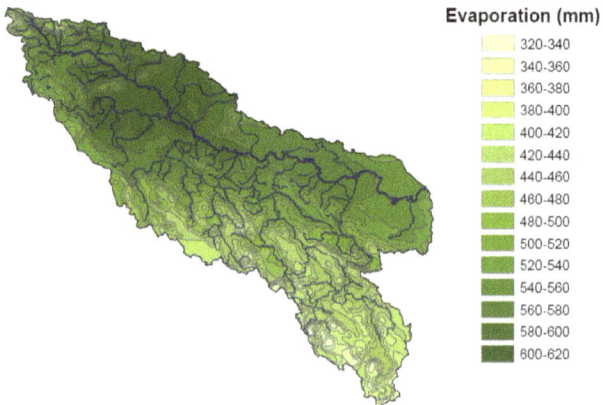

Fig. 4. Mean annual evapotranspiration in the Sava river basin (UNESCO, 2006)

The Sava river is formed by two mountainous streams: Sava Dolinka and Sava Bohinjka. From their confluence to its mouth to the Danube in Belgrade (Serbia), the Sava river is 945 km long, thus being the third longest tributary of the Danube. Together with its longer headwater, the Sava Dolinka river (Fig. 5), it measures 990 km. With its **average discharge** at the confluence of about 1,700 m³/s (Fig. 5), the Sava river represents the richest-in-water Danube tributary, contributing with almost 25% to the Danube's total discharge. The longitudinal presentation of annual discharges along the Sava river is given in Fig. 6.

Fig. 5. Source and mouth of the Sava river (Left photo: "Zelenci", the Sava Dolinka source, Author: Milan Vogrin. Right photo: "Mouth", Author: Vlada Marinković. Credit: ISRBC)

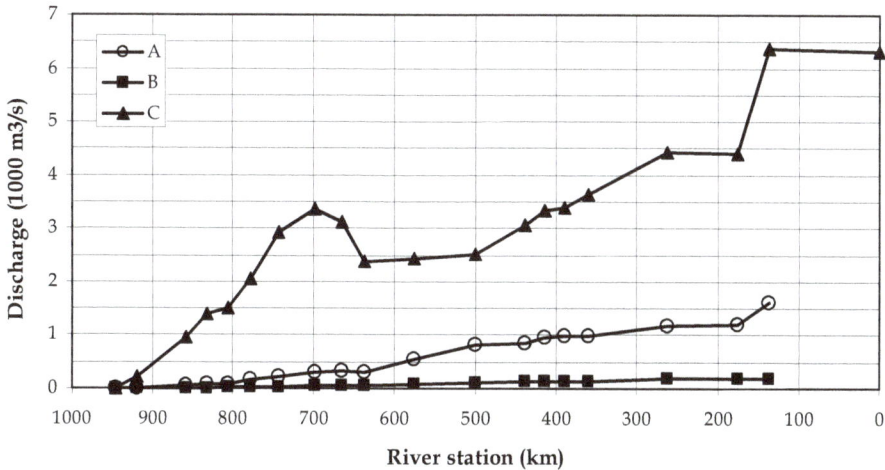

Fig. 6. Annual discharges along the Sava river: A – mean values; B – 100-year return period low flows; C – 100-year return period high flows (ISRBC, 2009d). Note: the river station is measured in the upstream direction (the zero station corresponds to the river mouth)

## 2.2 Environmental and socio-economic values of the basin

The Sava river basin is widely known for its high environmental and socio-economic values, associated not only with a natural beauty, an outstanding biological and landscape diversity (Fig. 7) and large retention areas along the river, but also with a high potential for development of economic activities, such as the waterway transport of cargo and passengers, hydropower generation, tourism and recreation, as well as other activities related to the use of water.

Fig. 7. Tributaries of the Sava river, Una and Drina (Left photo: "Una – Martin Brod", right photo: "Mouth of the Drina River". Author: Miroslav Jeremić. Credit: ISRBC)

The basin hosts the largest complex of alluvial wetlands in the Danube basin and large lowland forest complexes, being a unique example of a river basin with some of the floodplains still intact, thus supporting biodiversity and flood alleviation (Fig. 8). For illustration, the drop of the 100-year high flow, shown in Fig. 6, happens between the river stations 700 km and 640 km, which correspond exactly to the location of Lonjsko polje, being associated with hydraulic effects of this retention area. There are 167 protected areas in total, including six Ramsar sites, eight national parks, as well as numerous important bird and plant areas, protected areas at the national level, and Natura 2000 sites.

Fig. 8. Lonjsko polje – a nature park and retention area (Left photo: "Lonjsko polje – Kratečko", right photo: "Lonjsko polje – at *gmajna*". Author: Boris Krstinić. Credit: ISRBC)

The total annual **water use** in the basin is estimated at about 4.8 billion $m^3$. The overview of various types of the consumptive water uses is shown in Fig. 9. The non-consumptive uses include transportation, hydropower generation, recreation and fishing. The Sava river contributes to the Danube inland waterway transport network with 594 km of the waterway (Fig. 10), from Belgrade to Sisak (Croatia), and provides numerous sites and opportunities for different kinds of tourism and recreation along the whole watercourse of the Sava river (ISRBC, 2011c), as well as on the tributaries (Fig. 11).

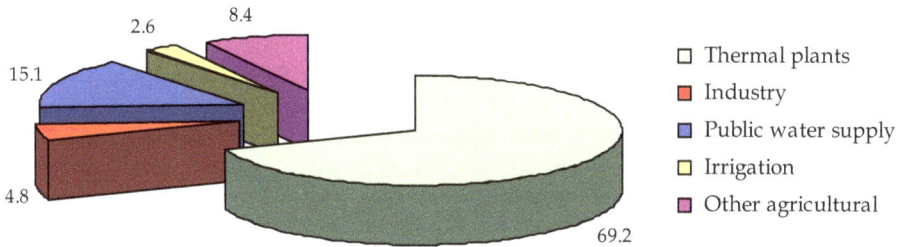

Fig. 9. Estimation of total water use in the Sava river basin (ISRBC, 2009d)

Fig. 10. Navigation on the Sava river (Left photo: Dragan M. Babović. Credit: ISRBC)

Fig. 11. Rafting on the Vrbas river, a Sava tributary (ISRBC, 2009d)

## 3. Legal and institutional framework for cooperation

After the political changes in the region in the early 1990-ies, the Sava river, which was the largest national river in the former country, has become an international river of a recognized importance. The establishment of the Stability Pact for South-Eastern Europe in 1999 provided a solid basis for triggering the cooperation of stakeholders in the region and, gradually, the creation of a new approach to water resources management in the Sava river basin. On these grounds, the four countries of the basin – Bosnia and Herzegovina, Federal Republic of Yugoslavia (later on Serbia & Montenegro, and then Republic of Serbia), Republic of Croatia and Republic of Slovenia, entered into a process of negotiations, with the primary aim to establish an appropriate framework for transboundary cooperation in the water sector, and thus foster sustainable development of the region.

As a key milestone of the process, the *Framework Agreement on the Sava River Basin* (*FASRB*) has been developed, as a unique international agreement integrating all aspects of water resources management and establishing the International Sava River Basin Commission (ISRBC) for implementation of the *FASRB*, with the legal status of an international organization. After signing the *FASRB* in December 2002, at Kranjska Gora (Slovenia), the Interim Sava Commission was formed to prepare all steps necessary for establishment of the permanent commission upon entry of the *FASRB* into force. Following the ratification of the *FASRB* by all Parties, and its entry into force in December 2004, the constitutional session of the ISRBC was held in June 2005, and subsequently, the permanent Secretariat of the ISRBC started to work in January 2006, with the seat in Zagreb (Croatia). Since then, the ISRBC has been a leader of cooperation of the Parties toward the *FASRB* implementation, the status of which is summarized in part 4.1.

### 3.1 *Framework Agreement on the Sava River Basin (FASRB)*

The *FASRB* is an international agreement that integrates all aspects of water resources management – different kinds of water use, the water and aquatic ecosystem protection, as well as the protection against harmful effects of water due to extreme hydrologic events and accidents involving the water pollution.

The overall objective of the FASRB is to establish and maintain the transboundary cooperation in order to provide conditions for sustainable development of the region within the Sava river basin. The particular objectives of the *FASRB* (*FASRB*, 2002) include the establishment of:

- international regime of navigation on the Sava river and its navigable tributaries;
- sustainable water management in the basin, and
- sustainable management of hazards in the basin (i.e. floods, droughts, ice, accidents).

The basic principles stipulated by the *FASRB* include:

- cooperation based on sovereign equality, territorial integrity, mutual benefit and good faith in order to achieve the goals of the *FASRB*, as well as based on regular exchange of information within the basin, cooperation with international organizations, and being in accordance with the *EU Water Framework Directive* (*WFD*) and other EU directives and UNECE conventions, and
- reasonable and equitable use of the water resources, applying measures aimed at securing the integrity of the water regime in the basin and reduction of transboundary impacts caused by economic and other activities of the Parties, and respecting the "no harm rule".

The *FASRB* implementation is being undertaken by the national institutions, officially nominated by the Parties, and is coordinated by the ISRBC.

The *FASRB* presents the first development-oriented multilateral agreement in the post-conflict period, concluded in the region of the former Yugoslavia after the *Dayton Peace Agreement* and the *Agreement on Succession*. By involving the whole water resources management and addressing both sustainability and development issues, the *FASRB* provides the ISRBC with the broadest scope of work among European basin organizations (i.e. river and lake commissions).

## 3.2 International Sava River Basin Commission (ISRBC)

The ISRBC is a joint institution established as an international organization with the international legal capacity necessary for exercising its functions.

In order to achieve the main goals of the *FASRB*, the following activities are coordinated by the ISRBC:

- preparation and implementation of joint or integrated plans for the basin (e.g. river basin management plan, flood risk management plan);
- preparation of development programs, e.g. for rehabilitation and development of navigation in the basin;
- establishment of integrated systems for the basin, such as geographical information system (GIS), river information services (RIS), flood forecasting and warning system, etc.;
- harmonization of national regulation with the EU regulation, and
- development of protocols for regulating specific aspects of the *FASRB* implementation.

In accordance with the mandate and responsibilities, the ISRBC is a central point in identification and implementation of projects of regional importance, aiming to strengthen the cooperation of the Sava countries and facilitate the fulfilment of the *FASRB* objectives.

The ISRBC is given the capacity for making decisions in the field of navigation (obligatory for the Parties) and providing recommendations on all other issues. Additionally, the ISRBC provides recommendations to the Meeting of the Parties, a ministerial-level body which makes decisions relating to strategic issues of the *FASRB* implementation and performs a general monitoring of the implementation process.

The ISRBC is composed of two representatives of each Party to the *FASRB*, one member and one deputy member of each Party, having one vote in the Commission. The Commission has a chairman who represents the ISRBC. The Secretariat is an administrative and executive body of the ISRBC.

In order to foster cooperation and ensure synergy in achieving its goals, the ISRBC has established permanent and *ad-hoc* expert groups, composed of delegated experts from each Party. There are four permanent expert groups, covering the key issues in the basin – river basin management, accident prevention and control, flood prevention, and navigation, as well as five *ad-hoc* expert groups, dealing with specific issues and tasks – legal issues, financial issues, hydro-meteorological issues, GIS and RIS.

## 4. Approach to sustainable development of the Sava river basin

Given the broad scope of the *FASRB*, the achievement of its principal objectives requires an integrated and sustainable approach, balancing the needs for development of economic activities such as navigation or tourism, against the needs of other water sub-sectors (i.e. other kinds of water use, protection against detrimental effects of water, and protection of water and aquatic ecosystem). The main features of the approach, as applied by the ISRBC, are illustrated in the following text, by reviewing the achievements and results, and presenting a vision of the future implementation of the *FASRB*.

## 4.1 Current status of the *FASRB* implementation

This part provides a brief summary of the achievements made, not only in the fields of navigation, other water uses, water protection and hazard management, but also with regard to „cross-cutting issues" (i.e. information management, hydrological and meteorological issues, cooperation, public participation and stakeholder involvement), which are dealt with in order to provide an overall support to the implementation process. Further information can be found elsewhere (Komatina and Zlatić-Jugović, 2010; Komatina, 2011a), or at the ISRBC web-site, www.savacommission.org, where majority of the documents, mentioned throughout the text, are available.

### 4.1.1 Navigation and other water uses

Following the economic decline in the former Yugoslavia in the 1980-ies, the armed conflict in the early 1990-ies has caused an additional decrease of transport and navigation on the Sava river, reducing the cargo traffic, which was around 10 million tons in 1982, and 5.7 million tons in 1990, to less than 1 million tons. A lack of investments into the waterway maintenance and infrastructure development resulted in unfavourable navigation conditions, characterized by a limited draft during long periods, a limited fairway width and a limited height for passages under some bridges, as well as insufficient marking. Navigability of the waterway, which used to be a class IV waterway in the past, was reduced to class III at many sections of the river. Given such an initial situation, ratification of the *FASRB* and establishment of the ISRBC provided a good basis for rehabilitation and development of navigation on the Sava river, which was further strengthened by simultaneous ratification of the *Protocol on Navigation Regime to the FASRB* (2004).

Since the beginning of the *FASRB* implementation, considerable efforts have been invested by the ISRBC and the Parties to provide conditions necessary for the Sava river to become an important, environment-friendly and navigation-safe lifeline for inland transport (ISRBC, 2009b; Komatina, 2011b). The undertaken activities have been focused on two major issues: (a) planning for rehabilitation and development of the Sava river waterway infrastructure, and (b) improvement of technical standards and safety of navigation, with the aim to prevent the environmental risks associated with navigation.

With regard to **rehabilitation and development of the waterway infrastructure**, the preliminary documentation has been developed, and future steps have been agreed upon by the Parties (ISRBC, 2011a, 2011b). Based on the assessment of transport demand (Fig. 12) and economic analyses, the upgrade of the whole waterway to class Va was shown to be feasible, the costs being 10% higher than the costs of rehabilitation of the whole waterway to class IV (ISRBC, 2008a). Nevertheless, in order to minimize negative environmental impacts of the project, the ISRBC has decided to develop the waterway to class Va only at 40% of the total length (section Belgrade – Brčko), while the rest part will be rehabilitated to class IV. For the same reason, no change of the present watercourse (i.e. no straightening of the river) has been planned, so that, in sharp bends, only one-way navigation is foreseen. Additionally, several other activities have been performed, including a full restoration of the waterway marking system after 20 years, removal of unexploded ordnances from the river banks, and the initial phase of establishment of the Sava RIS in accordance with the *EU RIS Directive*.

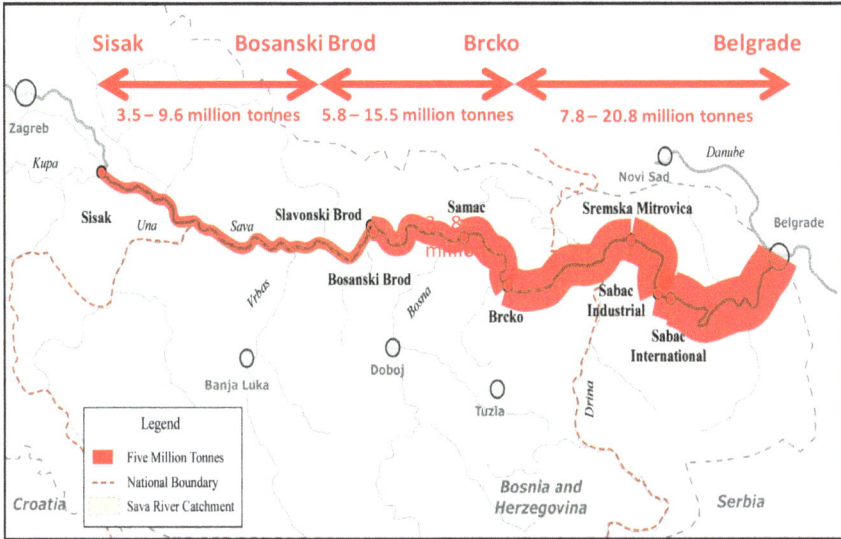

Fig. 12. Estimated margins of traffic volume on the Sava river for year 2027 (ISRBC, 2008a)

The **administrative and legal framework** has been strengthened by development of a set of rules and other documents related to technical issues and safety of navigation, harmonized with the corresponding EU and UNECE regulations (ISRBC, 2009b, 2009c). The *Protocol on Prevention of Water Pollution caused by Navigation to the FASRB* (ISRBC, 2009a) has been developed and signed, and is currently undergoing ratification. The *Protocol on Sediment Management to the FASRB*, aiming to regulate, *inter alia*, the sand and gravel exploitation in accordance with the *Sava River Basin Management Plan* (*Sava RBM Plan*), has entered the process of final harmonization by the Parties.

In order to ensure **environmental sustainability**, the issue of navigation development is considered as an integral part of the *Sava RBM Plan*, which is being developed in accordance with the *EU WFD*. A considerable attention has been paid, as well, to the improvement of technical standards and safety of navigation, through implementation of concrete projects (restoration of the waterway marking system, development of the Sava RIS) and strengthening of the administrative and legal framework, including the protocols to the *FASRB*, fully in line with the corresponding EU and UNECE regulations. Finally, the waterway planning has been based on a clear intention to minimize negative environmental impacts of the rehabilitation works, and accompanied with an active involvement of the ISRBC in the relevant processes on Danube and European levels (International Commission for the Protection of the Danube River [ICPDR] et al., 2008; *Manual on Good Practices in Sustainable Waterway Planning*, 2010).

Partly as a consequence of the above mentioned achievements, there are already several indicators of development in traffic and opening of new cargo flows on the Sava river, such as opening of transport of oil products from (Bosanski) Brod, new developments in Serbian ports (Sremska Mitrovica, Šabac), as well as the first passenger cruise along the whole Sava river waterway after 150 years.

In addition to navigation, efforts have also been made to develop **other economic activities** that can benefit from the use of river infrastructure. Being aware of great potentials for development of tourism in the basin in an environmentally friendly manner, the first *Nautical and Tourist Guide of the Sava River* has been developed in cooperation with regional chambers of commerce of the Parties (ISRBC, 2011c), while the preparation of a master plan for development of nautical tourism in the basin is planned to be undertaken as the next step. The preparation of another project, focusing on the contribution of small and medium enterprises to sustainable development of the Sava river basin, which has recently been initiated, targets not only river transport and tourism, but also other economic activities, including fish farming and shipbuilding.

### 4.1.2 Water protection and hazard management

The key activity in the field of **river basin management** has been the preparation of the first *Sava RBM Plan* in accordance with the *EU WFD*, given the commitment of the Parties to respect the *WFD*, although some of them (i.e. the non-EU member states) are not legally bound to do so. An important step in this regard was the development of the *Sava River Basin Analysis Report* (ISRBC, 2009d), a comprehensive document dealing with both water quality and quantity issues, hydrology and hydromorphology of the basin, and providing the first overview and thematic GIS maps of the basin (ISRBC, 2009e). To ensure an integrated approach from the very beginning of the *RBM Plan* preparation process, the Sava River Basin Analysis also included consideration of the flood management and navigation development issues. Further preparation of the first *Sava RBM Plan*, supported by the European Commission, is in progress. Following the drafting of the *Plan* in fall 2011, and the subsequent public consultation process, the *Sava RBM Plan* is expected to be finalized and adopted by the Parties in 2012.

As regional climate modelling suggests an overall reduction of around 15% to 30% in mean annual runoff in the Sava river basin by the middle of this century, which could be challenging for all investments made in the basin, the development of a *Climate Adaptation Plan for the Sava River Basin* has been undertaken by the World Bank. The main aim of this effort is to fill the knowledge gap on the climate change impact on the water sector in the basin and to show how to increase the climate resilience of critical water management infrastructure investments and of integrated water resource management in the region, by elaborating alternatives for adaptive management actions in water management sub-sectors, including navigation, hydropower, agricultural water use, flood protection and environmental protection.

In addition to these activities, the *Protocol on Sediment Management to the FASRB*, stipulating the preparation of a sediment management plan for the basin in accordance with the *Sava RBM Plan*, has been prepared and is undergoing final harmonization by the Parties, while the *Protocol on transboundary impact to the FASRB* is under development on the ISRBC level.

For the purpose of an efficient **accident prevention and control** in the Sava river basin, participation in testing of the existing Accident Emergency Warning System of the ICPDR is continuously being done, and efforts are being made to improve the work of the Principal International Alert Centres in the Parties to the *FASRB*, including the organization of training courses for the operational staff of the Alert Centres, in cooperation with the

ICPDR. The *Protocol on Emergency Situations to the FASRB*, aiming to enhance prevention, preparedness, response and mutual assistance of the Parties in case of emergency situations, has been drafted and entered the process of harmonization by the Parties. As an important future activity, development of a water contingency management plan for the basin is planned.

In the field of **flood management**, the *Flood Action Plan* for the Sava river basin (ICPDR & ISRBC, 2009) has been prepared in accordance with the *Flood Action Programme for the Danube River Basin* of the ICPDR, providing the first programme of measures for each Party to achieve the defined targets for flood management in its part of the Sava basin until 2015.

The *Protocol on Flood Protection to the FASRB* (ISRBC, 2010a), which aims to provide the legal basis for cooperation of the Parties in line with the *EU Flood Directive*, including the preparation of the *Flood Risk Management Plan* for the Sava river basin, has been developed and signed, and is currently under ratification. A number of preparatory activities toward the *Flood Risk Management Plan* have been performed so far, including an assessment of current flood management practices in the Parties, establishment of a database on the existing flood protection facilities, preparation of a GIS-based, indicative flood extent map for the whole Sava river (Fig. 13), development of a preliminary hydrological model of the Sava river basin and the hydraulic model of the Sava river, and launching an UNECE-supported project, aiming to assist linking the flood risk management planning and the climate change assessment in the basin.

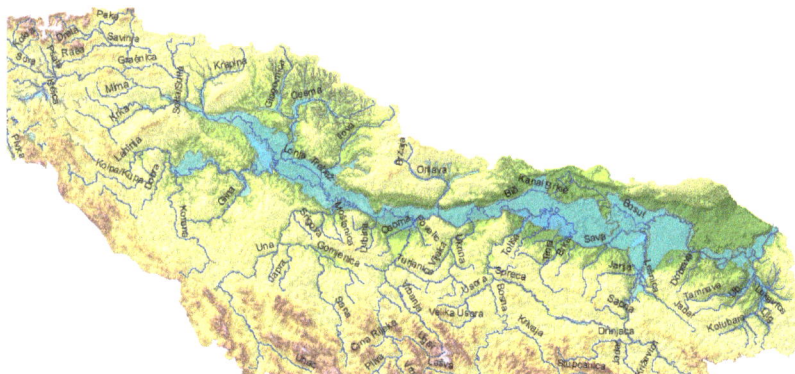

Fig. 13. Indicative map of important flood prone areas along the Sava river (ISRBC, 2009d)

### 4.1.3 Cross-cutting issues

In the field of **information management**, the *Sava GIS Strategy* (ISRBC, 2008b) has been developed, taking into account the *EU INSPIRE Directive* and the Water Information System for Europe. Subsequently, the implementing documents for the Sava GIS establishment have been prepared, the funding for the initial phase of the GIS establishment secured, and the initial phase launched.

As for the **hydrological and meteorological issues**, advances in the exchange of hydro-meteorological information and data within the basin have been made, including a revival of the *Hydrological Yearbook of the Sava River Basin* (ISRBC, 2010b) after more than 20 years.

Preliminary agreements have been made upon basic elements of a system for the exchange of hydrological and meteorological information and data within the basin. Preparatory activities have been undertaken toward the implementation of two important projects, namely a new *Hydrological Study for the Sava River Basin* (to be the first study for the whole basin since 1976), and the development and upgrade of the hydro-meteorological information and flood forecasting and warning system for the basin.

**Cooperation** of the ISRBC with a large number of international organizations and institutions has been established and maintained, with a special emphasis on those specified in the *FASRB*. The basis for cooperation with the ICPDR and Danube Commission has been strengthened by signing memoranda of understanding on cooperation with each of the two commissions. The support of the European Commission to the *FASRB*-related projects is becoming steady and their recognition of several priority projects of the ISRBC in the context of the *EU Strategy for the Danube Region* indicates a good will for a continued support. A good cooperation with the UNECE and their support to the projects of the ISRBC should be mentioned, as well. Cooperation with the institutions of the Parties responsible for the *FASRB* implementation has been established and maintained, as well as with other national institutions, such as agencies, offices, services, institutes and universities.

In order to ensure **public participation and stakeholder involvement** in major activities related to the *FASRB* implementation, cooperation with non-governmental organizations and other institutions and local actors from the Sava river basin has been established, a network of observers to the ISRBC has been created, and a number of mechanisms for information and consultation of stakeholders and/or wide public have been established, including the official web-site, the *Sava NewsFlash* bulletin, publications and promotion material of the ISRBC, celebration of the Sava Day (June 1), press releases, press conferences and media briefings, organization of consultation workshops, public presentations and other meetings with stakeholders by the ISRBC, or participation in ceremonies, conferences and other events, and contributions to bulletins and web-sites of other organizations/ institutions.

A good example of stakeholder involvement is the process of development and implementation of the *Joint Statement on Guiding Principles for the Development of Inland Navigation and Environmental Protection in the Danube River Basin* (ICPDR et al., 2008), led jointly by the ICPDR, Danube Commission and the ISRBC, where the issue is continuously discussed by a variety of stakeholders from navigation and environmental sector.

## 4.2 Vision of the future implementation of the *FASRB*

Since the beginning of the *FASRB* implementation, a wide range of activities have been undertaken or launched, as summarized in the previous text. However, in order to respond to a steady progress in the *FASRB* implementation during the last years, as well as to recent processes and initiatives on the Danube level (ICPDR et al., 2008; ICPDR, 2009, 2010) and European level (EC, 2010a, 2010b), relevant for the *FASRB* implementation, an updated *Strategy on Implementation of the FASRB* (ISRBC, 2011a) and the accompanying *Action Plan for the Period 2011-2015* (ISRBC, 2011b), have been developed to govern the future implementation. This part briefly outlines specific objectives in each priority area of the *FASRB* implementation, as well as measures for achievement of these objectives, in accordance with the *Strategy*.

Given the overall goal of the *FASRB* in the field of **navigation** (i.e. the establishment of the international regime of navigation on the Sava river and its navigable tributaries), as well as interests of the Parties related to other water uses, the future efforts should be oriented to:

- further unification and upgrading of the administrative and legal framework with the aim to increase navigation safety and remove administrative obstacles for navigation;
- rehabilitation, development and proper maintenance of the Sava river waterway with the aim to increase commercial traffic and improve navigation safety;
- establishment of an efficient system for the vessel waste management with the aim to protect the water against pollution from vessels;
- creation of a positive image of the inland navigation in general and promotion of the Sava river as an important regional transport corridor;
- development of nautical tourism in the Sava river basin, and
- consideration of other development activities in the basin (e.g. hydropower generation, water supply, agriculture, recreation, tourism), accompanied with careful analysis of their environmental sustainability, taking also climate change impacts into account.

Keeping in mind the overall goals of the *FASRB* in the fields of **water protection and hazard management**, i.e. the establishment of a sustainable water and hazard management in the Sava river basin, the future activities should be focused on:

- further efforts toward the achievement of the environmental objectives of the *EU WFD* in the Sava river basin, including the preparation and implementation of the *RBM Plans* and the accompanying *Programmes of Measures* for the Sava river basin, integration of water policy with other policies (i.e. navigation, climate change, hydropower generation, flood risk management), additional elaboration of the issues potentially important in future (sediment, water demand, groundwater quantity, etc.), as well as initiation of activities toward the establishment of sustainable management of sediment in the basin;
- establishment of a sustainable and efficient transnational system for management of accidental water pollution in the Sava river basin, through further improvement of functioning of the existing emergency warning system in the Parties and through development of a water pollution contingency management plan for the basin;
- establishment of a system for sustainable management of floods in the Sava river basin, by development of a common *Sava Flood Risk Management Plan* in accordance with the *EU Flood Directive* and in line with the *UNECE Water Convention*, as well as by adaptation of flood management to climate change.

In order to support the *FASRB* implementation and the achievement of its main objectives, the future activities related to the **cross-cutting issues**, should target:

- establishment of an integrated information system for the Sava river basin by development of the Sava GIS;
- strengthening of the platform for exchange and use of hydrological and meteorological information, including harmonization of national methodologies (e.g. related to data analysis), as well as development of the hydro-meteorological information and flood forecasting and warning system for the basin;
- facilitating the *FASRB* implementation related to navigation and other relevant economic issues by using statistical methods and techniques as tools, including

development of a system for data collection, processing and analysis in line with the Eurostat, regular data collection and processing, and dissemination of the data to relevant stakeholders;

- strengthening of the public participation process for the purposes of preparation and implementation of the *Sava RBM Plan*, targeting primarily the key stakeholders, i.e. the main water users, in the Sava river basin, and

- further improvement and broadening of stakeholder involvement in the *FASRB* implementation process, i.e. seeking a synergy of a top-down and a bottom-up approach, by exploring possibilities and elaborating options for the establishment of a multi-stakeholder platform that would facilitate, or further strengthen, the involvement of the civil, academic and business sectors, in addition to the existing involvement of stakeholders from the governmental and non-governmental sectors.

Accordingly, the main prerequisites for an effective further implementation of the *FASRB* include:

- further raising of awareness of benefits and the importance of the existing cooperation of the Parties in the framework of the *FASRB* implementation, not only in the institutions responsible for the implementation, but also in other national institutions;

- securing adequate human and financial resources in the Parties to follow up the activities coordinated by the ISRBC;

- providing adequate financial instruments for realization of respective activities and projects, especially those to be performed under the umbrella of the ISRBC;

- facilitating a free access to basic data needed for preparation of the studies coordinated by the ISRBC, with the special focus on the data owned by national institutions not officially nominated as responsible for implementation of the *FASRB*, and

- further developing the legal background and institutional arrangements in the Parties.

Additional strengthening of the capacity within the ISRBC framework, through a stronger support of the Parties to the members of the ISRBC expert groups in performing their obligations, as well as through further strengthening of the capacity of the ISRBC Secretariat, would certainly be of additional benefit for the *FASRB* implementation in future.

## 5. Conclusion

The so-far experience in the *FASRB* implementation indicates a number of advantages of the presented approach, showing that the approach is:

- cohesive, by providing conditions for the cooperation of the countries after a conflict, the implementation of joint, basin-wide projects, as well as the harmonization of national regulation, methodologies and procedures;

- integrated, not only in terms of the geographical scope (covering the whole basin and the related ecosystem), but also in terms of the scope of work (dealing with both sustainability and development aspects);

- transparent, as it is based on a number of public participation and stakeholder involvement activities;

- aligned with relevant EU and UNECE regulation;

- sub-regional, offering a "finer resolution" of results, that are complementary to those obtained on a regional (Danube) scale;

- pragmatic and practical, providing concrete "products" to the Parties, such as joint plans, development programs, protocols, harmonized regulation, integrated systems for the whole basin, etc.

The *FASRB* has proven to be a good platform for intensified contacts and an improved cooperation among the Parties, providing opportunities for exchange of experiences and an additional training of the experts from the region. It also provides for an improved inter-sectoral cooperation, especially among the competent authorities, within each of the Parties.

However, a number of challenges and (existing or potential) obstacles for the *FASRB* implementation have been identified. These are, generally, associated with:

- differences between the countries (i.e. EU membership status, eligibility for approaching funds, level of economic development, organizational structure in decision-making process, environmental awareness of the public);
- financing of major activities (e.g. priority projects, strategic studies, establishment of integrated systems for the basin), and
- resolving conflicts of interests of different users of water (on both transboundary and national levels), especially as they are likely to increase in future due to climate change.

Major obstacles and difficulties in the *FASRB* implementation are associated with a lack of human and financial resources of the Parties, as well as securing funds for implementation of the priority projects. The additional challenge is a limited access to basic data (topographic, hydrologic, etc.), needed for preparation of studies of common interest under the umbrella of the ISRBC, especially when the data are owned by national institutions not officially nominated as responsible for implementation of the *FASRB*.

Some challenges are associated with specific fields of the *FASRB* implementation. For example, on national level, the inland navigation is, although being the most efficient and environmentally-friendly mode of transport, generally underestimated in comparison with other modes of transport. Or, progress in the fields of water protection and hazard management, where requirements toward the Parties are based on recommendations and conclusions of the ISRBC (unlike the ISRBC decisions in the field of navigation, having a binding character for the Parties), is partly affected by a different perception of the requirements by the competent authorities of the Parties.

In some Parties, additional obstacles include lack of appropriate institutional arrangements and lack of harmonization of the legislation with the EU *acquis*. There is also a space for improvement of bilateral cooperation, where the ISRBC is perceived as a possible mediator.

The presented approach is considered as relevant to the processes on a wider (Danube and EU) scale, such as those associated with the *EU Strategy for the Danube Region (EUSDR)* and the *EU 2020 Strategy*, for several reasons:

- the overall objective of the *EUSDR* and *FASRB* is identical – sustainable development of the region they refer to;
- there is an obvious conformity of the ISRBC approach and its priority projects with the *EUSDR* priorities, and a high potential for synergy, as the implementation of the ISRBC projects within the *EUSDR* framework can contribute to implementation of both *EUSDR* and *FASRB*;

-   the sub-regional level, such as the Sava river basin level, is likely to be the most effective level from the viewpoint of the *EUSDR* implementation;
-   a majority of the ongoing activities of the ISRBC fully match the three main priorities of the *EU 2020 Strategy*, i.e. sustainable growth, smart growth and inclusive growth.

Given the existing interest of other regions (i.e. other parts of the South-Eastern Europe, Mediterranean region, Western Europe, Central Asia) in the Sava model of cooperation, the ISRBC approach seems to be an attractive example of good practice. In this context, a fundamental advantage of the *FASRB* is associated with the creation of a platform for transboundary cooperation, which is sufficiently broad to integrate all aspects of water management, and thus provide opportunities for specific interests of all Parties to be satisfied through the cooperation. It should be kept in mind, however, that providing the coordinating body with a twofold legal capacity (e.g. decisions in the field of navigation vs. recommendations in the fields of water protection and hazard management, as in the case of ISRBC) may, in some situations, challenge the efforts to find a right balance in satisfying the interests of all Parties.

Despite of all challenges, the *FASRB* is considered as a solid basis for the integrated water resources management in the Sava river basin. Although rather demanding in terms of the need for resources and continuous, joint efforts of the Parties, the *FASRB* implementation is perceived as a process providing multiple benefits, and making a steady progress toward the key objective – sustainable development of the region within the basin.

## 6. Acronyms and abbreviations

| | |
|---|---|
| AL | Albania |
| BA | Bosnia and Herzegovina |
| EC | European Commission |
| EU | European Union |
| Eurostat | EU Statistical Office |
| *EUSDR* | *EU Strategy for the Danube Region* |
| *FASRB* | *Framework Agreement on the Sava River Basin* |
| GIS | Geographical Information System |
| HR | Croatia |
| ICPDR | International Commission for the Protection of the Danube River |
| INSPIRE | Infrastructure for Spatial Information in the European Community |
| ISRBC | International Sava River Basin Commission |
| IWRM | Integrated Water Resources Management |
| ME | Montenegro |
| RBD | River Basin District |
| RBM | River Basin Management |
| RIS | River Information Services |
| RS | Serbia |
| SI | Slovenia |
| UNECE | United Nations Economic Commission for Europe |
| UNESCO | United Nations Educational, Scientific and Cultural Organization |
| *WFD* | *Water Framework Directive* |

# 7. References

Biswas, A.K. (2004). Integrated Water Resources Management: A Reassessment, A Water Forum Contribution, IWRA, *Water International*, Vol. 29, No. 2, June 2004, 248-256

EC. (2006). *NAIADES – An Integrated European Action Programme for Inland Waterway Transport*, Available from:
http://ec.europa.eu/transport/inland/promotion/promotion_en.htm

EC. (2010a). *Europe 2020 – A European Strategy for Smart, Sustainable and Inclusive Growth*, Available from: http://europa.eu/press_room/pdf/complet_en_barroso___007_-_europe_2020_-_en_version.pdf

EC. (2010b). *EU Strategy for the Danube Region*, Available from:
http://ec.europa.eu/regional_policy/cooperation/danube/index_en.htm

*Framework Agreement on the Sava River Basin*, 2002, Available from:
http://www.savacommission.org/basic_docs

Global Water Partnership. (2000). Integrated Water Resources Management, *Technical Advisory Committee Background Paper No. 4*, Global Water Partnership, Stockholm

Global Water Partnership & International Network of Basin Organizations. (2009). *A Handbook for Integrated Water Resources Management in Basins*, Global Water Partnership & International Network of Basin Organizations, ISBN: 978-91-85321-72-8, Stockholm

ICPDR. (2009). *Danube River Basin Management Plan*, Available from:
http://www.icpdr.org/icpdr-pages/danube_rbm_plan_ready.htm

ICPDR. (2010). *Danube Basin: Shared waters – Joint responsibilities*, Danube Declaration adopted by the Ministers of Environment on 16 February 2010, Available from:
http://www.icpdr.org/icpdr-pages/mm2010.htm

ICPDR, Danube Commission & ISRBC. (2008). *Joint Statement on Guiding Principles for the Development of Inland Navigation and Environmental Protection in the Danube River Basin*, Available from http://www.savacommission.org/basic_docs

ICPDR & ISRBC. (2009). *Flood Action Plan - Sava River Basin*, Available from:
http://www.savacommission.org/publication

ISRBC. (2008a). *Feasibility Study for Rehabilitation and Development of Navigation and Transport on Sava River Waterway*, Available from http://www.savacommission.org/project

ISRBC. (2008b). *Sava GIS Strategy*, Available from:
http://www.savacommission.org/basic_docs

ISRBC. (2009a). *Protocol on Prevention of Water Pollution caused by Navigation to the FASRB*, Available from http://www.savacommission.org/basic_docs

ISRBC. (2009b). *Rehabilitation and Development of Navigation in the Sava River Basin*, ISRBC, Zagreb

ISRBC. (2009c). Set of rules related to safety and technical issues of navigation, Available from http://www.savacommission.org/decision

ISRBC. (2009d). *Sava River Basin Analysis Report*, Available from:
http://www.savacommission.org/publication

ISRBC. (2009e). *Sava River Basin Overview Map*, Available from:
http://www.savacommission.org/publication

ISRBC. (2010a). *Protocol on Flood Protection to the FASRB*, Available from:
http://www.savacommission.org/basic_docs

ISRBC. (2010b). *The 2006 Hydrological Yearbook of the Sava River Basin*, Available from:
    http://www.savacommission.org/publication
ISRBC. (2011a). *Strategy on Implementation of the FASRB*, Available from:
    http://www.savacommission.org/basic_docs
ISRBC. (2011b). *Action Plan for the Period 2011-2015*, Available from:
    http://www.savacommission.org/basic_docs
ISRBC. (2011c). *Nautical and Tourist Guide of the Sava River*, Available from:
    http://www.savacommission.org/publication
Komatina, D. (2011a). The Framework Agreement on the Sava River Basin – a basis for
    sustainable development of the region, submitted for publication in *Danube News*
Komatina, D. (2011b). Development of navigation on the Sava River – an integrated
    approach, submitted for publication in *Danube News*
Komatina, D. & Zlatić-Jugović, J. (2010). Transboundary cooperation in the Sava River Basin
    in the field of water management. *Hrvatske vode*, Water Management Journal, Vol.
    18, No. 73, (September 2010), pp. 249-258, ISSN 1330-1144 (in Croatian)
*Manual on Good Practices in Sustainable Waterway Planning*, 2010, Available from:
    http://transfer.message.at/_ZJyVSK80VdDitR
*Protocol on Navigation Regime to the FASRB*, 2004, Available from:
    http://www.savacommission.org/basic_docs
Schmeier, S. (2010). *The Organizational Structure of River Basin Organizations – Lessons Learned
    and Recommendations for the Mekong River Commission (MRC)*, Technical Background
    Paper, Available from http://www.mrcmekong.org/download/free_download/
    MRC-Technical-Paper-Org-Structure-of-RBOs.pdf
*The Dublin Statement on Water and Sustainable Development*, 1992, Available from:
    http://www.wmo.ch/web/homs/documents/english/icwedece.html
UNECE. (1992). *Convention on the Protection and Use of Transboundary Watercourses and Inter-
    national Lakes (UNECE Water Convention)*, Available from:
    http://live.unece.org/env/water/text/text.html
UNESCO. (2006). *The Danube and its Basin – Hydrological Monograph*, Follow-up Volume VIII,
    *Basin-Wide Water Balance in the Danube River Basin*, Regional Cooperation of the
    Danube Countries in the Frame of the International Hydrological Programme of
    UNESCO and Water Research Institute Bratislava, Slovakia

# 4

# Assessing Environmental and Social Dimensions of Water Issues Through Sustainability Indicators in Arid and Semiarid Zones

Enrique Troyo-Diéguez[1], Arturo Cruz-Falcón[1], Alejandra Nieto-Garibay[1],
Ignacio Orona-Castillo[2], Bernardo Murillo-Amador[1],
José Luis García-Hernández[2] and Alfredo Ortega-Rubio[1]

*[1]Programa de Agricultura en Zonas Áridas,
Centro de Investigaciones Biológicas del Noroeste,
S.C. (CIBNOR, S.C.). La Paz, Baja California Sur;
[2]Facultad de Agricultura y Zootecnia (FAZ),
Universidad Juárez del Estado de Durango,
Gómez Palacio, Durango;
México*

## 1. Introduction

All humans need basic things or products for their material and physical well-being: air, water, food, shelter and others, all over the World, but in arid zones the prevailing condition is water scarcity. Besides the huge necessity of water for human consumption, agriculture and livestock demand large volumes of water. But sustainability for these activities is a highly fragile and complicated concept, difficult to be reached. Biomass production of a cash crop grown under drought will be lower than that grown under optimal soil moisture conditions; therefore in practical terms, it is not possible to obtain immunity against the effects of drought. In arid and semiarid zones, the environmental and agricultural issues depend on the availability of water and on its use efficiency, which is affected by high temperatures and elevated evapotranspiration rates. As a production dedicated activity, agriculture has changed dramatically especially since the end of the World War II. Food and fiber productivity were improved due to new technologies, mechanization, increased chemical use, specialization and government policies that favored maximizing production. These changes allowed fewer farmers with reduced labor demands to produce the majority of the food and fiber. But in this framework, one of the highest costs is the water extraction.

The objectives of this paper are to present a review about a method to assess the environmental and social dimensions of water issues through sustainability indicators in arid and semiarid zones, and to discuss the challenges for sustainability in terrestrial and agricultural ecosystems in a Mexican arid zone.

## 2. Hydrologic aspects of arid and semiarid zones

Several basic conditions can be distinguished between arid regions and other zones from a hydrological point of view:

1.  A few, often intensive, rain events with a low amount of overall precipitation, which causes most of the small rivers or seasonal streams to be active only a few months or weeks every year, and sometimes only once every few years.
2.  Loss of soil fertility and development of "desert crust" on the land surface.
3.  Commonly, the aquifer is characterized by a thick aeration (vadose) zone, the zone between the water table and the land surface, resulting in a deep water table, and
4.  Exacerbation of salinity-related problems, in both, groundwater and soils. When these factors are misunderstood, the lack of development and mismanagement of water cause environmental degradation and desertification (Sharma, 1998).

### 2.1 Low soil water retention, poor fertility and formation of desert crust

Vast areas of bare soils, low annual precipitation, and few high intensity rainfall events with high kinetic energy characterize arid zones. The parent material and the degree of weathering determines the level and availability of nutrients, and the type of clay and sandy minerals; these in turn largely determine cation exchange, base saturation, the adsorption, and sometimes fixation of Phosphorus, texture, permeability, moisture retention and the stability of soil aggregates. As a result of its chemical and physical properties, a soil may harden on drying, form a surface crust that restricts the entry of water and air and inhibits seedling emergence, develop a subsurface layer or pan that restricts root growth, be susceptible to erosion, become acid or alkaline on drying, or have a subsurface horizon that hardens irreversibly on exposure. Positive correlation between soil moisture in every layer and precipitation, and negative correlation between soil moisture and air temperature are common patterns (Zhuguo *et al.*, 2000). Such relationships are gaining more relevance in global change-related studies. In this sense, Seneviratne et al. (2006) reported that the impact of atmospheric circulation modifications might be indirect, by imposing changes in the seasonal cycle of soil moisture, which in turn can lead to modified soil-moisture-temperature coupling characteristics, although years ago, Calvet et al. (1998) pointed out that further studies are needed to investigate the soil water content retrieval by simple surface schemes using estimations of the surface soil moisture and temperature.

Soil constraints to plant growth may be summarized as:

a.  Inadequate moisture
b.  Deficiencies/imbalances in nutrients
c.  Low cation exchange capacity
d.  Low base saturation
e.  Low water retention in some cases, but under specific conditions, water logging impeded drainage and poor aeration
f.  Poor fixation of phosphorus
g.  Low pH (acidity) with high pH (alkalinity)
h.  Salinity
i.  Impermeability
j.  Shallowness
k.  Textural problems (crusting, hardening, stoniness)

l.   Physical loss of soil,
m.  Hardening of subsurface horizon, and
n.   High temperature on the soil surface.

In arid soils, dryness and salinity, and their interaction under mismanaged irrigation, are the main causes of soil degradation. Effluent irrigation results in increased soil sodicity, because of the medium-to-high salinity and sodium concentration (Balks et al., 1998).

Bare soils exposed to rainfall are subjected to physical and chemical processes that change the hydraulic properties of the soil near the surface (Arie and Resnick, 1996). When the soil is dry, a hard layer is formed in the soil surface that is often called "desert crust", commonly enriched with calcite or silica. Desert crust decreases the infiltration rate of soils, thereby increasing runoff and soil erosion, reducing the availability of water through the root zone, and impeding seedling and plant growth (Figure 1.A). Other kind of crusts are formed in agricultural plots, irrigated with saline waters (Figure 1.B). Understanding the formation and properties of such crusts, as well as developing engineering methods to break it, are essential to control the runoff-infiltration (groundwater recharge) ratio and to maintain successful and sustainable agricultural activities. When desert crusts are a result of microbial activity, these kinds of crusts could help to protect the soil surface (Campbell et al., 2009). Besides, arid soils typically possess within a predominant sandy soil, a very low organic matter content with a consequent low fertility. In this sense, crop productivity under dryland conditions is largely limited by soil water availability. Soil organic matter (SOM) contents have been found to be a reliable index of crop productivity in semiarid regions because it positively affects soil water-holding capacity (Diaz-Zorita, et al., 1999).

(a)                                         (b)

Fig. 1. Soil salinity symptoms and consequences. A: Photograph of soil fines and saline crusts; Wyoming, USA (http://www.powderriverbasin.org/assets/). B: Degraded unfertile soil because of salt accumulation in Chametla Baja California Sur, México

## 2.2 Vegetation and livestock control

Vegetation may be the most important control on water movement in arid soils. Because vegetation in arid regions is opportunistic, when the water application rate is increased, plant growth increases as it uses up the excess water. The opportunistic nature of desert vegetation is shown by a significantly higher concentration of vegetation in areas of increased water flow, such as in ephemeral streams and in fissured sediments or rock-beds.

Where the water supply is limited, plant activity decreases until the water-supply rate increases. The importance of vegetation on a local scale has been shown in several field studies elsewhere, including soil cover protection, maintaining the soil aggregation and other effects. As a negatively associated activity, the extensive production and maintenance of livestock generate overgrazing, lose of plant cover, soil exposure, lose of biodiversity and desertification, which turn to be economically irreversible.

## 2.3 Vadose zone studies

A number of studies of the vadose zone in arid environments have been conducted elsewhere primarily for water resources evaluation. In the last two decades of the twentieth century, however, emphasis shifted from water resources to waste disposal and the transport of salts and other contaminants. Arid areas are being proposed for low-level and high-level radioactive waste disposal. Most of the studies related to the vadose zone in arid settings were conducted in the western United States, in regions that are designated as waste facilities. Some of these sites include Hanford Washington, Sandia New Mexico, Ward Valley California, Eagle Flat Texas, Nevada test site and Yucca Mountain Nevada (Scanlon et al., 1997). The increasing interest in the desert environment for waste facilities, in general, and radioactive waste, in particular, raises the need to understand the importance of preferential flow in the subsurface. One could assume that a thick vadose zone combined with low precipitation promotes the safest possible environment for waste disposal. However, fast flow via fractures, cracks, and macropores had been suggested as a major mechanism leading to contaminant transport much faster than anticipated by models that predicted transport based on average soil properties. For hydrologic studies, dual-porosity models exist (i.e. HYDRUS) (Simunek, 2008), but they are difficult to parameterize for this kind of soils.

### Salt accumulation

Salinization is a significant issue to consider in arid environments is the salinization of both soils and groundwater. The low precipitation combined with high evapotranspiration and often-slow flow rates through the subsurface, result in higher concentrations of salts. Human-induced salinization has a long history. A major source of salts accumulating in the upper vadose zone is irrigation water, which is essential for sustaining agriculture in arid lands. More than one-third of the developed agricultural lands in arid and semiarid regions reflect some degree of salt accumulation. High salinity in agricultural lands imposes stress on the growing crops that can lead to decreased yield and in some cases complete crop failure. This problem emphasizes the need for careful management of desert land and water balance.

## 2.4 Water management issues

Despite the difficulties for plants, animals, and humans to live in desert regions, they are increasingly being utilized because of pressure from world population growth. This problem is expressed in the expansion of agricultural activities onto desert lands as well as by the formation and rapid growth of urban and industrial centers. These trends not only result in a growing demand for usable water, but also for the increased disposal of vast amounts of wastewater and solid wastes (e.g., radioactive wastes, hazardous wastes, and municipal solid wastes). In several cases, international conflicts have developed due to water rights in arid regions. Large rivers crossing desert regions are often the only potential source for water that is essential for agriculture, industrial use, and drinking water. For example, the rights to use the water of large rivers in Africa (e.g., the Nile) and in the Middle East (e.g., the Euphrates and Jordan) remain one of the major issues that govern the relations and conflicts between the countries upstream, where most of the river water discharges, and the countries that use the river water downstream.

## 2.5 Desalinization strategies

The process to separate salts from saline waters or desalinization of either deep saline groundwater or seawater is a feasible alternative source for water in arid regions. However, the cost of desalinization remains higher than most other alternatives. A complex infrastructure is required, and the need for a close source of saline water makes this alternative impractical in many arid environments. The world's largest desalinization projects are in the Arabian Gulf (Saudi Arabia, United Arab Emirates, Kuwait), United States, and Japan, all which are wealthy countries with long seashores; lately, in Northwest Mexico some desalinating plants are being installed, with uncertainty about operational costs in the near future.

## 3. Methods

Sustainability indicators were reviewed and applied to an arid region; a study case was analyzed by means of the application of selected indicators and indexes in an overexploited aquifer. Results were interpreted within the framework of sustainability of water resources. A water usage balance study was analyzed for the La Paz Watershed, Baja California Sur, in a semiarid zone of Northwest Mexico, in order to determine environmental and social dimensions of water issues through sustainability indicators. In this zone, conventional crops are a major user of irrigation water, because of its water-demanding nature, due to an average of the five to seven irrigations needed per year. Within the La Paz watershed, four micro basins were evaluated for water deficit: El Cajoncito, La Huerta, La Palma and El Novillo.

Three variables were assessed in order to estimate the index of water scarcity Iwsc (water availability indicator), a composed integrated index which takes into the regional hydrological account, the natural groundwater recharge, the extraction and the resulting balance. In order to understand the relationship 'availability-demand', the index of water scarcity (Iwsc), which combines information about water abstractions and water availability, is assessed at first. For this purpose, the regional water availability index (Irwa) is a measure of water available for socio-economic development and agricultural production. It is the accessible water diverted from the runoff cycle in a country, region or drainage basin, expressed as volume per person per year, $m^3/p/y$. The indicator Iwsc is defined by:

$$Iwsc = (W - S)/Q \qquad (1)$$

Where:

Iwsc   water scarcity index [-]
W      annual freshwater abstractions in $Mm^3$, (M: millions)
S      desalinated water in $Mm^3$
Q      the annual available water in $Mm^3$

$$Q = R + \alpha\,S\,Dup \qquad (2)$$

Where:

R      the internal water resources in the country in Mm$^3$
Dup  the amount of external water resources in Mm$^3$
α      ratio of the external water resources that can be used.

The factor α is influenced by the quality of the transboundary water, by the consumption of water resources in the upstream region, and the accessibility of water.

Critical values of Iwsc identify various ranges for water scarcity and its parameters; the most common range for Q oscillates between 1000 and 1700 m$^3$/p/y. A region is considered highly water stressed if Iwsc is higher than 0.4 (Alcamo et al., 2003), which is a reasonable although not definitive threshold value, because not all the renewable freshwater resources are used by human society. Data with shorter time scales will enable more detailed assessments considering the effects of seasonal variability in the hydrological cycles (Oki, 2006). These values are important because the World Bank and other aid organizations use them to prioritize and to direct aid to developing nations.

An indicator related to the 'efficiency of land cultivation' is the cultivation factor R.

The ratio of cultivated to non-cultivated land was defined by Ruthenberg (1976) as:

$$R = (C \times 100)/(C+F) \tag{3}$$

Where:
R    cultivation factor (years of cultivation as % total cycle)
C    length of cropping period, years
F    length of the fallow period, years

For the interpretation of R, Ruthenberg defined:
R < 30 as shifting cultivation;
R = 30 to 70 as semi-permanent cultivation;
R > 70 as permanent cultivation.

# 4. Application of sustainability indicators in Baja California Sur, México

## 4.1 Study region

Baja California Sur is one of the driest Mexican States, with an annual average 140 mm of precipitation along its 72,000 km$^2$ extension. La Paz, the capital city, is located near the southern tip of the Baja California Peninsula (Figure 2). With a population approaching 200,000, it is the third largest city on this peninsula, after Tijuana and Mexicali. The La Paz region is dominated by desert and arid ecosystems, with a low availability of water resources. The annual mean temperature reaches 24°C with a yearly total rainfall of 180 mm, but with several dry months with null precipitation (Figure 3), with much of water coming in the form of hurricanes (CNA, 1999).

## Application of sustainability indicators

Information on quantity and quality of natural resources is essential for sustainable development. In particular, information on freshwater resources, their availability and use is becoming increasingly important with the emergence of regional water shortages and the need to improve water use efficiency.

Assessing Environmental and Social Dimensions of Water Issues Through Sustainability Indicators in Arid and
Semiarid Zones

73

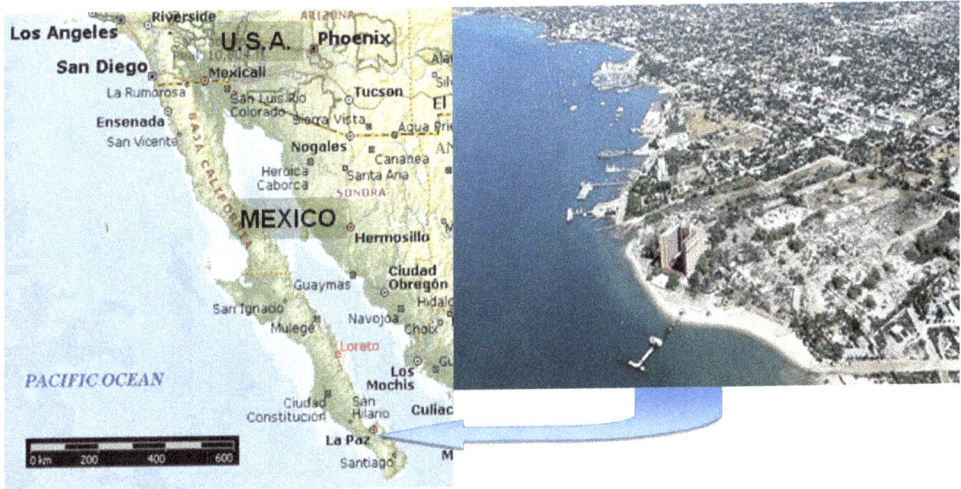

Fig. 2. Geographic location of the Mexican State Baja California Sur and La Paz, the capital city. Northwest México

For the application of sustainability indicators to the diagnosis of La Paz watershed, data on water uses, the natural groundwater recharge and the extraction were obtained from the National Water Commission (CONAGUA) reports and other previous studies (Cruz-Falcón, 2007; CONAGUA, 2008). Weather data (temperature, evaporation and precipitation) used in this study for La Paz B.C.S. (México) were obtained from División Hidrométrica de Baja California Sur, of the Comisión Nacional del Agua (CONAGUA, 2008), who collects this information from the La Paz Weather Station, located at 24°09'N and 110°20'W, 3 km south La Paz City. Our analysis indicate that agriculture is the major water consuming activity, in both, Baja California Sur state and the whole Peninsula of Baja California (Table 1).

The natural groundwater recharge was estimated by the method according to the groundwater lever fluctuation method, which is an indirect method of deducing the recharge from the fluctuation of the water table. The rise in the water table during the rainy season is used to estimate the recharge, provided that there is a distinct rainy season with the remainder of the year being notoriously drier (Cruz Falcon, 2007).

| Water use | Baja California | Baja California Sur | Peninsula (total) |
|---|---|---|---|
| Agriculture | 1830227 | 6450 | 1836677 |
| Domestic | 74 | 1914 | 1988 |
| Multiple (industries) | 164 | 425 | 589 |
| Livestock | 809 | 2403 | 3212 |
| Urban-public | 146640 | 421 | 147061 |
| Services (county) | 28 | 129 | 157 |
| Total | 1977942 | 11742 | 1989684 |

Source: Official data base from REDPA - National Water Commission of Mexico (CONAGUA, 2008).

Table 1. Synopsis of the water use in the Baja California Peninsula and related States, according to the consumptive use; unit: Millions m$^3$/yr

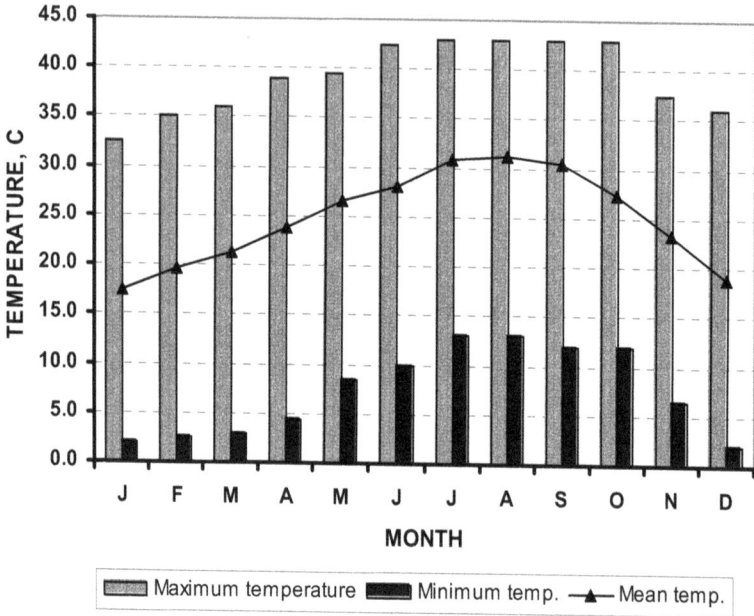

(a)

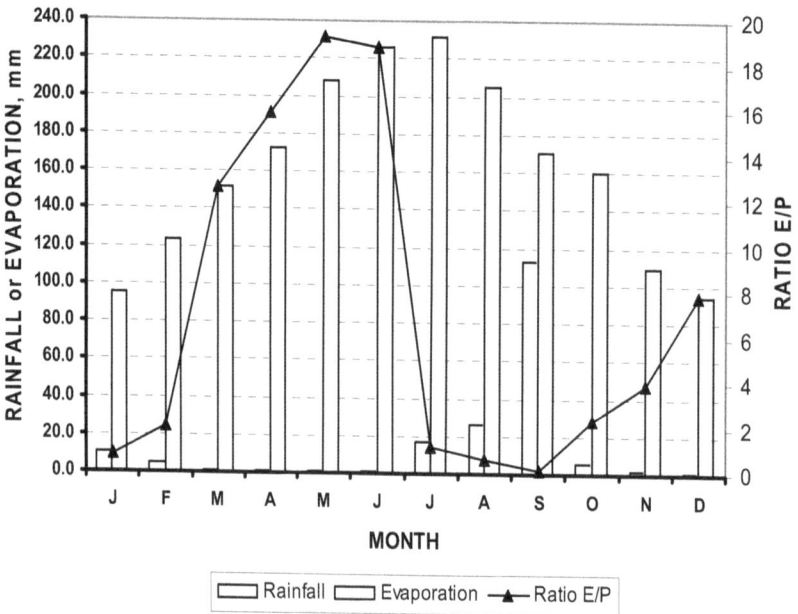

(b)

Fig. 3. Pattern of climatic variables for La Paz weather station, (A): maximum, minimum and mean temperature; (B): rainfall, evaporation and ratio E/P

## 4.2 Results and discussion

Factors affecting agriculture in Baja California Sur were found to be water deficit (evaporation dramatically exceeds rainfall): 2,380 mm – 180 mm = 2,200 mm of hydrological deficit, water scarcity (evidenced by absence of surface water with groundwater depletion), high temperatures: Temp avg = 24.5, Temp max = 42 C; Salinity (natural and caused), low fertility of soils, and socio-economical factors (long distance form main markets, complex marketing policies, others).

The oriented-extensive ground water extractions have caused notorious water depletion in two out of three contiguous watersheds, at La Paz municipality (Table 2).

| Name of Aquifer | Recharge (Mm³) | Extraction (Mm³) | Availability (Mm³) | Possible has with irrigation 100 cm depth |
|---|---|---|---|---|
| El Carrizal | 16.0 | 8.60 | 7.40 | 861.00 |
| La Paz | 27.8 | 36.95 | -9.15 | 3,200.00 |
| Los Planes | 8.5 | 9.57 | -1.10 | 957.00 |

Table 2. Hydrological balances for three contiguous watersheds in Baja California Sur, Northwest Mexico

Calculation of the index of water scarcity for La Paz watershed and its four microbasins rendered high values, from 1.11 (microbasin La Palma), to 2.74 (microbasin El Novillo) (Figure 1, Table 3). Results suggest that El Novillo faces a critical condition as a result of high extraction rates, over passing the natural groundwater recharge, with a notorious deficit, estimated in -4,450,068.75 m³, which affects the water availability for urban growth and development.

| Microbasin | Recharge | Extraction | Balance | Iwsc |
|---|---|---|---|---|
| | ================ m³ ================ | | | |
| El Cajoncito | 2,233,967.75 | 3,689,549.00 | -1,455,581.25 | 1.65 |
| La Huerta | 7,519,608.13 | 8,565,189.38 | -1,045,581.25 | 1.14 |
| La Palma | 16,524,756.81 | 18,420,338.06 | -1,895,581.25 | 1.11 |
| El Novillo | 2,559,500.39 | 7,009,569.14 | -4,450,068.75 | 2.74 |

Table 3. Values of Iwsc calculated for four microbasins of La Paz B.C.S. watershed, Northwest Mexico

The concept of 'sustainable development' as well as 'sustainable agriculture' integrates three main goals: environmental health, economic profitability, and social and economic equity. A variety of philosophies, policies and practices have contributed to these goals. People with many different capacities, from farmers to consumers, have shared this vision. In the case of agriculture, one of the most water-demanding activity, an agroecosystem must be viewed as a source of 'goods' and a sink of 'inputs' (i.e. water). For this activity and for the others, 'nature' (the ecosystem) is the main source of all we consume, but the ecosystem also serves as a sink for all wastes. For production systems, the main resources basically are: water, plants, grains, animals, energy from the sun, wind, and other nonrenewable: oil (from fossil). Human activity is necessarily focused for extracting resources and producing waste to produce, transport to consumer, and dispose of materials. As agriculture and the other socioeconomic activities in arid and semiarid zones depend on water, which mainly is obtained from an aquifer, it is crucial to analyze the concept of sustainability of an aquifer: Maintaining a balance between recharge and extraction, or seeking that average annual extraction does not exceed the annual recharge. Although extraction from some groundwater resources has been above the long-term sustainable yield, its use must be managed to the sustainable yield through the implementation of water sharing plans for groundwater (Scanlon, 1997). Current droughts have increased the demand on groundwater resources, causing localized stresses in parts of groundwater systems and overexploitation, with a high risk of water scarcity. A wide variety of ecosystems depend on groundwater for their continued survival. Significant changes in groundwater quality and quantity have the potential to degrade ecosystems and affect human uses of water.

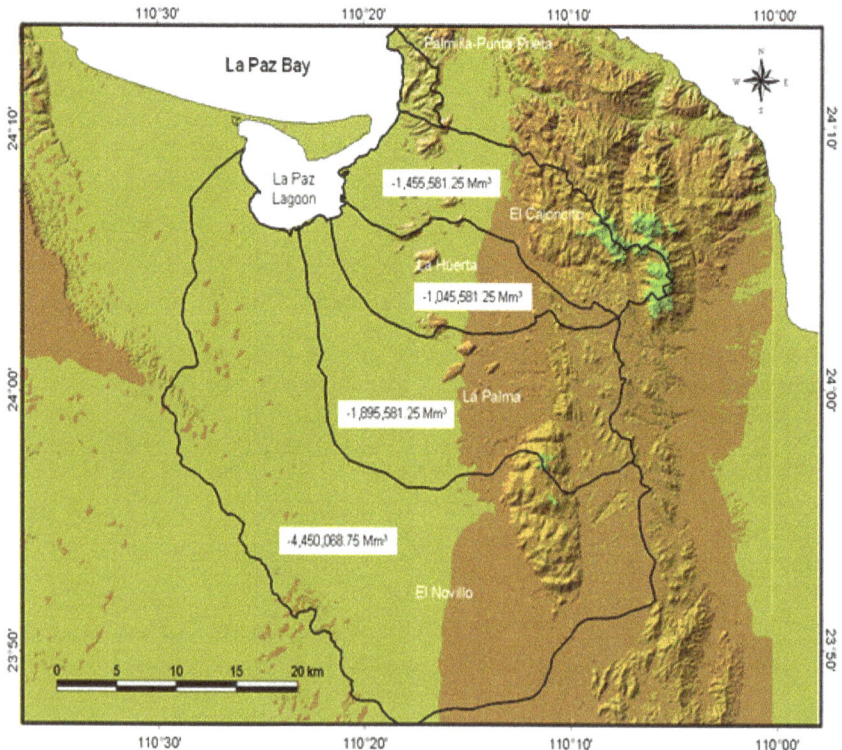

Fig. 4. Geographic and topographic configuration of four microbasins within La Paz BCS watershed, Northwest Mexico, with estimated water deficit, in millions (M) m³

Agriculture and livestock depend on water. Under natural conditions, water deficit is a common condition for agro-ecosystems and grazing-lands in arid zones. "Rainfall" is an important climatic parameter, but in arid and semiarid zones, its analysis scarcely explains the dryness intensity and the aridity pattern. At this stage in earth's history, it is believed that mankind can make 'productive efforts' to decrease or diminish the water depletion, such as artificial groundwater recharge, small dams, artificial infiltration ponds, desalination, increasing the water use efficiency, mitigating or remediating losses, promotion of native agro-forestries for water retention, ad others. In a sort or medium term, the hydrological cycle will be intensified in several zones, with more evaporation and more precipitation, but the extra precipitation will be unequally distributed around the globe. Some parts of the world may see significant reductions in precipitation, or major alterations in the timing of wet and dry seasons (Arnell, 1999). A challenge for scientists is to find appropriate models to diagnose the water deficit, which is the real parameter that impact livestock and agricultural ecosystems. Advanced extensive techniques are so varied and complex that only large areas are feasible and realistic. Most of adapted species to drought do not have yet a real important market, although there are possibilities to develop it. i.e.: *Salicornia bigelovii, Aloe vera, Opuntia spp.* An additional problem is the significant distance from suppliers and market.

## 5. Conclusions

The study reported here has applied a series of spatially-resolved data sets depicting the biogeophysical and socioeconomic properties of the South-Baja Californian communities in Northwest Mexico. We, as others, have found that Northwest Mexico is a dry zone. Associated with this dryness is a highly dynamic water cycle, providing a large degree of variability in terms of climate, runoff and discharge. A systems perspective is essential to understanding sustainability. The system is envisioned in its broadest sense, from the individual farm, to the local ecosystem, and to communities affected by this farming system, both locally and globally. An emphasis on the system allows a larger and more thorough view of the consequences of farming practices on human communities and the environment. The application and interpretation of sustainability indicators motivates, for the La Paz watershed case, the design and instrumentation of strategies in order to improve the water use efficiency, and to alleviate the water deficit of the aquifer.

A significant fraction of agricultural land and human population is located in the study region with low runoff hydrography at the center of the La Paz valley, with disperse small localities and villages at the high sections of the watershed, with high runoff and high variability. Hence, agricultural water demand defines the aggregate water use for the watershed. These characteristics of human-water interactions, in turn, provide challenges to the water infrastructure of the watershed, with evidences that the region may be experiencing curtailed use of water, relative to its high demands. Biogeophysical data sets, emerging rapidly from the local science community, can make important contributions to emerging water resource assessments.

On the base of available evidences, we conclude that both, bio-geophysical as well as socioeconomic indicators will be necessary to map the patterns and intensities of water scarcity. Interdisciplinary study is thus an important component of future research.

## 6. Acknowledgment

This work was financed by the Consejo Nacional de Ciencia y Tecnología (CONACyT), CONACyT-CIENCIA BASICA Fund, Project 134460 "Determinación y construcción de indicadores de la huella hídrica y desertificación como consecuencia de la sobreexplotación agropecuaria y del cambio climático en cuencas de zonas áridas", and by the Centro de Investigaciones Biológicas del Noroeste´ (CIBNOR). Thanks are due to la Comisión Nacional del Agua (CONAGUA), Dirección Local in Baja California Sur, for providing climate and geo-hydrological data of La Paz B.C.S.

# 7. References

Alcamo, J.; Döll, P.; Henrichs, T.; Kaspar, F.; Lehner, B.; Rösch, T.; Siebert, S. 2003. Global estimation of water withdrawals and availability under current and "business as usual" conditions. Hydrological Science, 48(3): 339-348.

Arie S.I., and Resnick, S.D. 1996. Runoff, Infiltration and Subsurface Flow of Water in Arid and Semi-Arid Regions. Water Science and Technology Library. Dordrecht, The Netherlands: Kluwer Academic Press.

Arnell, N.W. 1999. Climate change and global water resources. Global Environmental Change, 9 (Supplement 1): S31-S49.

Balks, M.R., Bond, W.J. and Smith, C.J. 1998. Effects of sodium accumulation on soil physical properties under an effluent-irrigated plantation. Australian Journal of Soil Research, 36(5): 821-830.

Calvet, J.-C., J. Noilhan, and P. Bessemoulin. 1998. Retrieving the root-zone soil moisture from surface soil moisture or temperature estimates: a feasibility study based on field measurements. Journal of Applied Meteorology, 37: 371-386.

Campbell, S.E., Seeler, J.S. and Golubic, S. 2009. Desert crust formation and soil stabilization. Arid Soil Research and Rehabilitation, 3(2): 217-228.

CNA (Comisión Nacional del Agua). 1999. Documento de Respaldo para la Publicación de la Disponibilidad. Acuífero B. C. S.-24 La Paz. Gerencia Regional de la Península de Baja California, Mexicali, B. C. 17 p.

CONAGUA (Comisión Nacional del Agua). 2008. Estudio para Actualizar la Disponibilidad Media Anual de las Aguas Nacionales Superficiales en las 85 (ochenta y cinco) Subregiones Hidrológicas de las 7 (siete) Regiones Hidrológicas 1, 2, 3, 4, 5, 6 y 7 de la Península de Baja California, Mediante la NOM-011-CNA-2000. México, D.F.

Cruz Falcón, A., 2007. Caracterización y Diagnóstico del Acuífero de La Paz B. C. S. Mediante Estudios Geofísicos y Geohidrológicos. Tesis de Doctorado, IPN-CICIMAR, Diciembre de 2007. 139 pp.

Diaz-Zorita, M., Buschiazzo, D.E., and Peinemann, N. 1999. Soil organic matter and wheat productivity in the semiarid Argentine pampas. Agronomy Journal, 91(2): 276-279.

Gleick, P. H. 2000. The World's Water: The Biennial Report on Freshwater Resources 2000–2001. Washington, D.C.: Island Press. USA.

Oki, T., and S. Kanae. 2006. Global hydrological cycles and world water resources. Science, 313(5790): 1068-1072.

Scanlon, B.R., Tyler, S.W., Wierenga, P.J. 1997. Hydrologic issues in arid, unsaturated systems and implications for contaminant transport. Reviews of Geophysics, 35(4): 461–490.

Seneviratne, S.I., Luthi, D., Litschi, M. & C. Schar. 2006. Land–atmosphere coupling and climate change in Europe. Nature, 443(14): 205-209.

Sharma, K.D. 1998. The hydrological indicators of desertification. Journal of Arid Environments, 39(2): 121-132.

Simunek, J., M. Th. van Genuchten and M. Sejna. 2008. Development and applications of the HYDRUS and STANMOD Software packages and related codes. Vadose Zone Journal; May 2008; v. 7; no. 2; p. 587-600.

Weisbrod, Noam et al. "Salt Accumulation and Flushing in Unsaturated Fractures in an Arid Environment." Groundwater 38, no. 3, 452–461.

Zhuguo, M., Helin, W., and Congbin, F. 2000. Relationship between regional soil moisture variation and climatic variability over East China. Acta Meteorologica Sinica, 2000-03.

# Part 2

# Water and Agriculture

# 5

# Integration Challenges of Water and Land Reform – A Critical Review of South Africa

Nikki Funke and Inga Jacobs
*Council for Scientific and Industrial Research (CSIR)*
*South Africa*

## 1. Introduction

The equitable utilisation of water in the real world is a very complex challenge involving a wide range of often competing actors and factors that need to work synergistically and be integrated if we are to effectively manage this valuable resource for productive land use. Additionally, the relationship between land and water is politically, economically and culturally complex and this complexity is expected to increase with the progression of growing populations, increasing water scarcity, growing demand for water, and food security concerns. This challenge is bound to gain global significance particularly in regions where communities are vulnerable to the profound impacts of global change. Integrated policy, planning and management of water and land resources can therefore provide improved benefits and create innovative opportunities for regional economic development by contributing to ecosystem stability, sustainable livelihoods and food security.

Water and land reform in South Africa is a special case highlighting the importance of integrated approaches. The last two decades have seen an abundance of comprehensive reforms the world over in the management of natural resources, with an emphasis on greater integration, the devolution of power and the decentralisation of government decision making. In the developing world, this phenomenon has been particularly prevalent in the water and agricultural sectors with new national development policies and action plans developed and harmonised to regional and/or international legal and institutional frameworks. Technocratic templates from developed countries in Europe and North America, such as the concept of Integrated Water Resources Management (IWRM), have also been suggested as best practice. However, not enough attention has been placed on factoring in local configurations, domestic policy, political identities, and social and cultural institutions, particularly in the African context (Jacobs, 2010).

In South Africa, water and land reform policies have been embedded within a complex socio-political and socio-economic environment, and yet have occurred largely independently of one another. South Africa presents an interesting example of the consequences of the non-integration of reform policy and yields lessons for countries in the rest of the world in terms of the challenges to successfully implementing land and water management reform programmes. The role of the South African government in providing a coordinating role is important. However, a concerted multi-stakeholder and multi-sectoral effort is required at all levels, from the local to the national, if integration is to be operational and implementable.

## 2. Motivation

Despite the fact that the interconnectedness of water and land and the relevance of these resources for sustainable development have been well-documented, both resources are still largely managed as isolated policy issues and only limited research focuses on the numerous links between them. There is still a weak link between land reform, agricultural support and water resource provision (Greenberg, 2010). In South Africa, many land reform farms have failed because of water not being available for production. The synchronisation between water allocation and land reform programmes in irrigation areas therefore has to be improved to ensure that beneficiaries hold secure land and water use rights once they have been allocated their land (Groenewald, 2004).

Integration is however easier said than done, and can only be achieved through the acknowledgement of a diverse multi-actor landscape and consequent diverging interests and perceptions. This can only be achieved if the current tendency by government departments and sectors to work in "silos", without much integration, is transformed from the programme level. Once this is achieved, we will be able to come to terms with the existence of multiple social and cultural norms that shape how water and land are managed.

## 3. Definition of terminology

Several integration approaches have been developed over time to better conceptualise the meaning of "integration" and how it applies to natural resource management processes, policy implementation, and theoretical frameworks.

### 3.1 The "Integrated" in IRM, INRM and IWRM

Firstly, the term Integrated Resource Management (IRM) is somewhat ambiguous and not always clearly defined, and, as such, is often operationalised in a variety of ways. Integrated Natural Resources Management (INRM) has been described as a conscious process of incorporating multiple aspects of natural resource use into a system of sustainable management to meet explicit production goals of farmers and other uses (e.g., profitability, risk reduction) as well as goals of the wider community (sustainability) (Sayer and Campbell, 2004). INRM is also described as an approach that integrates research about different types of natural resources into stakeholder-driven processes of adaptive management and innovation. The aim of this process is to improve livelihoods, agro-ecosystem resilience, agricultural productivity and environmental services at community, eco-regional and global scales of intervention and impact (Thomas, 2002). The focus is agriculture specific, which speaks to the chapter's focus on integrated water allocation and land reform.

A related term that is also of relevance to this chapter is that of Integrated Water Resources Management (IWRM), which the Technical Advisory Committee of the Global Water Partnership (GWP-TAC) defines as follows:

*"IWRM is a process, which promotes the co-ordinated development and management of water, land and related resources, in order to maximise the resultant economic and social welfare in an equitable manner without compromising the sustainability of vital ecosystems" (GWP-TAC, 2000).*

The distinction between "integrated" and "traditional" management of water and natural resources relies largely on the scope and sphere of operation of the two. "Traditional" management is typically sector-oriented (water supply, irrigation, agriculture, hydropower, etc.) and focused on satisfying the perceived demands within each sector. "Integrated" management, in contrast, attempts to take a cross-sectoral approach and focuses on the management of water and natural resources, as well as the demand, supply and use of water and natural resources (Gooch and Stålnacke 2006).

It is argued that the successful implementation of IWRM can prevent human health, economic and environmental losses that might hamper development and frustrate poverty reduction efforts. In addition, the participative processes that make up "good" IWRM can help developing countries to meet the millennium development goals (MDGs). The MDGs aim to address poverty, gender equality and health issues and also strive to attain environmental sustainability (Jonch-Clausen, 2004).

Operationally, and similar to INRM, IWRM approaches apply knowledge from several disciplines as well as multiple stakeholders to devise and implement efficient, equitable and sustainable solutions to water and development problems. As such, IWRM is a comprehensive, participatory planning tool that involves the coordinated planning and management of land, water and other environmental resources for their equitable, efficient and sustainable use (Calder, 1999). Key points here are process, coordination, and the relationship between sustainability and economic and social welfare. IWRM can be seen as consisting of five main characteristics that may cause complications and problems and necessitate action: Multi-functionality (e.g., fishing, farming, water supply), user interests and conflicts, multiple managers at different levels (e.g., local, regional, national), asymmetric power relations (e.g., up- and downstream users and managers), and technical complexity (Mostert, 1998).

## 3.2 Definition of integration

Having identified different possible approaches to integration of water and land management, it becomes important to establish a definition of "integration". This definition is based on the discussion of the international approaches above but also specifically applies them to the issue of water allocation and land reform in the South African context. Integration can therefore be defined as follows: the degree to which policies formulated in one government department are harmonised or coordinated with policies developed in other government departments, other sectors, or acknowledge the interconnectedness of various resources and the degree to which inter-departmental coordination and communication take place in the implementation of said policies.

Integration therefore refers to policy harmonisation and coordination across government departments and sectors as a result of the recognition of the interconnectedness of different natural resources. Furthermore, integration entails acknowledging and taking into account the diverse multi-actor landscape and consequent diverging interests and perceptions that make up the water allocation and land reform landscape in order to come to terms with the existence of multiple social and cultural norms that shape this landscape.

Integration as described here is important so that policies or programmes developed in one government department take into account the impacts on or of other sectors and do not operate in isolation from other sectors. In addition, coordination is not only imperative between different government departments and sectors, but also between different levels of government at the national, provincial and local levels.

## 4. Integration of land and water management in the context of a developing country

In terms of applying the concepts of IRM, INRM and specifically IWRM to developing countries, it is important to realise that no universal blueprint for IWRM exists. While certain basic principles are applicable universally, a number of factors affect their realisation and effective implementation in individual and specifically developing countries. These factors include the nature, character and intensity of water problems in individual countries, as well as human resources, institutional capacity, the relative strengths and characteristics of the public and private sectors, cultural setting, and the natural conditions present (GWP-TAC, 2000).

In addition, many of Africa's problems (and those of other developing countries) are uniquely "local", which may make it difficult for a "transplanted" solution to work. It is therefore important to ensure that the IRM, INRM and IWRM principles and specific practices that are implemented in an African country (or any other developing country) take sufficient account of local conditions to ensure they are sustainable and effective in the long-term (Ashton, 2007).

In terms of water reform in particular, which is of particular relevance for the South African context, it also seems to be difficult to overhaul water resources management and apply new legislation, strategies and institutions that are linked to paradigms such as IWRM in practice. These tasks often exceed the budgets and human resource capacities of most Southern African Development Community (SADC) states. It is also important to be aware of the largely political nature of water reform processes, such as proposing a profound re-alignment of decision-making power and decentralising management to the lowest possible level, in already fragile, underdeveloped states (Funke et al., 2007a). This statement is of particular relevance to the South African context not only in terms of water allocation reform, but also in terms of the land reform process and the shift in political and power dynamics that has played a part in the run up to and during the implementation of both of these processes. In addition, South Africa, similar to many other countries, has been struggling to implement IWRM for a number of reasons. These generally include an absence of relevant institutions (Catchment Management Strategies that are supposed to be the implementing agencies of IWRM), lack of coordination of available data, lack of capacity and skills, and lack of communication within the South African Department of Water Affairs (DWA) (Funke et al., 2007a). Therefore there seems to be a discrepancy between developing policy or paradigms that sound highly promising on paper and implementing these in practice.

# 5. The South African case study

South Africa is characterised by substantial socio-economic inequalities and inequitable access to water resources and land as a result of its historical legacy, coupled with challenging climatic conditions and problems of water management. It therefore makes for an interesting case study of the need for integrating approaches to water allocation and land reform as well as the consequences of non-integration.

## 5.1 Climatic conditions in South Africa

South Africa is a water scarce country. Although some parts of the country receive more rainfall than others, the country's average rainfall of 450mm per year is far below the global average which amounts to 860mm per year. In addition, factors such as climate change and international obligations to neighbouring countries with shared watercourses limit the amount of water that can be used (Claassen, 2010).

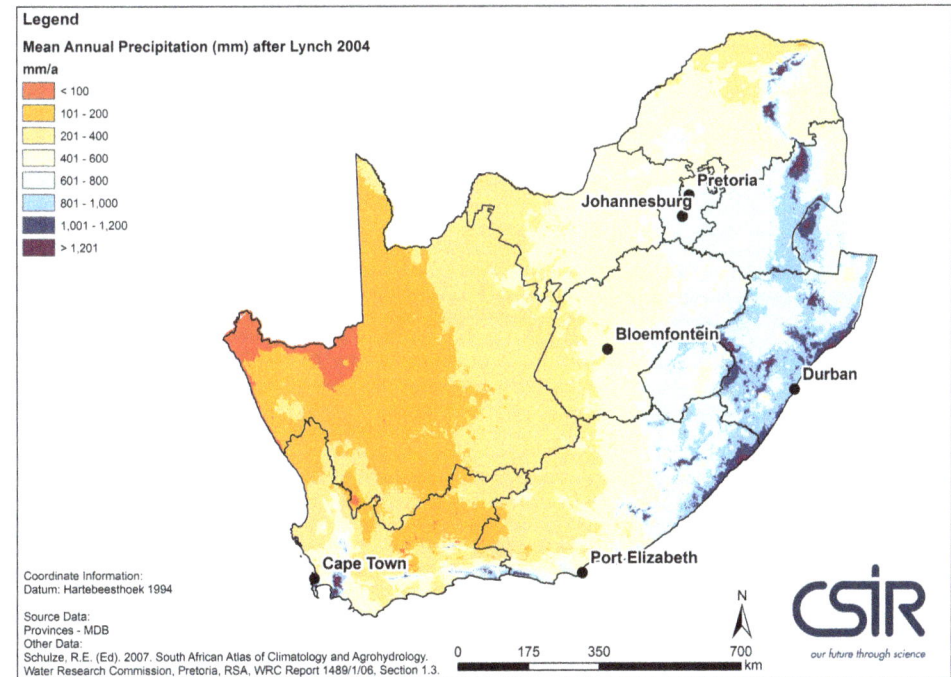

Fig. 1. South Africa's average rainfall (Maherry, 2010)

While South Africa has enough water to meet its needs in the immediate future, based on calculations of runoff, yield and water use, there is a growing demand for water, which is currently being met by the development of the country's surface water resources. South Africa's estimated mean annual runoff is 43 500 million cubic metres per annum (excluding the runoff from Swaziland and Lesotho), the total available yield is 13 227 million m3/a and for the year 2000 the total water use requirements were 12 871 million m3/a (Classsen, 2010).

In terms of water use, the water requirements of irrigated agriculture are an estimated 56% of the total annual water requirements of 22 045 million m³ surface and groundwater (Backeberg, 2007). Although the contribution of irrigation to total agricultural production varies according to crop type, most of this water is used for commercial food production in local and export markets. In South Africa, the total land area under irrigation is 1.3 million ha, of which 100ha are food plots and smallholder irrigation schemes. This land falls in various rainfall regions, with a highly variable average of 500mm per year. The two most important irrigation practices are permanent irrigation and the sprinkler method (Backeberg, 2006).

| Region | Rainfall [mm] | Total [ha] | Perma-nent [%] | Supple-mentary [%] | Occasional [%] | Flood [%] | Sprinkler [%] | Micro [%] |
|---|---|---|---|---|---|---|---|---|
| | | | | Type of irrigation | | | Method of irrigation | |
| 1 | <126 | 19174 | 92.5 | 0 | 7.5 | 66.6 | 8.3 | 25.2 |
| 2 | 126-250 | 161197 | 61.1 | 0.4 | 38.5 | 77.1 | 16.8 | 6.1 |
| 3 | (251-500) | 399278 | 86.7 | 7.7 | 5.7 | 42.8 | 43.6 | 13.6 |
| 4 | (501-750) | 488543 | 75.2 | 20.8 | 4.0 | 21.0 | 65.4 | 10.8 |
| 5 | (>750) | 221940 | 81.5 | 16.6 | 1.9 | 5.3 | 80.9 | 13.8 |
| Total | | 1290232 | 78.3 | 13.1 | 8.6 | 32.8 | 54.4 | 11.8 |

Table 1. Total areas, type and method of irrigation in different rainfall regions (WRC, 1996)

Despite just enough water being available for current use, including agriculture, South Africa's water resources face political, social and economic pressures. These include having enough infrastructure to secure water during low rainfall periods and supply areas of high demand, growing enough food to supply the growing population and meeting the water demands of energy, industry and mining (Claassen, 2010). In addition, due to increasing urbanisation and higher standards of living, competing demands are experienced for domestic, mining and industrial water use (Backeberg and Odendaal, 1998). At present, most of the country's water supply has already been allocated, and the only "supply options" available are linked to re-allocations between different water use sectors (De Lange, 2010).

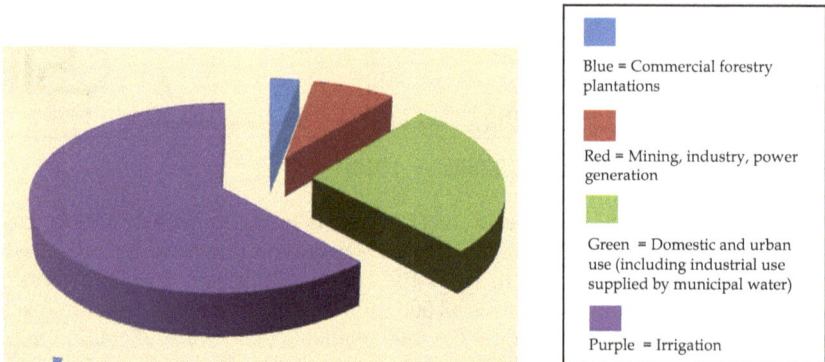

Blue = Commercial forestry plantations

Red = Mining, industry, power generation

Green = Domestic and urban use (including industrial use supplied by municipal water)

Purple = Irrigation

Fig. 2. Water use per sector (Strydom, 2010)

In addition to the above, the South African government also faces other challenges related to water governance. After coming to power in 1994, the post-apartheid South African government passed world class water legislation to address the backlog in water supply and sanitation, which it inherited from the apartheid government, and to manage South Africa's situation of water scarcity (Funke et al., 2007b). In combination, South Africa's Water Services Act of 1997 and National Water Act of 1998 were designed to "redress the inequalities of racial and gender discrimination of the past; link water management to economic development and poverty eradication; and ensure the preservation of the ecological resource base for future generations" (Schreiner et al., 2002).

However, to date, the implementation of this legislation has been slow and problematic (Funke et al., 2007b). Challenges include high staff turnover and lack of institutional capacity in numerous government departments, resulting in these departments being overburdened (Hattingh et al., 2004, Funke and Nienaber, in press); a disconnect between water supply and water resource management (more water is being supplied at the municipal level than is ecologically feasible) (Pollard and Du Toit, 2005); the inability of many municipalities to treat domestic sewage and industrial effluent to enable this to be safely discharged into rivers and streams (Ashton, 2010); a serious backlog in setting up South Africa's Catchment Management Agencies (Hattingh et al. 2004); and a focus on development at the expense of the conservation of freshwater ecosystems (Funke and Nienaber, in press).

In addition, the country is characterised by deteriorating water quality in its major river systems, water storage reservoirs and ground water resources, which results in social, economic and health risks to society (Ashton, 2010). Almost half of South Africa's 112 river ecosystems are currently at a level of critical endangerment.

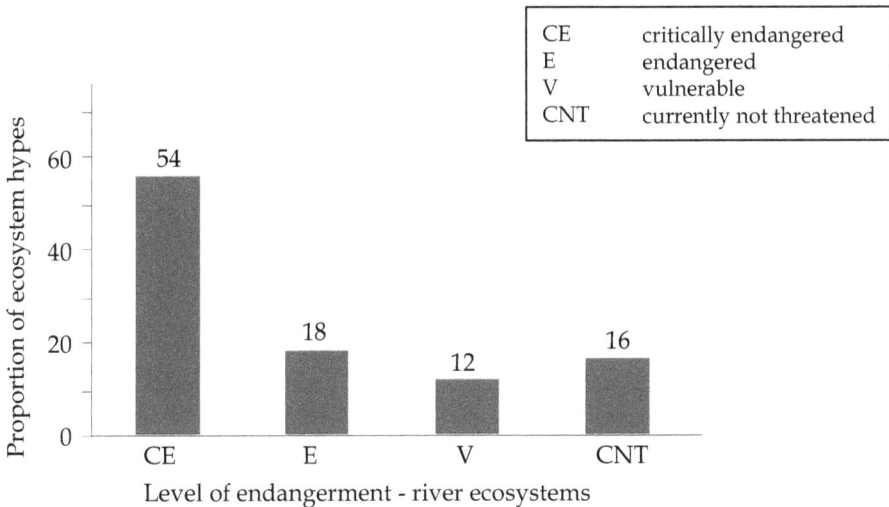

Fig. 3. Level of endangerment – river ecosystems (Nel, 2010)

One water governance related issue in particular that is of current interest and perceived national importance in South Africa is that of Acid Mine Drainage (AMD) (Hobbs, 2010). This issue has recently featured considerably in the country's newspapers and electronic media, where investigative journalists have flagged their concerns about it from various angles (Funke et al., in press). Acid mine water started decanting from abandoned underground mine workings close to Krugersdorp on the West Rand of the Gauteng Province in 2002. Now, the potential volume of AMD from the Witwatersrand Goldfield alone amounts to 350 Ml/day (Hobbs, 2010).

AMD, or the uncontrolled discharge of polluted water from defunct gold mining operations into surface and ground water resources, presents a serious threat to the receiving environment and has severe socio-economic and environmental impacts. Specifically, these impacts include the release of chemical contaminants into water resources, persistent environmental damage long after mine closure, and negative impacts on the health and safety of communities living in the vicinity of mining operations (Hobbs, 2010). There is no indication that the AMD threat will subside in the foreseeable future as mining operations remain active throughout South Africa as evident on the map below.

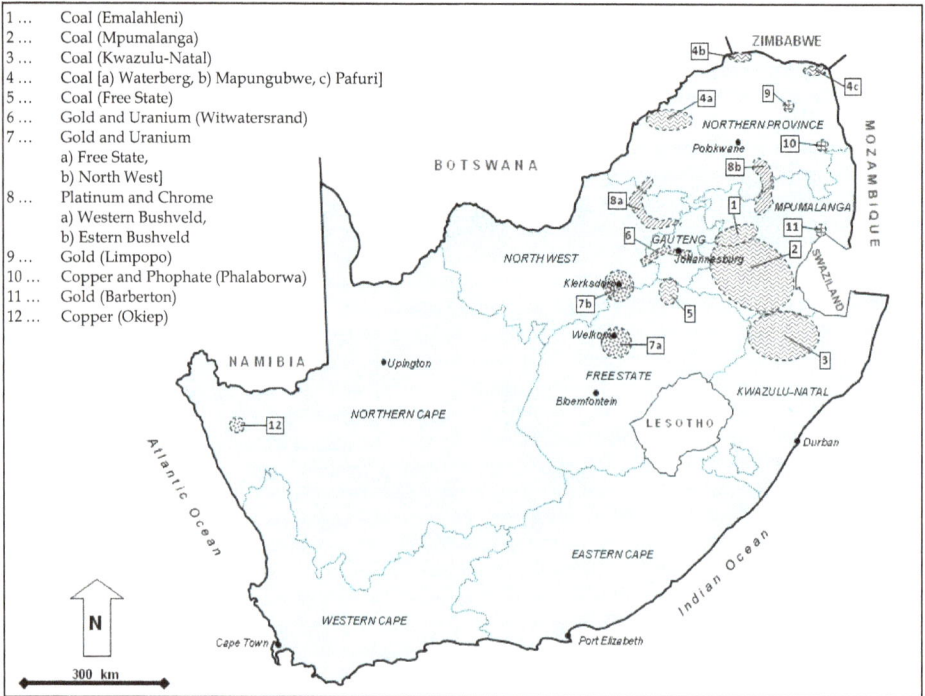

Fig. 4. Mining areas and minerals particularly susceptible to the formation of AMD (Hobbs and Kennedy, 2010)

Having sketched a picture of the climatic conditions in South Africa, an overview of water allocation and land reform in South Africa is presented to explain how the government chose to follow this reform path and what the results have been to date.

## 5.2 Overview of water allocation and land reform in South Africa

South Africa's political transformation formalised by the country's first democratic elections in 1994 brought with it a host of progressive reforms in the water and agricultural sectors. The Water Services Act was ratified in 1997 and the landmark National Water Act in 1998 (Republic of South Africa, 1998). The National Water Act is in line with other international reforms in water management. It prioritises decentralised water management and common property aspects of water; separates ownership of land from ownership of water; confirms the need to ensure that aquatic ecosystems receive sufficient water to function properly; stipulates the need to ensure that neighbouring states utilise shared water resources equitably; and prioritises the right of all South Africans to have adequate access to wholesome supplies of water (DWAF, 1997). The National Water Act is regarded, along with the EU Water Framework Directive (EU, 2000), as a pioneer of an international wave of reform and one of the most innovative and far-reaching water laws in the world, which has set the benchmark for new ways of managing water resources (Woodhouse, 2008; Ashton et al, 2008; Postel and Richter, 2003).

However, the necessary goal of redressing past racial and gender inequality means that South Africa's water reform is expected to deliver on changes in process (holistic, decentralised, participatory and economically cost effective), social outcomes (Woodhouse, 2008) as well as ensuring higher environmental standards as stipulated in the 1998 National Water Act. According to Woodhouse, "The prospect of redistribution from existing 'haves' to 'have nots' raises considerably the political risks and expectations attached to the implementation of reform" (Woodhouse, 2008: 3).

In line with the South African government's social redress priority, the land reform programme intends to transfer approximately 30% of white-owned commercial farms to "new" black commercial farmers by 2014 in an effective and sustainable manner (Cousins and Scoones, 2010). The land reform programme in the country has three different dimensions namely, land restitution, land redistribution and tenure reform. The restitution component of the policy aims to return land that was taken away forcibly from black people during apartheid, or to provide those affected with financial compensation. It targets both rural and urban lands. The land redistribution dimension aims to equitably share resources by transferring land from white to black people so that the land ownership share of black people is increased. This is considered necessary because black people make up the large majority of the South African population but have less land compared to the white population. For instance, in 1991 they held only about 13.9 % (17 million ha) of the national land (Lyne and Darroch, 2003). The land tenure reform dimension aims to enhance the tenure security of vulnerable people, such as workers and their families residing on private commercial farms as well as people living in the former homelands.

However, according to the Department of Rural Development and Land Reform's (DRDLR's) Strategic Plan for 2009-2012, by the end of the 2007/08 financial year the combined programme had only achieved 4.9 million ha. Cumulatively, from 1994 to the second quarter of 2008/09, the National Land Reform Programme had achieved just over 5.1 million hectares of land delivery. This means that from 1994 the yearly average output of 0.371 million ha has been less than one third of the expected 1.23 million required to meet the 2014 target. More importantly, it has been acknowledged that 90% of land reform projects on redistributed farming land have failed (Pressley, 2010).

In parallel to the land reform process, the water reform process has also been underway, with one of its central pillars being the Water Allocation Reform Strategy of 2008. Water Allocation Reform (WAR) aims to provide water for subsistence farming or for sustaining basic livelihoods, and to start a development path of commercial and competitive water use in support of broad based black economic empowerment. Thus the water allocation process must be undertaken in a fair, reasonable and consistent manner and existing lawful uses will not be arbitrarily curtailed (DWAF, 2006). Furthermore, the strategy aims to allocate 30% of available water to black people. By 2024 the target is 60%, half of which should be under control of black women.

Currently, 15% of water use licenses are allocated to historically disadvantaged individuals for irrigation purposes. By 2011/12 the Department of Water Affairs (DWA) plans to address the existing backlog of issuing licenses and is aiming to increase this target to 40% by 2013/14 (PLAAS 2009). DWA also wants to review progress towards integrated water, rural development and land reform by 2013/14 (PLAAS, 2009).

Compulsory licensing is an integral part of the Water Allocation Reform programme (DWAF, 2004). This allows for water currently allocated to users to be re-allocated to previously disadvantaged people. All commercial water users must now register their water use and will have to apply for a water use license (DWAF, 2004). In practice, however, not much re-allocation of water has occurred. In fact, the process of compulsory licensing has not yet started with only three pilot studies being carried out in various provinces. Similarly, of the 1212 ad hoc licenses for new water use that had been allocated by 2006, 98% were for non-historically disadvantaged individuals. Van Koppen et al. (2009) argue that for administration-proficient, larger-scale users, obtaining a license simply means submitting an application. The DWA appears to have very limited capacity to evaluate and judge each application on its own merits, check on-site or enforce the licensing process. Administrative pressure, and the proven threat that vested applicants can report any delays to the Water Tribunal, pushes officials towards allocating whatever is being asked for. The redistributive potential of water allocation reform risks fading away amid these legal complexities and to the detriment of small-scale users.

A parallel process to land and water sector reform is agricultural policy reform, which pays particular attention to irrigation policy. The overall objective of the agricultural policy reform process is to create more opportunities for smallholders and resource-poor farmers to improve productivity and contribute more to the mainstream economy. This notion was supported by the African National Congress (ANC)'s 2007 Polokwane conference, during which the importance was stressed of integrating smallholders into the formal value chain and linking them with markets. The problem is that there is insufficient support for the agricultural sector, which means that plans related to the agricultural sector cannot easily be carried out. This situation has not been helped by the fact that national government has reallocated resources from agriculture to other priority areas during the recent global economic depression. Agriculture is seen as a declining sector, as opposed to urban areas, which are seen as the future of the country (Greenberg, 2010).

So, despite efforts at socio-economic and political transformation, the legacy of apartheid policies has resulted in most available land and water for irrigation remaining in the hands of the large-scale commercial farming sector, low productivity levels of land transferred to beneficiaries of land reform, insufficient post-settlement support, very little knowledge by farmers of their use rights, and overall food security concerns to name a few.

Additionally, even though parallel processes of water allocation and land reform have been of high priority to the South African government, both have had less than satisfactory results. Water allocation and land reform processes have both had redistributive, socio-economic and social redress objectives, through which the South African government intended to make water and land vehicles for rural economic transformation. However, in many respects, the state has failed to live up to its reform objectives, facing backlogs, falling short of its targets, and contributing little to improving the productivity of beneficiaries of the water allocation and land reform programmes. A commonly cited reason for this failure is the uncoordinated nature of the land reform and water allocation reform policy formulation processes as well as the uncoordinated nature of their implementation. While the South African government has put in place several trans-sectoral instruments, procedures and principles to accommodate dual sectoral policy objectives, these two sectoral reforms still seem to operate in relative isolation of each other.

## 5.3 Problems with the land reform process

Problems with the land reform process include the fact that the land and water reform targets set at the national level have not been matched by meaningful implementation on the ground. The land reform programme in South Africa has been characterised by a slow pace of land redistribution and has failed to impact significantly on the land tenure systems prevailing on commercial farms and in the communal areas (Hall, 2009; Greenberg, 2010). In addition, the "willing-buyer-willing-seller" approach is only able to transfer modest amounts of land to a small minority of the rural population, while leaving the underlying structure of the agrarian economy largely intact (Walker, 2005). The perception exists that most of the land that has already been redistributed is not being used as productively as originally planned. There have also not been significant livelihood benefits for the majority of the beneficiaries (Lahiff, 2008; Cousins and Scoones, 2010; Greenberg, 2010). This is probably as a result of weak delivery systems and institutions, inadequate budgets, top-down implementation (with the high expectations of a passive citizenry) and very poor provision of agricultural support (Greenberg, 2010).

Another major shortcoming of the land redistribution process is the lack of resources made available for post-transfer support and adequate resources to beneficiaries (Turner and Ibsen, 2000; Cousins, 2005). Various studies have shown that beneficiaries experience severe problems accessing services such as credit, training, extension advice, transport and ploughing services, veterinary services, and access to input and produce markets (HSRC 2003; Hall 2004; Bradstock 2005; Lahiff 2007). Other challenges include the types of beneficiaries accessing the programme, drawn out transfer periods, lack of transparency and possible illegitimate activities of local government institutions, the often inappropriate models of land-use being imposed on beneficiaries, the general failure of post-settlement support and, ultimately, the generally disappointing performance of land reform projects (Aliber and Mokoena, 2000; CASE, 2006). In addition land may be transferred to groups who may not be interested in agriculture or have any agricultural experience, rather than motivated, interested and experienced individuals (De Lange et al., 2004). Farms are also often transferred in their entirety, rather than divided into smaller, more manageable units for small-scale farming purposes (Van Koppen, 2009).

In terms of training emerging farmers, a number of difficulties exist. Agricultural training colleges have been characterised by low student numbers, which has meant a shift from training extension officers to training farmers directly. The Agricultural Sector Education and Training Authority (AgriSETA) was established to provide work-based, functional training in agriculture. This institution is flooded with requests for training assistance from both farm workers and land reform beneficiaries. However, it only approves very few of these applications. For instance, in 2006/07 AgriSETA received 16245 applications for learnerships (only 400 were approved) and 59000 applications for skills programmes (only 475 were approved) (Greenberg, 2010).

In addition, the public agricultural extension office has declined over the past 15-20 years. In 2008, 2152 agricultural extension officers, who assist farmers and land reform beneficiaries, were active in South Africa. Of these 60% were working in the Eastern Cape and Limpopo. There is currently a ratio of one officer to 878 farmers (which is comparable to India, Zambia and Zimbabwe who face similar agricultural issues). The Department of Agriculture Forestry and Fisheries (DAFF) has an Extension Recovery Plan in place which is aimed at reviving public extension services by increasing numbers and reskilling public extension officers. This initiative is however not planned or budgeted for in all provincial departments. The potential role of community-based extension workers as auxilliaries can also be considered (Greenberg, 2010).

In terms of power asymmetries in the agricultural sector, an alliance of conservative landowners, agricultural economists, officials and analysts has been promoting the need for sustainable commercial viability among emerging farmers (Doyer, 2004). This orientation does not sufficiently capture and address the historical inequities of land and water ownership and rural poverty (Vink and Van Rooyen, 2009; Walker, 2005). The 2005 National Land Summit tried to address this problem by calling for land redistribution to be embedded within a wider agrarian reform process that focuses on poverty reduction and creating opportunities for smallholder farmers. This idea has however not been developed in more detail and it is not clear what this may mean for beneficiary selection, programme design, post-transfer support and agricultural policy in general (Lahiff, 2008). The weaknesses mentioned above reflect deep-seated structural and implementation shortcomings as well as inappropriateness of current redistribution models.

Another challenge is the role of traditional leaders, which has not been clearly defined. Traditional leaders continue to perform unregulated land administration functions outside of any legal framework. These functions would otherwise exceed the capacity of local government. As a result, functions of traditional leaders are not matched or aligned to the planning and development functions of elected local government, which in some cases is resulting in a stand-off between these two institutions. It is therefore important to find a solution to this issue (Manona, 2009).

## 5.4 Problems with the water reform process

In terms of water reform, gaps in access to water appear even wider than gaps in access to land. 95% of water for irrigation is primarily used by large-scale commercial farmers, while smallholders have access to the remaining 5%. New users therefore have to compete for the available water with well-entrenched users. Irrigated agriculture is the main user, taking up 72% of the available water resources. Water re-allocation is therefore one component of a wider mandate to address the inequalities of the past (Anderson et al., 2008).

In addition, very few water-use licenses have actually been awarded and taken up by emerging black commercial farmers. This means that farmers often have to put production on hold until a license is granted even though other infrastructure may be in place (Surplus People Project, 2007). Evidence is also increasing that many water and land reform projects are not leading to meaningful and efficient productivity on most of the "new" black-owned irrigated farms. The challenge is, amongst others, to synchronise reform programmes in irrigation areas and ensure that beneficiaries hold secure land and water use entitlements.

According to Van Koppen et al. (2009), in implementing land restitution and redistribution as part of the land reform programme, there was at first little collaboration between the former Department of Land Affairs (DLA) (now Department of Rural Development and Land Reform) and the former Department of Water Affairs and Forestry (DWAF) (now Department of Water Affairs). Riparian water rights were not always completely registered as part of the land entitlement. Also, in a few cases, water rights tied to land under claim were sold, leaving an asset of lesser value. Without readily available registers of land under claim, the DWAF could not easily track this problem. In the late 1990s, however, it introduced a policy that the trading of water rights of land under claim should not be approved. Further coordination has since been established between the DWAF, the provincial Departments of Agriculture and the provincial governments with the signing of a memorandum of understanding on collaboration on land and water reform in 2008.

There have also been arguments that question the wisdom of transferring land and water to beneficiaries who may not be able to use it productively. According to such narratives, attempts to address equity needs must be balanced with the consideration that many existing lawful water users are making productive, efficient and beneficial use and are contributing to socio-economic stability and growth (Adger et al., 2001; Forsyth, 2003). There is also an argument that if reallocations take place too quickly, the result is likely to be economic and environmental damage as emerging users struggle to establish productive uses of the reallocated water (Forsyth, 2003). These narratives have influenced government thinking and contributed to the maintenance of the status quo instead of rapid allocation of water use entitlements to the "new" farmers.

Another factor affecting the uptake of water use entitlements is that many emerging farmers are not sufficiently capacitated to understand their water needs, the scales and rates of payment for water rights, use and management of water or their roles and responsibilities on Water User Associations. On the other hand, large scale commercial farmers who have historically used water for productive purposes are more knowledgeable about administrative processes and can easily apply and obtain water licenses. It is therefore important that the capacity constraints experienced by emerging farmers be recognised and addressed if they are to begin to make a more meaningful contribution (Surplus People Project, 2007).

In cases where new farmers start irrigating their lands, they often do not properly determine the optimum irrigation potential of their farms. This means that chances for under-utilisation are high (Backeberg, 2005). In addition, Joubert and Kruger (2005) attribute the high failure rates of the new farmers to inadequate appraisal of farm potential (e.g. marginal farms that have been offered for sale), and unrealistic business plans designed by consultants who are only interested in maximising their commission paid by government, and which do not provide sufficient guidance to new farmers.

The apartheid government invested heavily in infrastructure (including dams, irrigation schemes, private pumps and farm dams) for white farmers as well as black smallholder irrigation. However, after 1994 state support to white irrigators declined, although at a much smaller scale than for smallholder irrigators, who suddenly lost almost all government support. Many smallholder schemes collapsed and the recent revitalisation efforts have not yet produced any results. The DWA and former DoA (now Department of Agriculture, Forestry and Fisheries) have undertaken some commendable efforts to promote water harvesting at homesteads for food security, but these efforts are still too marginal in numbers and volumes to really redress the problems that smallholder irrigators are facing. Without government champions to boost infrastructure development for small-scale water users, the prospects of achieving the WAR targets remain gloomy (Van Koppen, 2009).

Important competitors for water for smallholder irrigation are water for urban, energy and industry purposes, as well as water for the environment (the Ecological Reserve is provided for in the National Water Act). As a result, many urban-biased water resource managers tend to perceive the use of water for small-scale farming as an "unproductive" use (Van Koppen, 2009).

## 5.5 The importance of integrating water allocation and land reform for South Africa

Given the problems characterising the land and water reform processes, as discussed above, the integration of these processes has been supported by a number of authors.

Greenberg (2010) states that there is a realisation at the highest levels of government that the link between land reform, agricultural support and water resource provision is weak. There is thus a need to invest in irrigation, both for commercial and for resource-poor farmers, and also to link water provision to the land transfer process. It is essential to ensure that water is available to land reform farms and this must be built into the planning stages at the outset of the transfer process. Many land reform farms have failed as a result of water not being available for production. It is important to improve the synchronisation between water and land reform programmes in irrigation areas to ensure that beneficiaries hold secure land and water use rights, once they have been allocated the land (Groenewald, 2004). Derman (2005) argues that the distinction made between land and water in the reform programmes does not fit with local conceptions of livelihoods, or the increasing evidence of the importance of the land-water interface, including natural wetlands and irrigation systems.

The lack of linkages between water allocation and land reform policies has resulted in "dry", unsustainable land reform projects. There is therefore a need to align land and water reform programmes at both the policy and programme level, as both programmes are the cornerstones of the South African rural development strategy. Addressing this integration requires leaders in the land and water sectors to establish joint think-tanks aimed at finding workable solutions that enhance both programmes in pursuit of a sustainable rural development path (Greenberg, 2010).

Here it can be argued that while the integration of the land and water allocation reform processes is imperative, it may not be sufficient to ensure the successful functioning of the two processes. A number of other challenges have to be addressed, once the beneficiaries' basic needs and challenges have been identified. There may also be a need to develop a wider understanding and appreciation of water for productive uses at sustainable livelihoods levels and how this can impact on the quest for equity (Chikozho and Jacobs, 2010). Services that need to be provided include extension, training, credit/finance, marketing, inputs, infrastructure, management labour, capital equipment and provision of facilitation and strategic services that are appropriate to emerging farmers (Walker, 2005). Additional support structures that are needed are secure land title deeds, secure water/rights licenses, physical infrastructure such as water supply systems and roads, soft loans, markets, fertilizer, irrigation machinery, seeds, energy, information and research (Chikozho and Jacobs, 2010). There is also a need for the development of a coherent vision of equitable redistribution of water and sustainable economic transformation. This necessitates developing effective institutional mechanisms (that would need to differ from failed integration attempts in the past) that link water management to agriculture, land, finance and other support systems (Van Koppen et al., 2009).

Government departments and agencies have to create an enabling and supportive environment for new farmers in terms of infrastructure and institutional development. The greatest challenge in the reform processes is how to implement them and to ensure that the stated objectives and targets are met in a sustainable manner (Chikozho and Jacobs, 2010).

It is also important to consider not merely reproducing and expanding on the current commercial agricultural model in South Africa, but to take into account lessons from the past to build a more equitable agricultural model in South Africa which will not lead to a repeat of the mistakes of dispossession or environmental degradation (Chikozho and Jacobs, 2010). Transformation of the agrarian sector from its current extractive commercial form to a more equitable and sustainable form is key (Chikozho and Jacobs, 2010).

What form could an alternative agricultural model in South Africa take and importantly how could this realistically be implemented? While deracialisation of the agricultural industry is necessary, it is not sufficient. Ideas around the multifunctionalism of agriculture and food security suggest that food production is perhaps only one of the functions of agriculture. Other elements that are equally important and that complement the notion of needs-based smallholder agriculture are sustainable livelihoods, living landscapes and environmental integrity, which are all integral to rural sustainability (Greenberg, 2010).

## 5.6 Critical review of government policies to address the challenges to water allocation and land reform

South African government departments have attempted to increase the implementation of water allocation and land reform by coming up with a number of programmes. Here follows a critical review of each of these efforts followed by a summary describing to what extent they have been successful.

### 5.6.1 Comprehensive Agricultural Support Programme (CASP)

The Comprehensive Agricultural Support Programme (CASP) (2003) is the biggest sub-programme at the provincial level in all provinces except Gauteng and the North West Province (Greenberg, 2010).The CASP is designed to help black famers to participate in a market that is dominated by white agri-business, but without altering the logic of the market or production system. The money that is awarded as part of the CASP is used mainly for infrastructural development i.e. warehouses, access roads, irrigation systems.

Money is also spent on training and capacity building and marketing. Farmers apply on a yearly basis and grants are awarded for a five year period (DAFF, 2011a). The CASP is therefore a potentially very valuable support programme as it is meant to supply emerging farmers with much needed infrastructure.

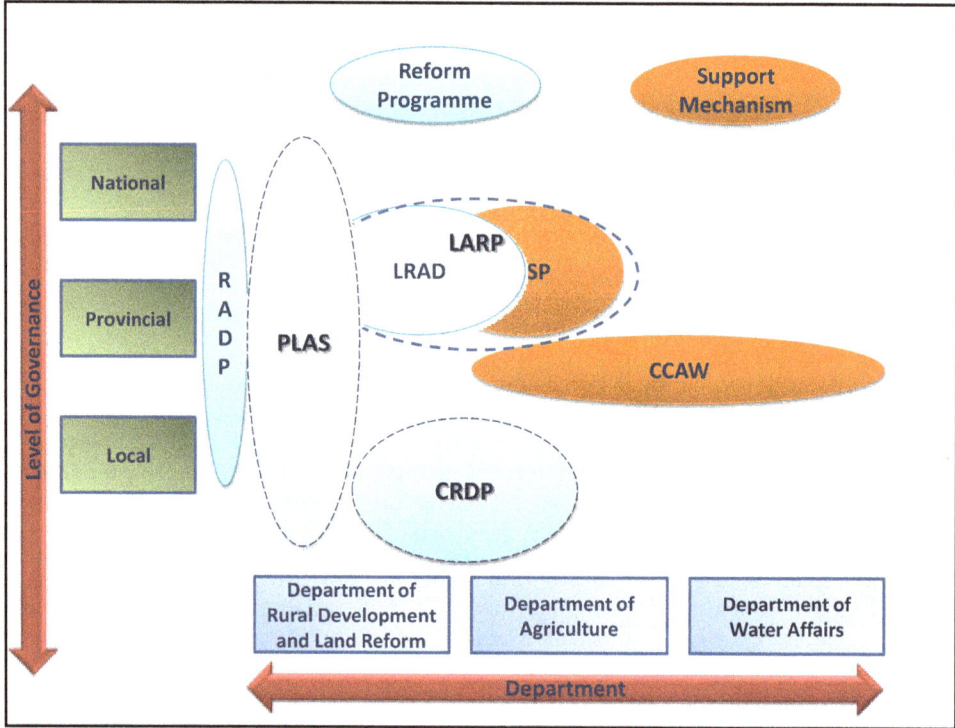

Fig. 5. Conceptualisation of transsectorality of reform programmes and support programmes

The South African Department of Agriculture, Forestry and Fisheries (DAFF) monitors the success of the impact of the CASP by checking whether the infrastructure that has been promised has been completed, and whether farmers are using it for the right purpose. While it is important to monitor whether infrastructure is being provided, no impact assessment studies have been conducted to establish the success of the CASP (DAFF, 2011a). When the former Department of Agriculture and the former Department of Land Affairs (now the Department of Rural Development and Land Reform) fell under one minister, it was easier to ensure that the land reform financial support programmes of both departments were coordinated. This has become more difficult now that the two competencies are no longer governed by the same minister (DRDLR, 2011a). Therefore, there used to be a greater level of transsectorality between the agriculture support specific and land reform specific programmes, which has now been compromised.

To date, the success of implementation of the CASP programme has been uneven, although most provincial farmer support programmes have been expanded (Greenberg, 2010). It appears that currently the CASP needs between three and four times its current budget in order to function effectively. Other difficulties include farmers not being aware of the different funding options they qualify for, the complicated nature of the government procurement process and the non-alignment of funding between different government departments (DRDLR, 2011a). In addition, it can take a number of years for a CASP application to be successful. This is problematic for new beneficiaries who want to start farming immediately and cannot wait years for financial support to set up infrastructure or use the money for other purposes (Raholane and Baloyi, 2011).

### 5.6.2 Land and Agrarian Reform Programme (LARP)

The government has attempted to integrate the CASP (the agricultural support programme) with the Land Redistribution for Agricultural Development (LRAD) programme (which focuses primarily on land reform) in the form of the Land and Agrarian Reform Programme (LARP), which was established in 2008 (Greenberg, 2010). The LARP is meant to offer collaboration on delivery and collaboration on land reform and agricultural support "to accelerate the rate and sustainability of transformation through aligned and joint action of all involved stakeholders". The idea is to have "one stop shop" service centres in close proximity to farms and beneficiaries (LARP, 2008). Implementation of the LARP has been slow and to date there is little evidence of any significant change in practice (Greenberg, 2010). This programme again demonstrates an attempt at coordinating different sectors – agriculture and land – and services from these sectors to serve beneficiaries of land reform with a large focus on stakeholder involvement.

### 5.6.3 Comprehensive Rural Development Programme (CRDP)

A subsequent attempt by the South African government to integrate agricultural support, land reform and broader rural development without putting more money into rural areas has been the Comprehensive Rural Development Programme (CRPD) (CRDP, 2011). Again, this is an attempt at transsectoral coordination between agriculture and land, this time with an additional focus on broader rural development. The programme aims to achieve "co-ordinated and integrated broad-based agrarian transformation, an improved land reform programme and strategic investments in economic and social infrastructure in rural areas" (CRDP, 2011). It is likely that this programme will run into difficulties as it relies on the weak institutions of the former Department of Land Affairs (which now is the Department of Rural Development and Land Reform with a bigger mandate but not a bigger budget). It also appears that the approach to planning and implementation is rushed with a focus on immediate delivery at all costs. The consequences of this approach are poor quality and lack of sustainability. Policy-making continues to be dominated by agri-business, which exerts a strong influence on the agricultural sector (Greenberg, 2010).

### 5.6.4 Pro-active Land Acquisition Strategy (PLAS)

An important component of revising the land reform programme has been the Proactive Land Acquisition Strategy (PLAS). This programme currently involves approximately 1000 farmers. As part of the programme, land is leased out to beneficiaries for a trial period of three to five years during which they have to prove that they can productively use the land for agricultural purposes. This programme has been in place since April 2010 (DRDLR,

2011b). One of the potential benefits of this programme is that it moves away from handing over land ownership rights to beneficiaries, which has often led to failure in terms of productivity in the past, and instead requires beneficiaries to prove that they are able to be productive by leasing land to them for a limited trial period. Beneficiaries have complained that this programme sets them up to fail and that the absence of a title deed makes it impossible for them to get financial assistance from banks. A reference group has been formed to further investigate the matter. It is still to be decided whether beneficiaries will eventually be able to own the land that is leased to them (DRDLR, 2011b).

As part of PLAS, grants to the value of 25% of the value of the land are to be awarded. This award will be once-off for now, but it is planned that in future it will be invested over a period of five years. This will take the form of a pyramid scheme with most of the money being awarded in Year One, and then less and less with farmers co-investing more of their own money over the next five years. The idea is not to give aid to the emerging farmers/beneficiaries but to teach them to farm on their own (DAFF, 2011a).

In addition, to aid emerging farmers, it is planned that they will team up with strategic partners, namely established commercial farmers, who will share in the profits and risks of the new enterprises. This partnership is regulated by means of an agreement between the strategic partner (farmer), the beneficiaries and the DRDLR (DAFF, 2011a).

Strategic partners are supposed to oversee activities on farms, ensure that the sowing and harvesting happens when it is supposed to, repair infrastructure, ensure that water allocations are paid for etc. The involvement of strategic partners has worked well in some cases as these partners often have in-depth knowledge of the ins and outs of commercial farming and are therefore able to ensure that productive farming takes place. This approach can however also be problematic as it can in some cases engender an over-reliance of beneficiaries on the strategic partner, as the partners often take over the management of the farm completely and also often leave the farming operation after their five-year contract has expired. This leaves the beneficiary with little new knowledge of how to manage a farm and also deprives them of the independence of managing the farming operation on their own (DAFF, 2011b).

### 5.6.5 Recapitalisation and Development Programme (RADP)

In response to the implementation challenges of CASP as well as the Settlement and Implementation Support (SIS) Programme, the DRDLR introduced the RADP to address programmatic weaknesses, such as the lack of monitoring for example. The RADP applies to all emerging farmers needing support and future land transactions, and aims to ensure increased production and food security; to graduate small farmers into commercial farmers; to create employment opportunities in the agricultural sector; to promote capacity building through training and mentorship; and to establish rural development rangers (DRDLR, 2010). The programme is to be sustained by the Recapitalisation and Development Fund (RDF), created from 25% of the baseline land reform budget per annum (DRDLR, 2010), and replaces the following land reform grants:

- The 25% PLAS Operational Budget
- The 25% Household Development Grant
- The 25% Restitution Development Grant

- The Restitution Settlement Grant
- Commonage Infrastructure Grant

Additionally, it places an emphasis on compliance to strict monitoring criteria. In this regard, it "will issue stringent conditions for those who qualify to benefit from it so as to avoid creating a culture of entitlement from unscrupulous individuals who are in it, for personal gain (DRDLR, 2010: 4).

### 5.6.6 Coordinating Committee on Agricultural Water (CCAW)

CCAW is a non-statutory cooperative government structure that serves as a provincial mechanism for joint effort between the Departments' of Water Affairs, Agriculture and DRDLR. Its objective is to ensure that government-funded projects are sustainable from a water utilisation, agricultural engineering and economic perspective. Projects submitted have to be evaluated to determine their feasibility and sustainability. Ultimately, the CCAW should also be responsible for the evaluation of any water use license application that is submitted to the DWA, however, the status and effectiveness of each provincial CCAW varies from functional to non-existent. Some have therefore not taken on this evaluating task, in which case it falls to the DWA.

### 5.6.7 Evaluation of "Integration" programmes

Each of the above efforts has been an important attempt at integrating the land and water allocation reform processes more closely. It is imperative for the different government competencies to work more closely together, and also to find ways of adjusting the current land and water allocation reform models to try and address some of the shortcomings of past attempts at effective water and land reform. Initiatives such as leasing out land until farmers can show that they are able to be productive with sufficient government support in terms of infrastructure and cooperation with strategic partners are potentially very valuable.

Unfortunately, however, to date none of the programmes seems to have been functioning ideally. The problem has been that attempts at integration between the water allocation and land reform processes have been fraught with difficulties, often linked to the design of the different programmes as well as the organisational weaknesses at the governmental level.

Such difficulties include budgets for water allocation and land reform programmes being housed in different departments and funding not being coordinated, underfunding for certain programmes, a lack of monitoring capacity by government departments to establish how well the programmes are being implemented and whether they are successful, a lack of awareness among emerging farmers about which funding options are available to them, a "quick fix" approach with not sufficient attention being paid to quality programmes and quality implementation, an overreliance by the government and emerging farmers on strategic partners.

In addition, different government departments generally do not communicate effectively with each other, do not know who they need to be speaking to in their sister departments and have no clear idea of what activities other government departments are engaging in. In addition to challenges of inter-departmental cooperation and coordination, there is also a problem with intra-departmental communication as decisions that are made at ministerial level, specifically in this instance those regarding cooperation with other departments, do not filter down to the lower levels of government. The top-down ways in which decisions are made therefore impact negatively on cooperation between operational managers in different government departments. Conversely, if government officials at the operational level wish to collaborate more closely, it becomes difficult for them to obtain the approval for such cooperation from their superiors, given the substantial amount of bureaucratic red tape that South African government departments are characterised by.

## 6. Conclusion

Given that integration of water allocation and land reform is very important but at the same time also a problem in South Africa, not only in terms of policy development but also in terms of implementation, what recommendations can be given for more effective integration of these two processes in future? In addition, what lessons does this chapter provide to other countries in terms of the impacts of non-integration and the challenges to successfully implementing integrated reform programmes?

With the Departments for Water Affairs and Rural Development and Land Reform (DWA and DRDLR) jointly acknowledging the importance of joint water allocation and land reform (Kleinbooi, 2009), there is a renewed onus on the South African government to achieve higher levels of integration between these two processes. The question now remains how this can best be achieved.

Integrated water allocation and land reform needs to go beyond quick-fix attempts to try and merge different existing programmes, and instead has to focus on identifying the root causes of why existing programmes are not working and how these causes can best be addressed. It is of course also important to ground any water and land reform integration programmes in the context of the South African legislative framework to ensure that the ethos of the country's progressive legislation is adhered to. Noticeably, existing trans-sectoral coordination efforts seem to have focused mostly on collaboration between the departments (DAFF, DRDLR and DWA). Other government departments that might also have an important role to play, such as the Department of Environmental Affairs (DEA), in terms of the environmental sustainability of agricultural practices, and the Department of Cooperative Governance and Traditional Affairs (DCGTA), perhaps need to be more involved.

In addition, three important premises can be identified to achieve more effective integration between the water allocation and land reform processes in South Africa? Firstly, it is important to acknowledge the multiplicity of the actor landscape and the presence of different stakeholder perspectives, linkages and interdependencies with other resources and sectors as a starting point. This may involve bringing on board other key stakeholders such as macro- and micro- lending institutions (for example, the Land Bank), commercial farmers who function as "strategic partners", irrigation boards, water user associations, land reform beneficiaries, and members of civil society, to try and find more innovative and inclusive solutions to address the need for integration. By determining the needs of stakeholders on the ground, it may be easier to establish how coordination between different parties may function more effectively. What should be key for government departments when involving a range of stakeholders is knowing when to solicit whose inputs and doing so strategically to prevent themselves from being overwhelmed by too many inputs all at once as this could be counter-productive.

Secondly, different government structures need to stop working in silos and need to start cooperating in terms of budget allocation and promoting integration.

Thirdly, there needs to be an acknowledgement that increased levels of integration and communication can take a long time, which necessitates patience, endurance and a long term vision on behalf of those who are seeking to improve integration.

Other countries can also benefit from this analysis by taking note of some of the impacts of non-integration of water and land management related programmes and the challenges to successfully implementing integrated reform programmes. The impacts of non-integration include governments having to deal with the effects of failed programmes and stakeholder collaboration, and the simultaneous manifestation of a disjuncture between policy and practice. Another impact is that failed integration efforts cause promising paradigms such as IWRM to lose credibility, both at the national and international level. It seems so difficult to implement integration focused water and land management programmes because of the bureaucratic culture of managing projects and programmes in silos. Implementation is furthermore impeded by government structures traditionally being hierarchical and compartmentalised and making it difficult for information to flow freely and easily between different units within government. Therefore there is a need in South Africa and elsewhere, when promoting greater levels of integration between water and land management related programmes, to try to ensure that different government structures work together both horizontally across sectors, and hierarchically from the national to the local level. In addition, a multi-stakeholder and multi-sectoral effort is required at all levels, from the local to national if integration is to be operational and implementable.

## 7. Acknowledgment

The authors would like to thank the South African Water Research Commission (WRC) for the role they have played in soliciting and funding a research project on which this paper is loosely based. The case study and contextual background study is based on work conducted in this project, "An Investigation of Water Conservation in Food Value Chains by Beneficiaries of Water Allocation Reform and Land Reform Programmes in South Africa" (K5/1958/4). More information on these and similar projects can be found at the WRC's website: http://www.wrc.org.za/

## 8. References

Adger, N. W., Benjaminsen, T. A., Brown, K. and Svarstad, H. (2001). Advancing a political ecology of global environmental discourses. *Development and Change* Vol. 32 No. 4, pp. (687-715), ISSN 0012-155X.

Aliber, M. and Mokoena, R. (2000). *The land redistribution programme and the land market.* Department of Land Affairs, Unpublished paper, Pretoria.

Anderson, K., Kurzweil, M., Martin, W., Sandri, D., Valenzuela, E. (2008). Measuring distortions to agricultural incentives, revisited. World Bank Policy Research Working Paper 4612, April 2008, Washington.

Ashton, P.J., 2007. The Role of Good Governance in Sustainable Development: Implications for Integrated Water Resources Management in Southern Africa. In: Turton, A.R., Hattingh, J., Maree, G.A., Roux, D.J., Claassen, M. and Strydom, W. (eds.), *Governance as a Trialogue –Government-Society-Science in Transition*, pp. 78-60, Springer-Verlag, ISBN 978-3-540-46265-1, Berlin, Germany.

Ashton, P.J., Hardwick, D., and Breen, C. (2008). Changes in Water Availability and Demand Within South Africa's Shared River Basins as Determinants of Regional Social and Ecological Resilience. In: *Exploring Sustainability Science : A Southern African Perspective*, M.J. Burns and A.v.B. Weaver (Eds), Stellenbosch University Press, ISBN 978-1-920109-51-6, Stellenbosch.

Ashton, P.J. (2010). The road ahead. In : A CSIR perspective on water – 2010. CSIR Report No. CSIR/NRE/PW/IR//2011/0012/A. ISBN 978-0-7988-5595-2, Pretoria, CSIR.

Backeberg, G.R. (2005). Water institutional reforms in South Africa. *Water Policy* Vol. 7 (2005), pp. (107–123), 1366-7017.

Backeberg, G.R. (2006). Reform of user charges, market pricing and management of water: problem or opportunity for irrigated agriculture? *Irrigation and Drainage* Vol. 55 (2006), pp. (1-12).

Backeberg, G.R. (2007). Allocation of water use rights in irrigated agriculture: experience with designing institutions and facilitating market processes in South Africa. Paper presented at USCID 4th International Conference on Irrigation and Drainage; 5 October 2007; Sacramento, California, USA; Copyright © 2007, US Committee on

Backeberg, G.R. and Odendaal, P.E. (1998). Water for agriculture: a future perspective. Proceedings of the 39th Ordinary General Meeting of the Fertilizer Society of South Africa (FSSA). 24 April 1998. Sun City Hotel. Pilansberg. FSSA Journal. Pretoria. Irrigation and Drainage.

Bradstock, A. (2005). *Key experiences of land reform in the Northern Cape Province of South Africa.* Policy and Research Series, FARM-Africa, ISBN: 1 904029 02 7, London.

Calder, I.R. (1999). *The Blue Revolution: Land Use and Integrated Water Resources Management.* Earthscan, ISBN 1-85383-634-6, London, U.K.

CASE (Community Agency for Social Enquiry) (2006). Assessment of the status quo of settled land restitution claims with a developmental component nationally. Research conducted for the Monitoring and Evaluation Directorate, Department of Land Affairs.

Chikozho, C. and Jacobs, I. (2010). The key factors and variables influencing water allocation and land reform programmes in South Africa. Position paper for WRC project KC/1958/4.

Claassen, M. (2010). How much water do we have? In :*A CSIR perspective on water – 2010.* CSIR Report No. CSIR/NRE/PW/IR//2011/0012/A, ISBN 978-0-7988-5595-2, Pretoria, CSIR.

Cousins, B. (2005). Agrarian reform and the 'two economies': transforming South Africa's countryside. In: *The land question in South Africa: the challenge of transformation and redistribution*, Hall, R and Ntsebeza, L (Eds), HSRC Press, ISBN, 0796921636, London.

Cousins, B. And Scoones, I. (2010). Contested paradigms of viability in redistributive land reform perspectives from southern Africa, The Journal of Peasant Studies, Vol. 37, No. 1 (2010), pp. ( 31-66), ISSN 0306-6150.

CSIR (2010). *A CSIR perspective on water in South Africa – 2010*. CSIR Report No. CSIR/NRE/PW/IR/2011/0012/A, ISBN 978-0-7988-5595-2.

De Lange, A. Swanepoel, F., Nesamvuni, E., Nyamande-Pitso, A. and Stroebel, A. (2004). The evaluation of empowerment policies, strategies and performance within the agricultural sector: Executive summary. University of the Free State, Bloemfontein.

De Lange, W. (2010). The water situation in South Africa: some inconvenient truths. In: A CSIR perspective on water – 2010. CSIR Report No. CSIR/NRE/PW/IR//2011/0012/A, ISBN 978-0-7988-5595-2, Pretoria, CSIR.

Department of Water Affairs and Forestry (DWAF) (1997). White Paper on a National Water Policy for South Africa, Government of the Republic of South Africa, Pretoria.

Department of Water Affairs and Forestry (DWAF) (2006). A strategy for water allocation reform in South Africa. DWAF, Pretoria.

Derman, B. (2005). The incredible heaviness of water: Water policy and water reform in the new millennium in Southern Africa. In: *Globalization, water and health: resource management in times of scarcity.* Pp. 209-230. Whiteford, S., and Whiteford, L. (Eds), James Currey Publishers, ISBN 0-85255-974-7, Oxford.

Doyer, T. (2004). BEE: A call for calm. Farmer's Weekly, 13 August 2004: 8.

European Union (EU) (2000). Water Framework Directive, 20 July 2011. Available from: <http://europa.eu.int/comm/environment/water/water-framework/index_en.html>

Forsyth, T. (2003). *Critical Political Ecology: The politics of environmental science*, Routledge, ISBN 0 415185637, New York.

Funke, N., Oelofse, S.H.H., Hattingh, J., Ashton, P.J. and Turton, A.R. 2007a. IWRM in Developing Countries: Lessons from the Mhlatuze Catchment in South Africa. *Physics and Chemistry of the Earth*, Vol. 32, pp. 1237-1245, ISSN 1474-7065.

Funke, N., Nortje, K., Findlater, K., Burns, M., Turton, A., Weaver, A. and Hattingh, H. 2007b. Redressing inequality: South Africa's new water policy. Environment, Vol. 49, No.3, (April, 2007), pp. (12-23).

Funke, N. and Nienaber, S. (in press). Promoting uptake and use of conservation science in South Africa. Accepted in *Water SA.*

Funke, N., Nienaber, S. and Gioia, C. (in press). Interest groups at work: environmental activism and the case of Acid Mine Drainage on Johannesburg's West Rand. To be published in a Handbook on Activism by the University of Pretoria.

Global Water Partnership, Technical Advisory Committee (GWP-TAC) (2000). GWP-TAC. 2000. 'Integrated Water Resources Management', Background paper No 4. Global Water Partnership – Technical Advisory Committee, 71 p, Global Water Partnership, ISBN: 91-630-9229-8, Stockholm, Sweden.

Gooch, G.D. and P. Stålnacke. (Eds.) (2006). *Integrated Transboundary Water Management in Theory and Practice: Experiences from the New EU Eastern borders.* IWA Publishing, ISBN: 9781843390848, London.

Greenberg, S. (2010). *Status Report on Land and Agricultural Policy in South Africa*, 2010. Research Report 38, Programme for Land and Agrarian Studies, University of the Western Cape, ISBN 978-1-86808-703-7, Cape Town, South Africa.

Groenewald, J.A. (2004). Conditions for Successful Land Reform in Africa. *South African Journal of Economic Management Science*, Vol. 7, No. 4., Juta & Co. ISSN 1015-8812

Hall, R. (2004). Land and agrarian reform in South Africa: A status report 2004. Research Report No. 20, Programme for Land and Agrarian Studies (PLAAS), University of the Western Cape, ISBN: 1-86808-600-3, Cape Town.

Hall, R. (2009). Land reform in South Africa: Successes, challenges and concrete proposals for the way forward. Proceedings of the Workshop on 'Land Reform in South Africa: Constructive Aims and Positive Outcomes – Reflecting on Experiences on the Way to 2014'. ISBN: 978-0-9814032-6-7, Roode Vallei Country Lodge, Pretoria, 26–27 August 2008.

Hattingh, J., Maree, G., Oelofse, S., Turton, A. and van Wyk, E. (2004). Environmental governance and equity in a democratic South Africa, paper presented at the AWRA/IWLRI International Conference on Water Law Governance in Dundee, Scotland.

Hobbs, P. (2010). Water and sustainable mining. In: A CSIR perspective on water – 2010. CSIR Report No. CSIR/NRE/PW/IR//2011/0012/A, ISBN 978-0-7988-5595-2, Pretoria, CSIR.

Hobbs, P. and Kennedy, K. (2010). Acid mine drainage: addressing the problem in South Africa. Report No.: CSIR/NRE/WR/IR/2011/0029/A. Pretoria, CSIR.

HSRC (Human Sciences Research Council) (2003) Land redistribution for agricultural development: Case studies in three provinces. Unpublished report. Integrated Rural and Regional Development Division, HSRC, Pretoria.

Jacobs, P. (2003). Support for agricultural development, *Evaluating land and agrarian reform in South Africa: An occasional paper series*. Hall, R (Ed), Institute of Poverty, Land and Agrarian Studies, University of the Western Cape, ISBN: 1-86808-588-0, Cape Town.

Jacobs, I.M. 2010. Norms and transboundary co-operation in Africa: the cases of the Orange-Senqu and Nile Rivers. PhD Dissertation: University of St. Andrews.

Jonch-Clausen, T., 2004. *Integrated Water Resources Management (IWRM) and Water Efficiency Plans by 2005 Why, What and How?* Global Water Partnership, ISBN 91-974559-5-4, Stockholm, Sweden.

Joubert, R. and Kruger, G. (2005). Land reform and redistribution implementation in the Mpumalanga Province sugar industry, South Africa: A process approach. RICS Research Paper Series, Vol. 5, No. 10, Royal Institute of Chartered Surveyors, London.

Kleinbooi, K. (2009). Resources and rights: water and land in rural development. PLAAS Umhlaba Wethu 9, November 2009.

Lahiff, E. (2007). "Willing buyer, willing seller": South Africa's failed experiment in market led agrarian reform. *Third World Quarterly*, Vol. 28 No. 8, pp. (1577-98), ISSN 0143-6597.

Lahiff, E. (2008). Land reform in South Africa: A status report 2008. Research Report 38, Programme for Land and Agrarian Studies (PLAAS), University of the Western Cape, ISBN 9781868086849, Cape Town.

Lyne, M. and Darroch, M. (2003). Land redistribution in South Africa: Past performance and future policy. In: *The challenge of change: Agriculture, land and the South African economy*, Nieuwoudt, L. and Groeneveld, J. (Eds.), pp. (65-86), University of Natal Press, ISBN 186914032X, Pietermaritzburg.

Maherry, A. (2010). Map of South Africa's annual rainfall. In :*A CSIR perspective on water – 2010*. CSIR Report No. CSIR/NRE/PW/IR//2011/0012/A, ISBN 978-0-7988-5595-2, Pretoria, CSIR.

Manona, S. (2009). Key policy challenges for rural development: land, water and traditional leaders. PLAAS Umhlaba Wethu 9, November 2009.

Mostert, E. (1998). River Basin Management in the European Union; How It Is Done and How It Should Be Done. *European Water Management*, Vol. 1, No. 3, (1998), pp. (2635), ISSN 1994-8549.

Nel. J. (2010). Sustainable water ecosystems. In: *A CSIR perspective on water – 2010*. CSIR Report No. CSIR/NRE/PW/IR//2011/0012/A, ISBN 978-0-7988-5595-2, Pretoria, CSIR.

PLAAS. (2009). Umhlaba Wethu 9: A bulletin tracking land reform in South Africa. November 2009, Institute for Poverty, Land and Agrarian Studies, Date of access 20 July 2011, Available from:
http://www.plaas.org.za/pubs/ne/uw/Umhlaba_Wethu_9.pdf/

Pollard, S. and Du Toit, D. 2005. Achieving Integrated Water Resources Management: the mismatches in boundaries between water resources management and water supply, paper presented at the International Workshop on African Water Laws: Plural Legislative Frameworks for Rural Water Management in Africa in Johannesburg, South Africa, 2005.

Postel, S. and Richter, B. (2003). *Rivers for Life: Managing Water for People and Nature*. Island Press, ISBN 1-55963-444-8, Washington D.C.

Pressley, D. (2010). Farming reform a failure for 9 out of 10. Business Report. Pretoria News, Wednesday, March 3, 2010.

Republic of South Africa (1998). National Water Act (Act No. 36 of 1998), Government of the Republic of South Africa, Pretoria.

Sayer J.A and Campbell, B. (2004). *The Science of Sustainable Development: Local Livelihoods and the Global Environment*. Cambridge University Press, ISBN 0-521-82728-0, Cambridge.

Schreiner, B., van Koppen, B. and Khumbane, T. 2002. From bucket to basin: a new paradigm for water management, poverty eradication and gender equity. In: *Hydropolitics in the developing world: a Southern African perspective*, A.R. Turton and R. Henwood, pp. (127-140), African Water Issues Research Unit, ISBN 0-620-29519-8, Pretoria.

Surplus People Project (2007). Input on poverty alleviation: Emerging farmers and forestry and water licensing to previously disadvantaged individuals. SPP, Pretoria.

Strydom, W. (2010). Map of water use per sector. In: *A CSIR perspective on water – 2010*. CSIR Report No. CSIR/NRE/PW/IR//2011/0012/A, ISBN 978-0-7988-5595-2, Pretoria, CSIR.

Thomas, R. J. (2002) Revisiting the Conceptual Framework : Project Sites and Results of Assessment Methodology 97 for INRM Developed in Penang and Cali. *Proceedings of Putting INRM into Action*. F. Turkelboom, R. LaRovere, J. Hageman, R. El-Khartib and K. Jazeh (eds.), 4th INRM Workshop held at ICARDA, Aleppo, Syria, 16-19 September, 2002.

Turner, S. and Ibsen. I. (2000). Land and agrarian reform in South Africa: a status report. Research Report No. 6. Cape Town: Institute of Poverty, Land and Agrarian Studies, University of the Western Cape.

Van Koppen, B. (2009). Widening gaps in water reform. PLAAS Umhlaba Wethu 9, November 2009.

Van Koppen, B., Sally, H., Aliber, M., Cousins, B. and Tapela, B. (2009). Water resources management, rural redress and agrarian reform. Development Planning Division Working Paper Series No.7, DBSA, Midrand.

Vink, N. and Van Rooyen, J. (2009). The Economic Performance of Agriculture in South Africa since 1994: Implications for Food Security. Development Planning Division Working Paper Series No.17, DBSA, Midrand.

Walker, C. (2005). The limits to land reform: Rethinking 'the land question'. *Journal of Southern African Studies*, Vol. 31, No. 4, (December 2005), pp. (805-824), ISSN 0305-7070.

Water Research Commission (1996). Policy proposal for irrigated agriculture in South Africa. Discussion document prepared by Backeberg, G.R., Bembridge, T.J., Bennie, A.T.P., Groenewald, J.A., Hammes, P.S., Pullen, R.A. and Thompson, H. Report No. KV96/96, Pretoria.

Woodhouse, P. (2008). Water Rights in South Africa: Insights from Legislative Reform. BWPI Working Paper 36, Brooks World Poverty Institute, ISBN 978-1-906518-35-6, Manchester.

Department of Agriculture, Forestry and Fisheries (DAFF). 2011a. Personal communication with government official. 7 February 2011, Pretoria, South Africa.

Department of Agriculture, Forestry and Fisheries (DAFF). 2011b. Personal communication with government official. 22 February 2011, Pretoria, South Africa.

Department of Rural Development and Land Reform (DRDLR). 2011a. Personal communication with government official. 21 February 2011. Pretoria, South Africa.

Department of Rural Development and Land Reform (DRDLR). 2011b. Personal communication with government official. 27 January 2011. Pretoria, South Africa.

# 6

# Paddy Water Management
# for Precision Farming of Rice

M.S.M. Amin, M.K. Rowshon and W. Aimrun
*Smart Farming Technology Centre of Excellence*
*Department of Biological and Agricultural Engineering, Faculty of Engineering,*
*Universiti Putra Malaysia (UPM), 43400 UPM Serdang, Selangor DE,*
*Malaysia*

## 1. Introduction

Irrigation is the largest water user in the world, using up to 85% of the available water in the developing countries [1]. A lot of irrigation water is used in the production of rice as the staple food of more than half the world population. However, despite the constraints of water scarcity, rice production must be raised to feed the growing population. Producing more rice with less water is therefore a formidable challenge for food, economic, social and water security.

Asia is relatively well endowed with water resources, but water resources per inhabitant are only slightly above half of the world's average. Countries like India and China are approaching the limit of water scarcity. About 84% of water withdrawal is for agriculture, with major emphasis on flooded rice irrigation. There has been a rapid increase in irrigation development. Most countries have achieved self-sufficiency in rice. Schemes are designed primarily to secure rice cultivation in the main cropping season. Some countries design new irrigation schemes for year-round irrigation. Rice represents about 45% of irrigated areas and 59% of the rice land is irrigated. Average cropping intensity is 127%. The 28 million hectares under intense irrigation producing two to three crops per year suffer from declining productivity.

Growth in irrigated areas has declined in recent years. Groundwater draw-down has reached alarming levels in many areas. Declining prices of rice, higher marginal development costs, environmental concerns, and poor performance of existing schemes are among the main factors of the decline in irrigation growth and investment both by governments and farmers.

Increased competition for water between sectors already affects agriculture. Poor operation and maintenance of large public schemes has led to irrigation management transfer or increased participation of users through water users' associations. Socio-economic changes and water scarcity call for a transformation of irrigation by the adoption of measures to modify water demands and maximize efficiency in water use, to improve its economic, technical, and environmental performance, together with diversification of produce and cropping patterns, changes in management systems and structures, financial and fiscal sustainability. But rehabilitation programs are assuming increasing importance. Progress in modernization is slow when compared with other regions.

Scenarios for growth in water demand suggest that because of the projected increases in food demand, irrigated food production will need to increase significantly. Demand from other sectors will also increase because of economic development and increase in population. Nearly all countries in the region will need to invest considerable efforts and resources in a mixture of improved demand management of the water sector and interventions on the supply side to achieve the very considerable improvements in water use which are required. But approximately 1 billion people would live in regions of absolute water scarcity.

Therefore there is a need to improve water productivity as well as water use efficiency. Land preparation, land soaking for maintaining water level in the paddy fields and soil saturation require more water than plant transpiration. System and farm irrigation efficiency is quite low (in the range of 30 to 40%). A river basin perspective should be adopted, defining the boundaries of intervention (farm, system, river basin), paying attention to managing the return flows and to water quality. However, practices which minimize irrigation inflow are of a direct interest to farmers who receive less water and more costly water. In the long run, sustainability of irrigated agriculture will ensure sustainable environment for all human beings.

Environmental sustainability is very synonymous with precision farming or site-specific management. Precision farming requires quick soil spatial variability description for decision-making on the right input at the right place, at the right time and in the right amount or site-specific zone management. VerisEC sensor is used widely for spatial variability description and it relates to soil properties such as salt concentration, texture and cation exchange capacity (CEC), an indication of the soils potential to hold plant nutrients. It is a sensor to measure the ability of soil in conducting electrical current using rotating discs as electrodes, which penetrate 6 cm into the soil, while pulled through the field by a tractor and locations determined by GPS. For upland crops, farmers use VerisEC to measure $EC_a$ at shallow and deep depths for Nitrogen management and hardpan depth determination.

Land management zone delineation using soil electrical conductivity (MAZDEC) shortens the time taken to determine paddy soil variability, can be utilized for zoning of paddy fields, and helps rice farmers in site-specific application of their inputs. MAZDEC can complement a detailed soil series map and can be used as an estimator for soil physical and chemical properties. MAZDEC allows directed soil sampling to replace grid sampling, allows topping-up of the required nutrients at the needed rate at the right place and time. Each farmer will be able to quickly determine the soil management zones for variable application rates of seed, fertilizers, and water. Making this information available on-line to the farmers will be a major step in making available the benefits of new technologies.

## 2. Irrigation water management

### 2.1 Research on rice irrigation water management

In Malaysia, about 70% of the available water resources are consumed for rice production. Due to rapidly growing population and water competition among different sectors it is imperative that the available irrigation supplies be used efficiently. A small improvement of water use in rice production would result in significant water savings for other sectors. Traditionally rice is grown under continuous submergence or intermittent or variable ponding conditions depending on the farmer's choice and also on the available water resources. Continuous submergence with 5 to 7 cm of water is probably the best for irrigated rice considering all factors and extremely deep water resulted in poor growth and yield [2]. To evaluate the effect of ponding water depth on rice yield, the 9 cm of ponding water depth in Wagner's pots (a growth chamber 25 cm diameter and 30 cm height, filled with soil up to 15 cm depth) gave the optimum rice growth and yield [3]. Therefore, the importance of controlled water supply and monitoring is indispensable for the sustainability of rice production, which varies enormously from region to region.

Performance assessment has been prioritized as the most critical element to improve irrigation management [4]. Various analytical frameworks, criteria and indicators are available to understand the factors that influence the performance of irrigation system and to quantify water delivery performance. Some of them are suitable to identifying reasons for poor performance and prescribing management and physical interventions to improve the performance. All performance indicators have their own strengths and weaknesses. Many performance indicators fail to quantify reliability, adequacy and equity aspects of water distribution although many performance indicators are useful for quantifying the water delivery for irrigation seasons. Many performance indicators are useful for post-season evaluation but they are not useful as management tools which can be used to keep track of irrigation delivery performance during a crop growing season. If a particular indicator is useful both as a management tool and as an indicator of the overall irrigation delivery performance for any given irrigation interval or season, its credibility is undoubtedly superior to the other performance measures.

Detailed reviews for the advantages and disadvantages for various performance assessment methodologies [5] emphasized the simplest indicator of evaluating water delivery performance and how tightly adequate water can be delivered to the fields. The available water supply and the water demand are the most crucial factors in irrigation planning, design and operation of an irrigation system. A performance indicator [6], which is called Relative Water Supply (RWS), is simply the ratio of irrigation supply and demand. Detailed application and weakness have been described for monitoring and assessing of irrigation water delivery performance for rice irrigation scheme using the RWS concept [7 & 8]; and further illustration on the use of the RWS concept based on field research [9 & 10]. The RWS concept has been widely used to assess the performance of irrigation systems. Indeed for paddy irrigation, quantifying of the upper bound value of RWS for oversupply condition is a difficult task due to the many variables that influence the performance of irrigated agriculture. Therefore, it is essential to have an appropriate tool with feasible options to improve the performance of the irrigation supplies for rice production.

The water demand for rice irrigation is completely different from upland crops. To replenish the field water level up to the Maximum Ponding Water Depth ($WSmax_j$) for a particular period, the amount of water for the difference between the Maximum Ponding Water Depth ($WSmax_j$) and the Present Ponding Water Depth ($WS_j$) is gradually delivered to the paddy fields. The RWS concept incorrectly characterizes irrigation delivery performance for not considering this amount of water ($WSmax_j$ - $WS_j$) in the denominator. Due to this, RWS gives a wrong scenario to monitor irrigation water delivery performance [11]. To overcome this conflict, new indicators known as Rice Relative Water Supply (RRWS), Cumulative Rice Relative Water Supply (CRRWS) and Ponding Water Index (PWI) are introduced to evaluate the irrigation delivery performance especially for the paddy irrigation system.

To improve irrigation management with variable irrigation supplies, GIS is an essential element for modern information techniques and acts as the interface with the user. To promote more efficient ways of managing water in irrigated areas, modern GIS technique can be employed to collect, store, and process enhanced information on water use for crops, and to disseminate reliable and validated procedures. The modern GIS technique coupled with model can quickly guide the management in decision-making since the temporal and spatial dimensions could be studied at once. The GIS approach is particularly appropriate as it is the most efficient tool for spatial data management and utilization that allows understanding of the spatial variance [12]. GIS has been applied effectively for bringing spatial variability of soil, crop, water supply and environment in dealing with the complex problems for irrigation and water management [13]. GIS is one of the most simple and straightforward ways of providing a management tool for planning of water allocation policy in irrigation system. GIS together with its powerful spatial data management and analysis capabilities is therefore used to extend the scope of the estimation of irrigation delivery performance and its proper evaluation techniques for paddy irrigation system.

## 2.2 Study area

The Tanjung Karang Rice Irrigation Scheme is located at about $3^0\,25'$ to $3^0\,45'$ N latitude and $100^0\,58'$ to $101^0\,15'$ E longitude in the state of Selangor Malaysia (Fig. 1). The total command area of the scheme is about 18,848 ha. Rice is grown two times in a year mainly from August to January (main- or wet season) and February to July (off- or dry season). The ponding water depth of up to 10 cm is maintained depending on the crop growth stage, farmer's attitude and available water resources. A unique feature of this irrigation scheme is that it is a run of the river type with no reservoir or dam to store water. The Bernam River is the only source for the irrigation supplies which is diverted at the Bernam River Headwork (BRH) into the feeder canal. Then water is conveyed into Tengi River and thence to the intake point of the main canal at Tengi River Headwork (TRH). The distance from BRH to TRH is about 36 km.

Irrigation water is delivered directly from the main canal to tertiary canals, which are spaced 400 m apart along the main canal. A standard irrigation block has a net command area of about 150-200 ha. Irrigation blocks receive water in their paddy plots direct from two tertiary canals. A pump house constructed in 1962 on the lower reaches of the Bernam River at Bagan Terap provides water supply for the northern portion of approximately 1000 ha. The command areas under pumping condition are not considered in this study. In order to get better utilization of available water resources, the scheme is divided into three irrigation service areas (ISA) where water delivery is staggered by one month starting from August in main season and February in off season. In this way, pre-saturation of the whole project area is completed within three months. The detailed features of the irrigation distribution networks and irrigation compartments under each irrigation service areas namely ISA I (Sawah Sempadan and Sungai Burong), ISA II (Sekinchan, Sungai Leman, Pasir Panjang and Sungai Nipah) and ISA III (Panchang Bedena and Bagan Terap) for the scheme are illustrated in Fig. 1.

The design discharge at the BRH at the elevation of full supply level (FSL) of 9.6 m is 30.6 m³/s. The average annual rainfall is about 1800 mm [14]. The highest annual evaporation in the area was found to be 1600 mm during 1990 to 2003. The highest amount of rainfall normally occurs in March-April and October-November for the off season and main season, respectively. The excess water throughout the drainage network is drained out to the sea. This condition is often found for excess rainfall in the main season.

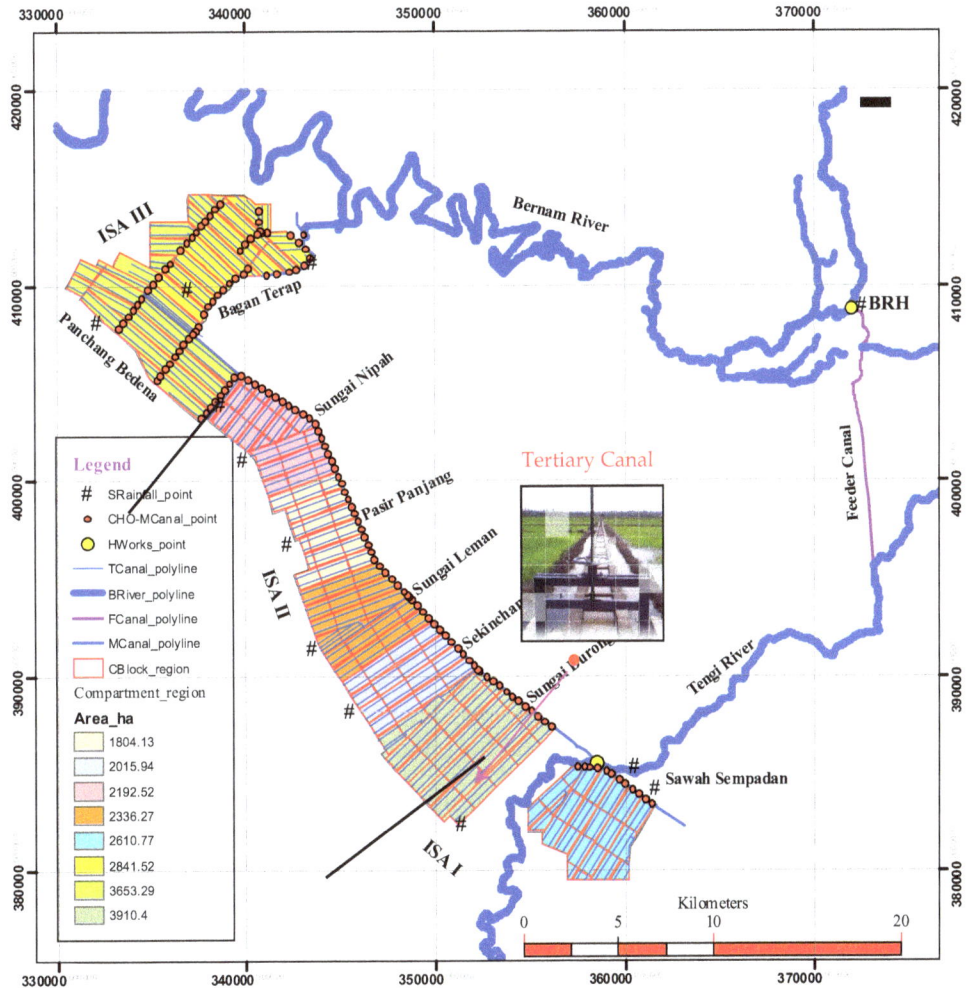

Fig. 1. Irrigation Distribution Network in the 18,000 ha Tanjung Karang Rice Irrigation Scheme (TAKRIS) Malaysia

## 2.3 Data collection and GIS database development

Many years of reliable climatic data records are required to estimate different parameters for a proper irrigation water management. Data and related information were obtained from relevant government agencies such as the Tanjung Karang Rice Irrigation Scheme Authority (IADA) for different ISAs, the Department of Irrigation and Drainage (DID), Department of Agriculture (DOA), Department of Survey and Mapping Malaysia (JUPEM) and Malaysia Meteorological Department (MMD). The detailed configuration of the irrigation canals, irrigation head regulator, Constant Head Orifice (CHO) offtake structures and specifications, stage and discharge data for the main canal were obtained from the Irrigation and Drainage Authority of the Scheme and also from the DID Headquarters, Malaysia. Database development is the crucial task to bring all the information obtained into a GIS database. All the data were properly registered and assembled in GIS platform.

## 2.4 Water demand estimation

Water demand estimation is the primary considerations for planning, design and evaluating of the irrigation scheduling of a scheme. In Malaysia, the recommended design presaturation and supplementary irrigation requirements for the rice irrigation systems are 2.31 l/s/ha (20 mm/day) and 1.16 l/s/ha (10 mm/day), respectively. The total water requirement for rice production is about 1000–1500 mm depending on characteristics of the schemes. A quantitative estimation of the major components of field water balance provides management decisions on how the scheme ought to be operated to ensure better distribution of irrigation water and the delivery performance.

### 2.4.1 Presaturation irrigation requirements

A huge amount of water is consumed to inundate fields for presaturation before planting of the crop. The water required during presaturation period can be determined as follows:

$$SAT = \frac{IR_{LS} + EP_S + SP + SW}{IE} \tag{1}$$

Where,
SAT = water requirement during presaturation period (mm/day)
$IR_{LS}$ = water requirement to saturate the soil (mm/day)
$EP_s$ = evaporation loss from saturated soil surface (mm/day)
SP = seepage and percolation losses (mm/day)
SW = additional supply to maintain the initial depth of flooding (mm/day)
IE = overall irrigation efficiency

### 2.4.2 Normal irrigation requirements

The required irrigation water during the normal irrigation period shall be allocated on the basis of the equation (2):

$$GIR_j = \frac{(ET_O)_j \times K_c + SP_j - ER_j}{IE} \tag{2}$$

where,
$GIR_j$ = gross irrigation water requirement (mm/day)
$(ET_0)_j$ = reference crop evapotranspiration (mm/day)

$SP_j$ = seepage-percolation loss (mm/day)
$ER_j$ = effective rainfall (mm/day)
$K_c$ = crop coefficient
IE = overall irrigation efficiency, which is assumed to be 45% ,[14].

For presaturation water depth, the DID recommendation of 20 mm/day is used. This would help to maintain the standing water depth of 100 mm for the normal irrigation period. The minimum standing water depth ($SW_{min}$) is maintained at 50 mm. The seepage-percolation (SP) rate of 2-3 mm/day is considered throughout the growth period [14 and 15].

If part of the water requirement is met by utilization of rainfall during crop growing period, then the net irrigation requirement on a particular day is determined as:

$$NIR_j = ET_j + SP_j - ER_j + SW_j - SW_{j-1} \qquad (3)$$

where, $ET_j$ is $ETo_j * kc$, $SW_j$ is the required standing water depth for a particular day, and $SW_{j-1}$ is field water level at the beginning of irrigation supply on (j-1)-th day. The $NIR_j$ is determined as in Equation 3 when paddy fields remain in the condition $SW_j \geq SW_{j-1}$. However, this condition is rare during peak water demand and it is possible only by storing a significant amount of rainfall in the paddy fields. The inequality between $SW_j$ and $SW_{j-1}$ leads to different water balance scenarios as well as water allocation rules, which are determined mainly by $SW_{j-1}$ that falls short or exceeds the required standing water depth. The conditions for the estimation of the net irrigation requirements are summarized in Table 1.

| Water balance condition | Availability of irrigation supply condition with respect to demand | Net irrigation requirements ($NIR_j$) |
|---|---|---|
| $SW_j = SW_{j-1}$ | $Q_{rs} \leq Q_{av}$ | $NIR_j = ET_j + SP_j - ER_j$ |
|  | $Q_{rs} > Q_{av}$ and $ER_j = 0$ | $NIR_j = 0$ and $SW_j = (SW_{min} \leq SW_{adj} < SW_j)$ |
| $SW_j < SW_{j-1}$ | $Q_{rs} \leq Q_{av}$ | $NIR_j = 0$ |
|  | $Q_{rs} > Q_{av}$ and $(ET_j + SP_j) \neq ER_j$ | $NIR_j = 0$ |
| $SW_j > SW_{j-1}$ | $Q_{rs} \leq Q_{av}$ | $NIR_j = ET_j + SP_j - ER_j + SW_j - SW_{j-1}$ |
|  | $Q_{rs} > Q_{av}$ | $NIR_j = ET_j + SP_j - ER_j$ and $SW_j = (SW_{min} \leq SW_{adj} < SW_j)$ |
| $SW_{j-1} = SW_{adj}$ | $Q_{rs} \leq Q_{av}$ | $NIR_j = ET_j + SP_j - ER_j + SW_j - SW_{j-1}$ |
|  | $Q_{rs} > Q_{av}$ | $NIR_j = ET_j + SP_j - ER_j$ and $SW_j = (SW_{min} \leq SW_{adj} < SW_j)$ |
| $Q_{rs}$ = Recommended demand and $Q_{av}$ = available discharge at the intake point of the main canal | | |

Table 1. Net Irrigation Requirements for Different Water Balance Scenarios

## 2.5 Assessment of the irrigation delivery performance for rice

Indicators and measures of irrigation water delivery performance are best when those can be used to evaluate the irrigation delivery performance and as management tool to keep track of the water delivery performance as the season progresses. In this regards, the RWS concept is appropriate and can be applied for paddy rice and upland rice or other crops. This discussion however is restricted mainly to paddy rice for characterizing the irrigation delivery performance using the RWS concept.

## 2.5.1 Relative water supply (RWS) concept

The available water supply and the water demand are the two most crucial factors for planning, design and operation of any irrigation system. The ratio of supply and demand constitutes an important concept called the Relative Water Supply [6]. This concept is actually the inverse of the engineering irrigation efficiency, output over input. The irrigation supply is the supply measured at the point of interest. The total water supply is defined to include both the irrigation supply and the effective rainfall during the period being considered. The effective rainfall is the fraction of the total rainfall over the irrigation command area that potentially could substitute for the irrigation supply. The total water demand is considered from losses due to crop evapotranspiration and seepage-percolation for the same duration. The RWS is mathematically expressed as follows:

For land preparation period or pre-saturation period,

$$RWS_j = \frac{IR_j + ER_j}{EP_j + SP_j + LS_j} \tag{4}$$

For normal crop growth period,

$$RWS_j = \left( \frac{IR_j + ER_j}{ET_j + SP_j} \right) \tag{5}$$

where, $LS_j$ = Land soaking water requirement in cm. The lower bound of RWS = 1.0 and the higher bound of RWS = 1.15 or more considered as the management strategy to maintain adequate water supply [7]. For any period, the value less than 1.0 represents undersupply and the value more than the upper bound represents oversupply. In fact, the upper bound value is not standard for characterizing the irrigation delivery as it depends on many factors involved in the irrigation systems. The RWS value can be maintained at lower bound level considering the expected stochastic rainfall in the next irrigation period. If no rainfall occurred, the paddy fields will remain undersupplied.

The RWS concept has proven to be a useful tool for understanding the performance of the irrigation systems and the impact on performance of the behavior of the major participants (the irrigation engineers and the farmers) in the irrigation process. It is useful for analysis and interpretation of irrigation performance for different time intervals and for different locations at system and sub-system levels. The computation flexibility in terms of time and space of the RWS makes it easy to use for evaluating the irrigation delivery performance. In addition to the potential for evaluating irrigation performance, the RWS concept is useful to evaluate the relative equity of water service with a system. It has also proven useful in understanding the decision rules that characterize system operation.

Traditionally rice is grown under continuous submergence or intermittent or variable ponding conditions depending on the farmer's choice and also on the water resources. This is the basic difference for the rice irrigation management system from other crops. Some additional water is required to maintain the standing water depth in the field due to the difference between the Desired or Maximum Standing Water Depth (WSmax$_j$) and the Present Standing Water Depth (WS$_j$). The widely used Relative Water Supply (RWS) concept gives incorrect determination to characterize an oversupply condition on irrigation deliveries for not considering this additional water supply for rice production.

## 2.5.2 Rice Relative Water Supply (RRWS)

The Rice Relative Water Supply (RRWS$_j$) is defined as the ratio of the total supply as Irrigation requirement (IR$_j$) and Effective Rainfall (ER$_j$) to the total demand as the sum of the difference between Maximum Ponding Water Depth and Present Ponding Water Depth (WSmax$_j$-WS$_j$) for a particular irrigation period; Evapotranspiration (ET$_j$) and Seepage-Percolation (SP$_j$) in the service areas for a duration being considered. It can distinctly characterize the oversupply for RRWS$_j$ > 1.0 and undersupply for RRWS$_j$ < 1.0 for any given period as the season advances. The value of RRWS$_j$ = 1.0 indicates irrigation supply is perfectly matched with the field water demand. Incorporating depleted ponding water (WSmax$_j$ - WS$_j$) into eq. (6) is the modification for the RWS concept given by [6] especially useful for evaluating irrigation delivery in rice-based systems. The RRWS is expressed as follows:

$$RRWS_j = \frac{IR_j + ER_j}{\left(WS\,max_j - WS_j\right) + ET_j + SP_j} \tag{6}$$

The oversupply and undersupply can be simply identified for any given irrigation period with the actual RRWS value compared to the RRWS = 1.0. For a particular period, irrigation supply is gradually increased with the amount of depleted ponding water until it reaches the maximum level in the field. Without considering this amount, RWS gives higher values, which are normally characterized as the false oversupply condition. In fact, it is not necessarily an oversupply. A value of RWS = 0.8 may not represent a problem, rather it may provide an indication that farmers are practicing deficit irrigation supply to maximize returns on water [16]. This remark can be adopted for operating irrigation system even at RRWS = 0.5 for a particular period to overcome water shortage and could be helpful to store more rainfall if WSmax$_j$ is retained.

## 2.5.3 Cumulative Rice Relative Water Supply (CRRWS)

The Cumulative Rice Relative Water Supply (CRRWS) is defined as the accumulated value of RRWS, which is the ratio of supply to the demand computed over short intervals of time starting from a particular time of the season. The advantage of CRRWS is similar throughout the season like CRWS. In addition, CRRWS can overcome the weakness of RWS and CRWS. It can be defined for a particular period as follows:

$$CRRWS_j = \sum_{j=1}^{n} \frac{IR_j + ER_j}{\left(WS\,max_j - WS_j\right) + ET_j + SP_j} \tag{7}$$

The slope and the trend of the CRRWS concept provide useful management inferences simpler than CRWS. The values of CRRWS for daily, weekly or any other short interval of cropping season with time interval can be plotted along the x-axis and CRRWS value along the Y-axis. This plot carries with the curve designated as CRRWS = 1.0. If computed CRRWS line follows the CRRWS = 1.0 line, it means that irrigation deliveries are entirely matched with the field water demand for a particular irrigation period. An increasing slope of the actual CRRWS curve with CRRWS = 1.0 means that irrigation supply can be slightly curtailed in the next period. On the other hand, if the slope is downwards, supply has to be increased.

### 2.5.4 Ponding Water Index (PWI)

The Ponding Water Index is the relation between present and the maximum desired ponding water depth for the duration being considered. It indicates the state of the water deliveries for a specific period with respect to the desired ponding water depth in the field during the crop growing season. It can be expressed as the following equation:

$$PWI_j = \frac{WS_j - WSmax_j}{WSmax_j} \tag{8}$$

Where
$WS_j$ is the measured ponding water depth on j-th day
$WSmax_j$ is the maximum desired ponding water depth on j-th day

The $WSmax_j$ is normally varied from 2 to 10 cm during the crop growth periods. For achieving good irrigation water delivery performance and efficient water management, the $PWI_j$ is recommended to be maintained at $-1.0 > PWI_j \leq 0$. Ponding Water Index gives the following interpretation over periodic irrigation supply at the end of a period:

$$PWI_j > 0 \text{ Over supply condition}$$

$$PWI_j = 0 \text{ Well-watered condition}$$

$$PWI_j = (-ve) \text{ Under supply}$$

$$PWI_j = -1.0 \text{ Critical or saturation condition.}$$

The $PWI_j$ gives distinct scenarios whether irrigation supply could be supplied properly or not during the crop growth periods. It gives the useful information for irrigation management decisions to managers. The indicator can also quantify the amount of undersupply with respect to maximum desired ponding water depth at the field level for the week being considered.

### 2.6 ArcGIS user-interface and operation procedures

Rice Irrigation Management Information System (RIMIS) with delivery performance assessment tool was developed for the Tanjung Karang Rice Irrigation Scheme as a user-friendly interactive system. RIMIS is an ArcGIS-VBA user-interface with many sub-modules and functions within the powerful ArcGIS environment. On activation of ArcGIS Software, the menu "RIMIS" appears directly on the Menu Bar in the ArcMap Window (Fig. 2). On clicking the Command Button, "Irrigation Delivery Performance" in Fig. 2, the Dialog Wizard for irrigation delivery performance will appear. RIMIS allows the irrigation manager to run the day-to-day operation and management activities as the season advances. The program is used to characterize the daily irrigation water allocation scenario and to evaluate the irrigation delivery performance using the selected performance indictors as the season advances.

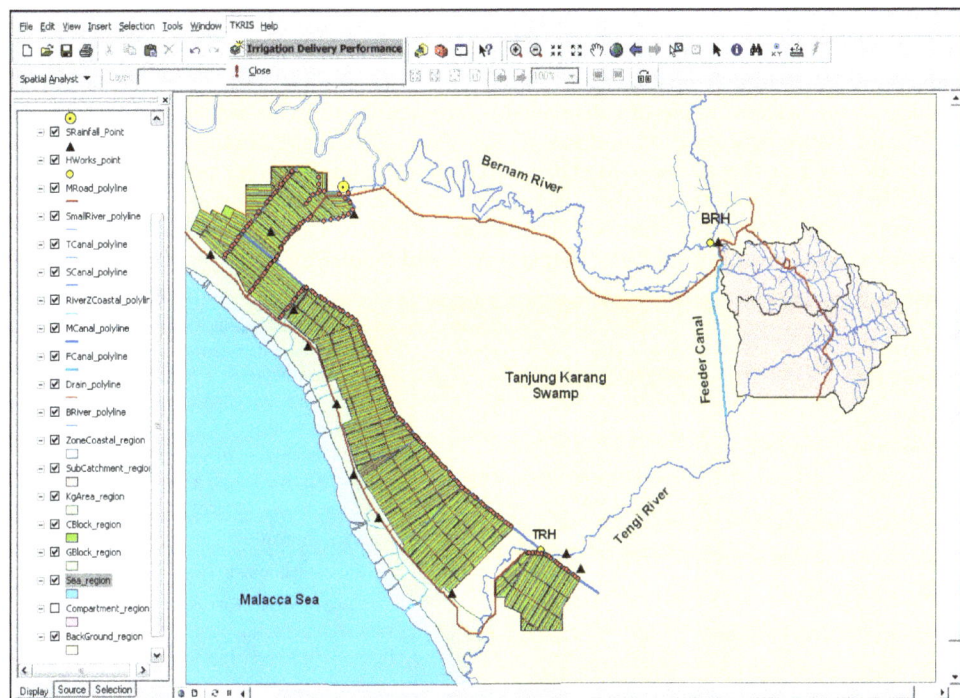

Fig. 2. The RIMIS Menu for the Tanjung Karang Rice Irrigation Scheme (TAKRIS) in ArcGIS

The user needs to enter the required inputs and choose the appropriate command button to view the results. Inputs can be directly fed into the dialog window by clicking on the command button "Data from Input Files" or manually (Fig. 3). All required inputs stored in Input Files are instantly retrieved into TextBoxes by each irrigation service area i.e. "ISAI: Input Daily Information (I)". Command Buttons for new irrigation delivery performance indicators, Rice Relative Water Supply (RRWS), Cumulative Rice Relative Water Supply (CRRWS) and Ponding Water Index (PWI) together with the Relative Water Supply (RWS) concept are shown in Fig. 3. Users may feed accurate information from other sources manually if available and reliable. To compute the performance indicators correctly, users should enter the actual information of rainfall and reference crop evapotranspiration on the day as they are available at the end of the irrigation day. A sub-routine was developed to compute the daily potential crop evapotranspiration as shown in Figures 4a and 4b. This sub-module allows an irrigation manager to compute the daily Reference Crop Evapotranspiration (ETo) using the Internationally recommended FAO Penman-Montieth Equation based on the available climatic information for a particular day. However studies by Hassan (2006) show that ETo calculated by the Penman-Montieth Equation gives an underestimation of ETo by 10% compared to ETo from micro-lysimeters installed in the paddy fields.

The system allows for storing of all inputs and outputs into the MS Access database simply by clicking on the Command Button "Save". It has also capabilities of retrieving, editing, deleting records from database for a particular day upon selection. The Command Button "Backup" helps to keep all files as backup any time. The dialog wizard linked with the Command Button "View Inputs & Outputs" in Fig. 3 allows interactive viewing for all input and output databases. It enables to users viewing data in table and graph formats. The records of the selected table are displayed. The records from the selected Table can be viewed for a specified duration and the selected fields. The easily obtained comprehensive information allows the manager to characterize the irrigation delivery performance faster. The dialog wizard for exploring the scheme's information can also be viewed on clicking the command button "Detailed Scheme Information" in Fig. 3. This sub-module gives salient features of the scheme in the form of maps, graphs and tabular formats.

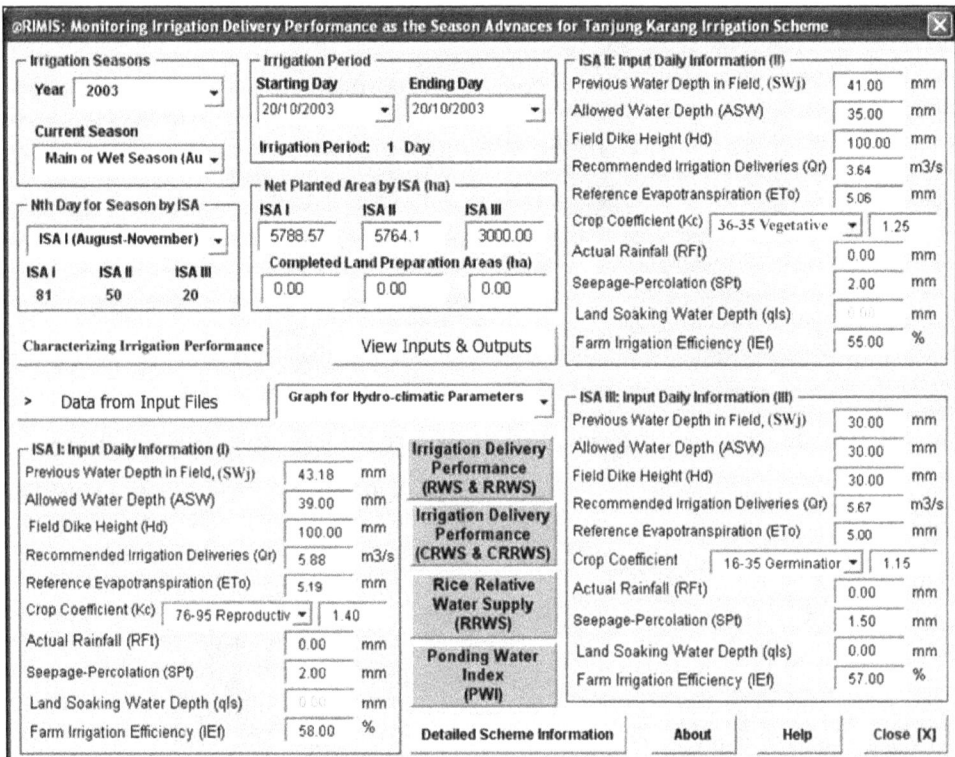

Fig. 3. Dialog Wizard for Characterizing Irrigation Delivery Performance for the Main Season

## 2.7 Results and discussions

To adopt proper irrigation supplies, the appropriate information system is a prerequisite to make the right decision on water allocation as well as possible remedial actions as the season advances. Typical results produced using the GIS based assessment tool is presented in the following sections for visualization in colored maps, charts and tables as the season

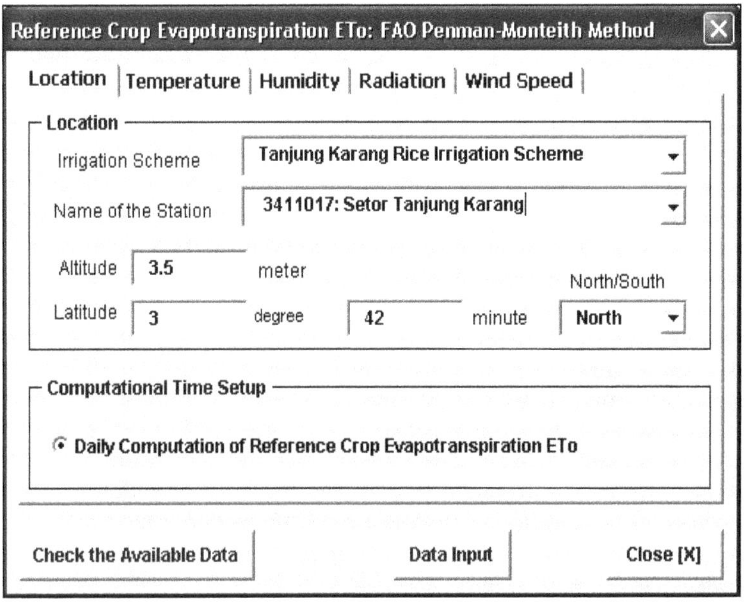

(a) Input Selection Dialog Wizard

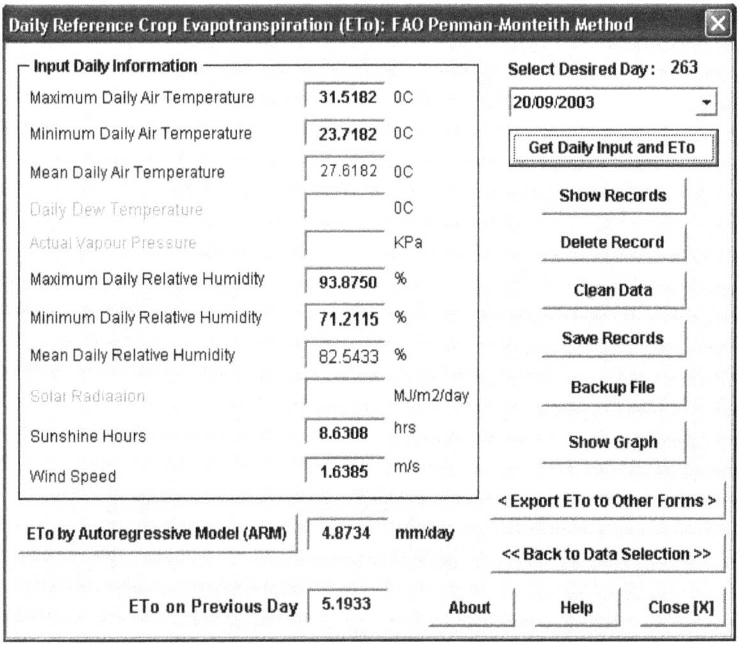

(b) Input Dialog Wizard for Daily Meteorological Data

Fig. 4. Computing Daily $ET_o$ using FAO Penman-Monteith Method

advances. The system provides useful information of the actual field condition with respect to the water demand as the season progresses and helps allocate the right amount of irrigation supplies for the next day or period.

### 2.7.1 Characterizing irrigation delivery performance

The characterization is essential to evaluate irrigation deliveries among tertiary canals and for their remedial measures as the season advances. A dialog wizard like that in Fig. 3 appears by clicking on the Menu Item "Irrigation Delivery Performance" in the dialog wizard shown in Fig. 3. To get the output, the user needs to enter the required inputs and the appropriate options. Inputs are directly fed from the input database by clicking on the Command Button "Data from Input Files" or manually. After that, the results for a particular day can be computed and viewed on a screen instantly like that shown in Fig. 5 by clicking on the Command Button "Monitoring Irrigation Performance" in Fig. 3. New performance indicators RRWS and PWI are used for evaluating and characterizing irrigation delivery performance and compared the advantages and disadvantages with the RWS concept especially for paddy irrigation system. The 18,000 ha irrigation scheme is divided into 3 Irrigation Service Areas (ISA) with a staggered supply of one month.

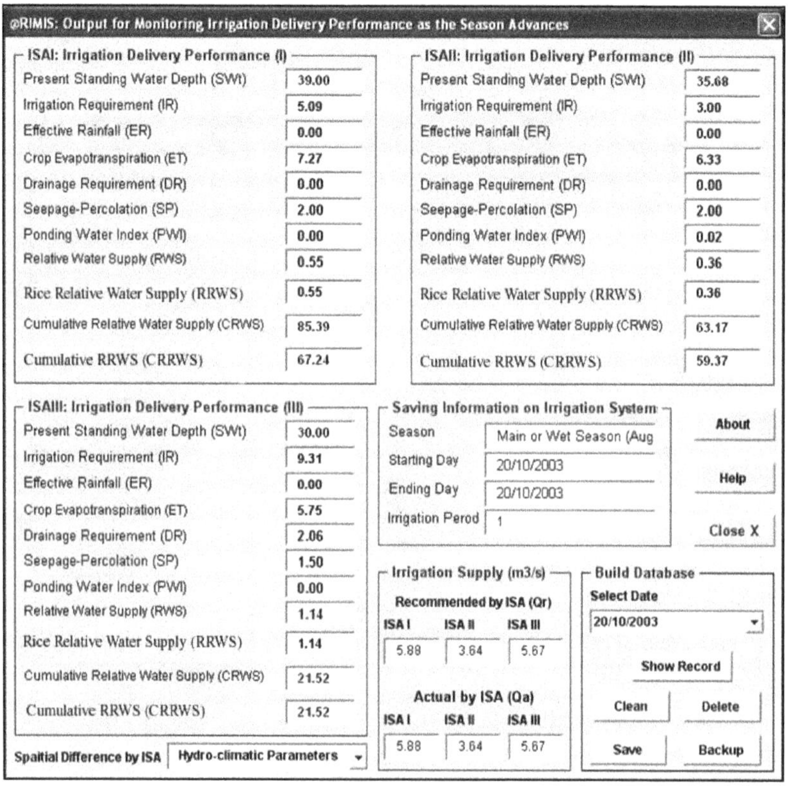

Fig. 5. Output Dialog Window for Characterizing Irrigation Delivery Performance on 20 October 2003

The irrigation performance characterizes as over supply for RRWS > 1.0, undersupply for RRWS < 1.0 and good performance for RRWS = 1.0. Therefore, the irrigation manager can easily quantify how irrigation water ought to be allocated with the available water resources for the next day or irrigation period.

## 2.7.2 Rice Relative Water Supply (RRWS)

After clicking on the Command Button "Rice Relative Water Supply (RRWS)" in Fig. 3, irrigation delivery performances from the beginning to the current days for the current irrigation season are plotted as shown in Fig. 6. The irrigation performance characterizes as over supply for RRWS > 1.0, undersupply for RRWS < 1.0 and good performance for RRWS = 1.0. Therefore, the irrigation manager can easily quantify how irrigation water ought to be delivered with respect to the available water resources for the next day.

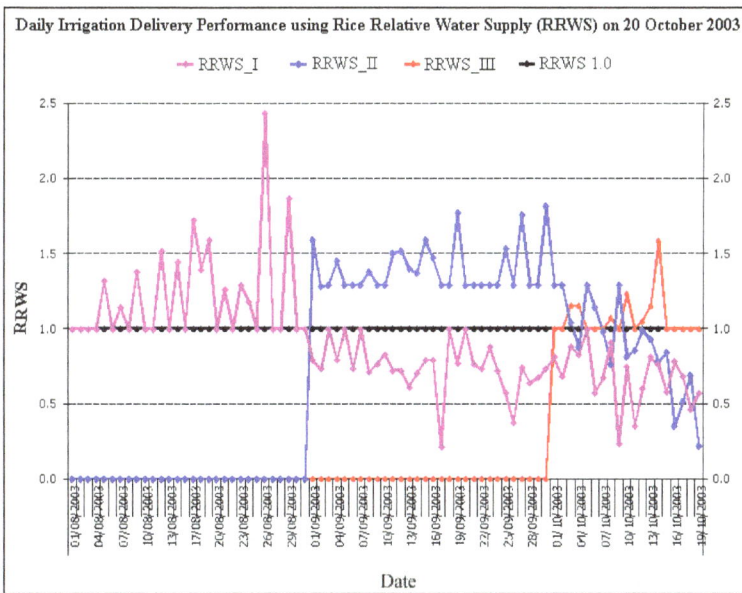

Fig. 6. Characterizing Daily Irrigation Delivery Performance using RRWS for Main Season 2003

The RRWS_I, RRWS_II and RRWS_III represent Rice Relative Water Supply for Irrigation Service Areas I, II and III where irrigation supply is staggered by one month, respectively. From Fig. 6, the following decisions can be drawn (Table 2):

| Month | Irrigation Delivery Conditions by ISA | | |
|---|---|---|---|
| | ISA I | ISA II | ISA III |
| August | Oversupply | No Supply | No Supply |
| September | Under Supply | Oversupply | No Supply |
| October | Oversupply | Under Supply | Under Supply |

Table 2. Characterization of Irrigation Delivery Conditions

From this scenario, it is noticed that oversupply conditions were found only during the presaturation period in the first month for each ISA. Undersupply condition was found during the normal irrigation supply for all ISAs. If irrigation supply would follow the RRWS guideline, then it could be possible to overcome the oversupply and undersupply conditions. If irrigation supplies would be curtailed for the ISA II and increase for the ISA I then irrigation delivery might have maintained good conditions as shown in the middle of the plot in Fig. 6.

### 2.7.3 Cumulative Rice Relative Water Supply (CRRWS)

The Cumulative Rice Relative Water Supply (CRRWS) values were plotted for both seasons of 2003/04 as shown in Figs. 7a and 7b. These plots have three other curves designated as CRRWS = 1.0 staggered by one month. The relative merits and demerits between the plots CRWS and CRRWS have brought the robustness of using the new indicators for evaluating irrigation delivery performance as the season advances. The slope of the actual CRRWS curve provides useful management information, which can enhance the decision-making for the irrigation delivery. If there is an increasing slope of CRRWS line with respect to the CRRWS = 1.0 line then irrigation supply could be slightly curtailed in the next irrigation period. On the other hand, if the slope is downwards and is reaching the lower CRRWS = 1.0 line, the supply could be increased for the next irrigation period. The coinciding between actual CRRWS and CRRWS = 1.0 lines characterizes well water distribution. The CRRWS plot should be maintained at the CRRWS = 1.0 to improve irrigation delivery performance.

The CRRWS_I, CRRWS_II and CRRWS_III represent Cumulative Rice Relative Water Supply (CRRWS) for Irrigation Service Area (ISA) I, II and III, respectively. In Fig. 6a, the CRRWS values obtained higher than 1.0 during the land preparation periods in ISA I, ISAII and ISA III but lower at normal irrigation periods in ISA I. The slope of CRRWS line of ISA I decreased when land preparation was started in ISA II and the slope of CRRWS line of ISA II decreased after starting the land preparation in ISA III due to water shortage in the off-season. The plot for all irrigation service areas are shown normally under supply condition after completing land preparation in ISA I from May.

In the Fig. 7b, the CRRWS values obtained were higher than 1.0 during the land preparation periods in ISA I, ISAII and ISA III but lower at normal irrigation periods in ISA I. The slope of CRRWS line of ISA I started to decrease when land preparation was started in ISA II and the slope of CRRWS line of ISA II decreased after starting the land preparation in ISA III in the main season. The irrigation supplies for all irrigation service areas were not shown under supply condition throughout the irrigation season. The slope of the CRRWS line for ISA III shows better irrigation delivery performance than ISA and ISA II. The plot shows that available water resources could not meet field water demand at the design ponding water level 10 cm during normal irrigation periods. The utilization of rainfall plays an important role to maintain target standing water depth in the fields. The oversupply conditions were normally found on rainy days.

### 2.7.4 Ponding Water Index (PWI)

A daily analysis of Ponding Water Index (PWI) was plotted for the main and off seasons in 2003/04 as shown in Fig. 8. The PWI_I, PWI_II and PWI_III represent Ponding Water Index

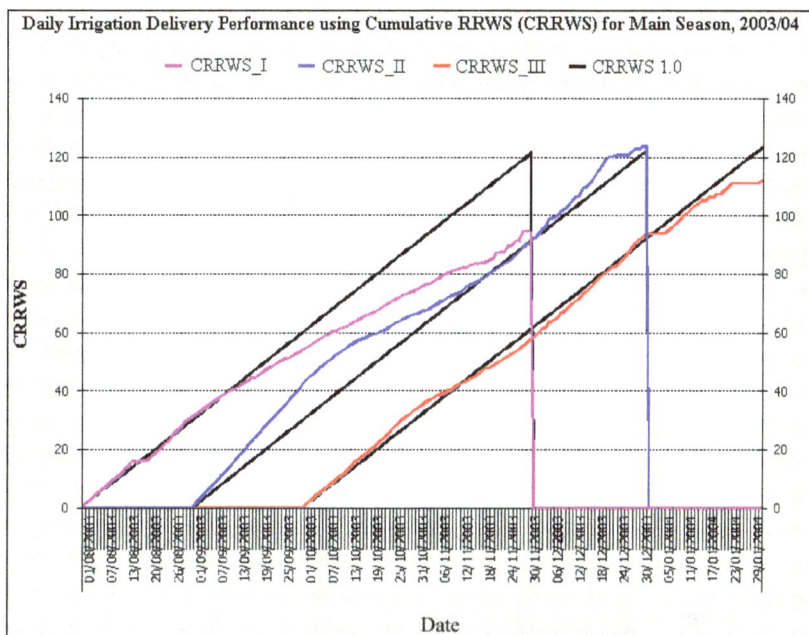

(a) Main Season in 2003/04

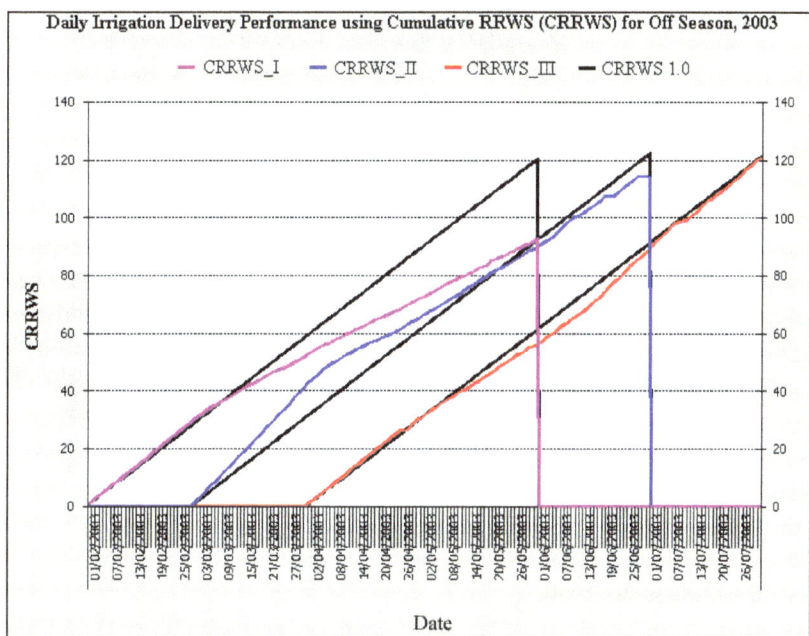

(b) Off Season in 2003/04

Fig. 7. Characterizing Daily Irrigation Delivery Performance using CRRWS

(PWI) for Irrigation Service Area (ISA) I, II and III, respectively. The undersupply condition was obtained due to shortage of water resources. The extreme values computed due to continuing irrigation deliveries while sufficient water remained in the fields. The oversupply condition is shown during land preparation period and especially rainy days. The peak negative values are due to severe shortage of water.

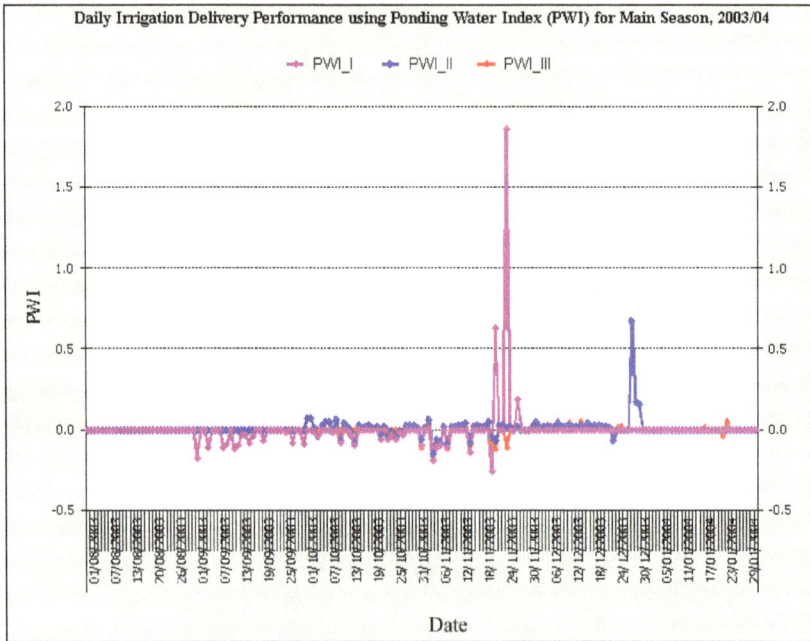

Fig. 8. Characterizing Daily Irrigation Delivery Performance using Ponding Water Index (PWI) for the Main Season in 2003/04

Irrigation deliveries have shown oversupply due to more rainfall from 20-24 November and end of the season in the main season. The days on the x-axis at zero value shows the well-watered condition. Irrigation deliveries will be considered as over supply or under supply if PWI is more or less than zero, respectively. The irrigation manager can simply identify and quantify the performance of irrigation deliveries for a given day and what decision has to be made for the next period. Both the PWI and RRWS can quantify and identify irrigation deliveries simultaneously. The average variation of the PWI values is higher in the off season than in the main season. The PWI also represents under supply condition in the off season as more severe than in the main season. Besides, the values of PWI are very close to PWI = 0 for the main season. The peak values are shown due to heavy rainfall.

## 2.8 New water management tool

This book chapter illustrates new irrigation performance indicators known as the Rice Relative Water Supply (RRWS) and Cumulative Rice Relative Water Supply (CRRWS) to evaluate irrigation water delivery performance of the paddy rice irrigation systems. The weakness of using the Relative Water Supply (RWS) and Cumulative Relative Water Supply (CRWS) and its adverse implications on irrigation management and operation was also highlighted. Each performance indicator has its own strengths and weaknesses, which may be relevant under particular conditions. The RRWS and CRRWS parameters instead of RWS and CRWS were found to be more useful for the irrigation managers and water users to characterize the water delivery performance for the paddy irrigation systems.

The following conclusions can be drawn from the study on performance indicators:

The RRWS and CRRWS can simply and distinctly characterize irrigation delivery performance with respect to the RRWS = 1.0. It gives the following interpretation over periodic irrigation supply at the end of a period:

$$RRWS_j = 1.0 \text{ Good irrigation delivery}$$

$$RRWS_j > 1.0 \text{ Oversupply condition}$$

$$RRWS_j < 1.0 \text{ Undersupply condition}$$

- The $RWS_j$ concept shows the over supply condition for not considering the depleted water depth ($WSmax_j - WS_j$) of the denominator in RWS given by Levin (1982) when $WS_j < WSmax_j$ in the paddy fields.
- The utility of RRWS and CRRWS justify the weakness of RWS and CRWS to evaluate the irrigation water delivery for paddy irrigation systems.
- It recommends that irrigation supply should be curtailed for increasing the slope of the actual CRRWS line than the CRRWS = 1.0 line to improve the irrigation delivery performance.
- Irrigation system can be operated even at RRWS = 0.5 to irrigate wider areas due to water shortage if the Ponding water depth is retained more than the field water demand for the next period.

The Ponding Water Index (PWI) also can simply characterize irrigation delivery performance. It gives the following interpretation over periodic irrigation supply at the end of a period:

$$PWI_j = 0 \text{ Good irrigation delivery condition}$$

$$PWI_j > 0 \text{ Oversupply condition}$$

$$PWI_j < 0 \text{ Undersupply condition}$$

The new indicators have successfully been evaluated for various water management scenarios in the study area. Therefore, they can be adopted to evaluate irrigation delivery performance and proper decision for water allocation for irrigated paddy. GIS interface coupled with new performance indicators has explicitly helped in integrating spatial and temporal information for evaluating the daily irrigation delivery performances for paddy cultivation. The new irrigation performance indicators are able to provide useful information and can be adopted to evaluate irrigation delivery performance and proper decision for water allocation in the paddy irrigation system. The availability of such a quantitative tool for irrigation systems operation can have a powerful impact on the overall water management strategy in an irrigation project area.

## 3. Precision farming of rice

### 3.1 Saving resources through precision farming

Precision farming (PF) is considered the best practicable approach to achieve sustainable agriculture. Precision farming is an integrated, information- and production- based farming system that aims to raise efficiency, productivity and profitability of long term, site-specific and whole farm production while avoiding the undesirable effects of excessive chemical loading to the environment or insufficient input application.

The role of PF in crop production technology is recognized worldwide, but so far, it is applied mostly on large farms. Implementation of PF should be followed in three main steps of information gathering in terms of variability, data processing to evaluate the significance of variation and employ new management strategy to apply farming inputs. Fig. 9 demonstrates some equipment and technologies in a typical precision farming crop growing cycle. Some describe precision farming as applying the right inputs, at the right place, at the right time, right amount, and in the right manner.

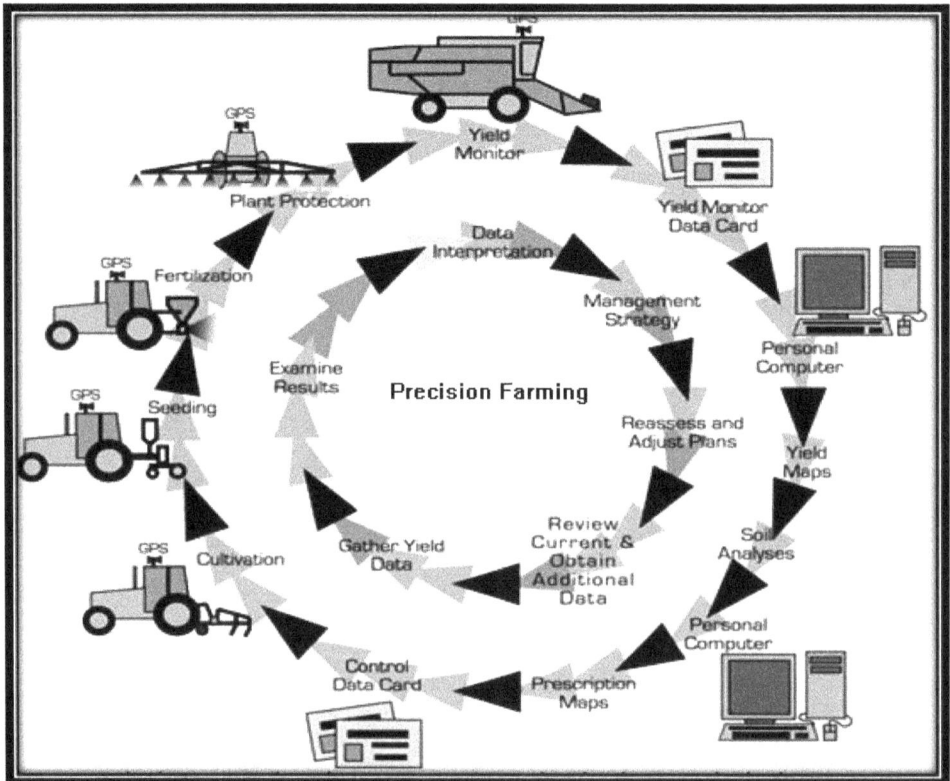

Fig. 9. Precision Farming Cycle (Grisso, 2009)

Implementation of management strategy based on precision farming concept is the vital factor to achieve a desired outcome for the farm. Managers should make out their own strategies that allow them to manage variability precisely. Blackmore (1999) stated the three types of variability that have been identified are spatial variability, which can be found through changes across the field, temporal variability which means changes over time and predictive variability, that identifies the difference between predicted and what actually happen in the field.

One of the precision farming approaches to manage spatial variability is site specific crop management (SSM). In order to match application of farm practices with soil and crop requirements, zone management was suggested. Zone management represents sub-fields with similar characteristics including soil properties, topography, slope, nutrient levels and so on.

## 3.2 Paddy soil variability

Soil variability in paddy fields is well recognized. The spatial description is an important component of the precision farming cycle for zone management practices. Precision farming requires topping up of only the nutrients that are lacking in the soil to attain the optimum yield with the least inputs. Manual soil sampling and the consequent laboratory analysis are expensive, labour intensive and requires a long time. The use of an on-the-go electrical conductivity (ECa) sensor can replace the traditional way of acquiring data in a more efficient way. Research results have confirmed the usefulness of the ECa data as a summary indicator for zoning paddy soils to facilitate water and fertilizer management.

Soil $EC_a$ measurements can provide information on soil texture, in addition to estimating soil water content. Maps of soil physical properties and yield maps have shown visible correlation. Soil $EC_a$ can serve as a proxy for soil physical properties such as organic matter (Jaynes, et al., 1994), clay content (Williams and Hoey, 1987), and cation exchange capacity (McBride, et al., 1990). These properties have a significant effect on water and nutrient-holding capacity, which are major drivers of yield (Jaynes, 1995). The relationship between soil $EC_a$ and yield has been reported (Kitchen and Sudduth, 1996; Fleming, et al., 1998). Sudduth et al. (1998) found that within field variation in soil properties could be explained with soil conductivity measurements. They found a significant relationship between soil conductivity and topsoil depth and Fraisse et al. (1999) added to this work by using soil electrical conductivity for zone delineation. Both of these works concentrated on using soil $EC_a$ to characterize local spatial variability. Lund et al. (1998) show that sampling according to soil management zones identified with a soil conductivity map can be more effective than grid sampling.

## 3.3 Materials and methods

This study uses an on-the-go $EC_a$ sensor for producing $EC_a$ map and to use it for soil nutrients assessment. The field soil salinity readings can be obtained through this soil-to-instrument contact device that permits rapid soil $EC_a$ measurement without requiring a permanently buried detector. The study was conducted in paddy fields at Tanjung Karang, Selangor, Malaysia. The study site has 118 plots covering 144 ha with an average plot size of about 1.2 ha (Figs. 10 and 11). The EC sensor was pulled by a tractor at a speed of about 15 km h$^{-1}$ in a U-shape pattern 15 m apart. The data was later transferred to a notebook computer for generation of $EC_a$ maps using Surfer 7.0 software and ArcGIS 8.3 with Spatial and 3D Analyst extensions. A total of 63,578 data points were obtained. For comparison, a total of 236 soil samples were collected and analyzed in the laboratory for their chemical and physical properties.

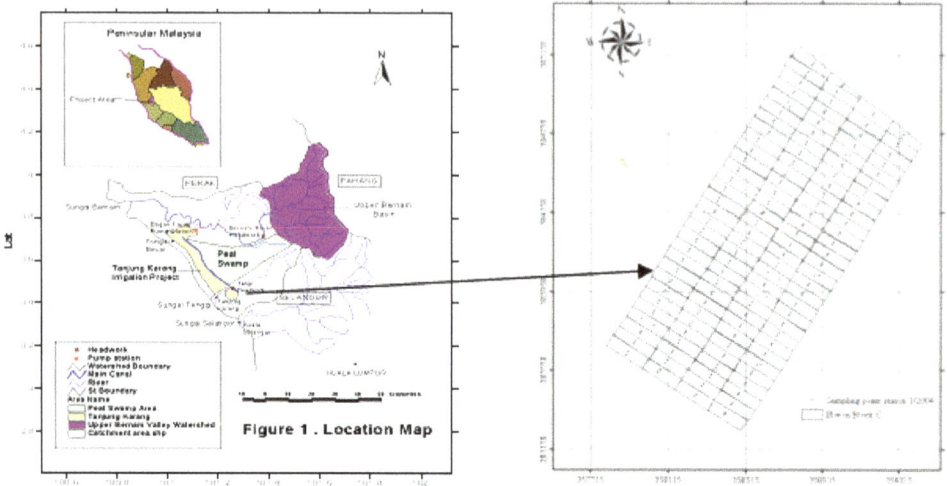

Fig. 10. Map of the Soil Chemical and Physical Sampling Points in Block C

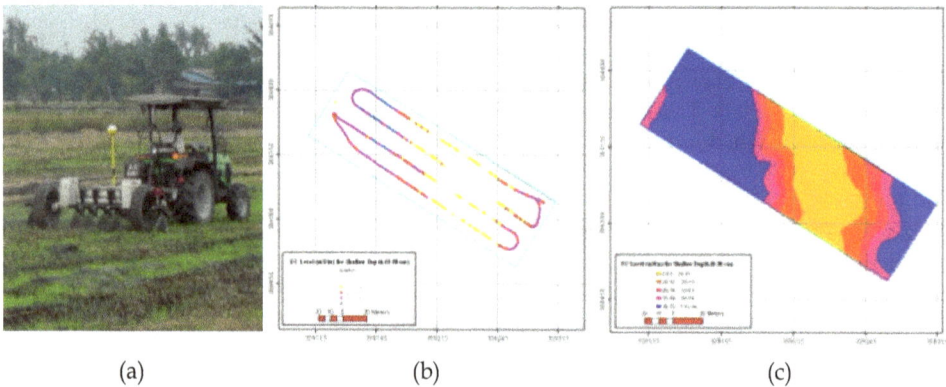

(a)                              (b)                              (c)

Fig. 11. (a) The EC sensor pulled by a tractor installed with DGPS in a paddy field, (b) results of 4 passes spaced 15m apart in a typical 1.2 ha plot, and (c) krigged map of ECa.

## 3.4 Results and discussion

Soil $EC_a$ could provide a measure of the spatial differences associated with soil physical and chemical properties, which for paddy soil may be a measure of soil suitability for crop growth, its water demand and its productivity. The $EC_a$ maps indicated that it is similar to some soil nutrient maps. It was found that the technique could identify the zone of a former river located within the study area while detailed soil series map alone could not have found it. The relation of $EC_a$ to soil P, K, Mg and CEC in the paddy fields indicates that their

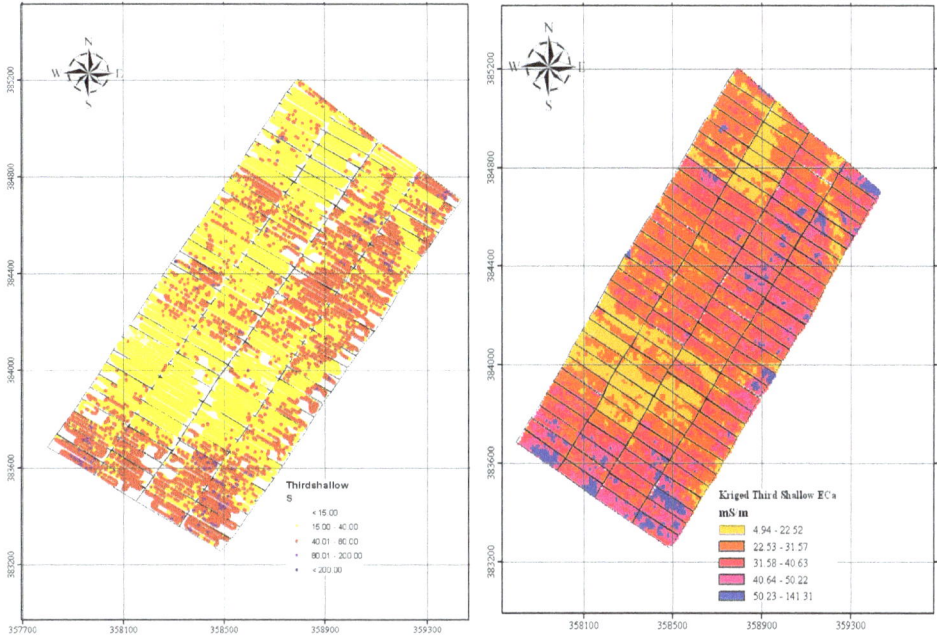

Fig. 12. Kriged Map for the Shallow $EC_a$ (mS m$^{-1}$) Classified by Smart Quantiles

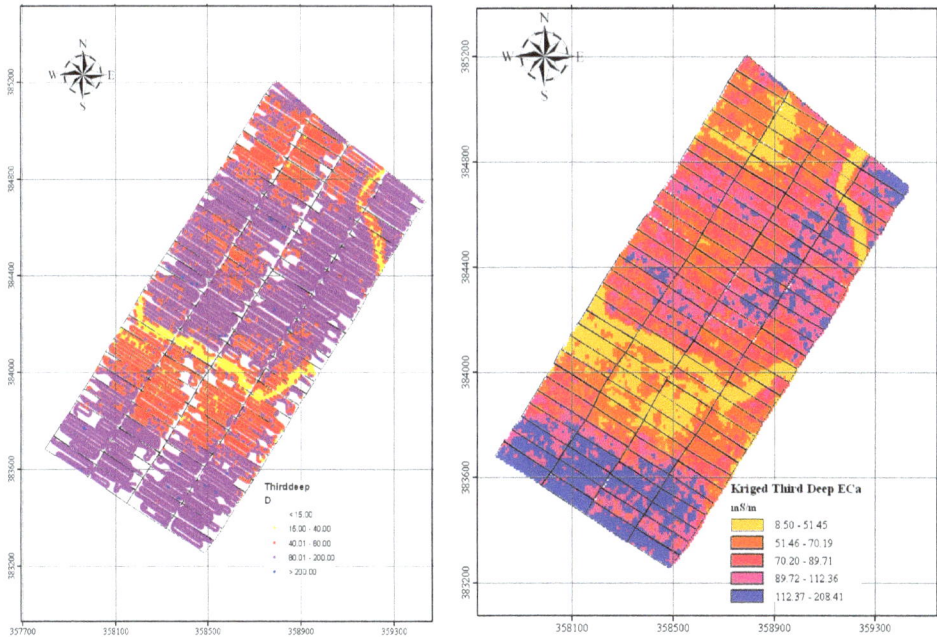

Fig. 13. Kriged Map for the Deep $EC_a$ (mS m$^{-1}$) classified by Smart Quantiles

concentration can be estimated. Hence, quick nutrients determination can be done through the $EC_a$ sensor detection. The average values of $EC_a$ are significantly different between shallow (0-30 cm) and deep depths (0-90 cm) signifying differences in soil structure and nutrient status. The sensor can measure the soil $EC_a$ through the field quickly for detailed features of the paddy soil, and can be operated by just one worker.

The study area was divided into 5 manageable zones by smart quantiles method (ESRI, 2001). Fig. 12 shows the shallow $EC_a$ and Fig. 13 shows the deep $EC_a$. The map for the deep $EC_a$ shows the distribution clearly, especially for very low and low $EC_a$ levels. Fig. 13 shows the pattern of a former river clearly as a continuous line about 45 m wide at the northern and central regions of the study area.

The on-the-go EC sensor can be used to replace the traditional way of acquiring soil data by intensive sampling technique and laboratory analysis, which is usually time consuming and laborious. The resulting $EC_a$ maps are useful in showing the management zones for improving crop productivity with minimum inputs. The delineation by $EC_a$ showed that some soil properties significantly differ from zone to zone. A total of 21 parameters were significantly predicted by using $EC_a$ which shows that the EC probe can predict multi-variables, hence reduces time for sampling and analyses.

### 3.5 Matrix correlation of soil properties

Pearson's 2-tailed test for soil chemical, physical and $EC_a$ correlation showed that shallow $EC_a$ has positively significant correlation to pH, EC, CEC, Mg, Fe, clay and deep $EC_a$ and negatively significant correlation to Al, fine sand and sand, at 99%. It has positively significant correlation to P, K and total cation, at 95%. The highest r value was 0.70** for deep $EC_a$ and followed by pH (r=0.39**). Deep $EC_a$ has positively significant correlation to pH, EC, P, CEC, Mg, K, Na, Fe, total cation, clay, fine sand, sand and shallow $EC_a$ and negatively significant correlation to Al, at 99%. The highest r value was 0.70 for shallow $EC_a$ and followed by Mg (r=0.46**). Eltaib (2003) found that laboratory EC has highest correlation to Mg (r = 0.79**, n = 36) for this study area.

### 3.6 Shallow $EC_a$ zoning characteristics

The mean value for shallow $EC_a$ slightly increased from zone to zone and significantly isolated between zones. But, some mean values of soil properties (i.e. EC, OM, C, S, N, CEC, Ca, Na and etc.) within the shallow $EC_a$ zones did not show linear trend as per mean shallow $EC_a$. The hypothesis for $EC_a$ zone establishment was that the soil properties within the zone were significantly different from zone to zone which indicated that soil $EC_a$ is a good zone delineator, a new classification approach for paddy soil properties.

The results indicate that the mean soil pH values within shallow $EC_a$ zone 1, 2 and 3 were not significantly different, but significantly higher than that in zone 4 and 5. However, zone 1 has significantly high OM, C, total S, total N, ESP, fine sand and sand, and significantly low Ca, total cation, BS and clay. Soil moisture, silt and coarse sand were not significantly different for all zones, at 0.05. The shallow $EC_a$ proved that zone 1 (former river) contained higher OM as compared to other parts of the study area. Therefore, to manage that area based on shallow $EC_a$, it should be managed differently according to organic matter content. Mean shallow $EC_a$ for 5 zones has significantly negative correlation to soil pH (r = -0.95) and total S ($r = -0.93$) at 95% level. This indicates that soil pH and total S decreased when shallow $EC_a$ increased. Hence a good water management is to apply more irrigation water to increase the soil pH in zones 4 and 5.

## 3.7 Deep $EC_a$ zoning characteristics

The stratification of total N, BS and coarse sand by deep $EC_a$ were homogenous between the zones when their mean values within the zone were not significantly different at 0.05. Mean soil OM, C and total S in zone 3 were significantly higher than those in other zones and they were different to shallow $EC_a$ zone where it indicated that zone 1 has significantly high OM, C and total S. Mean soil pH, EC, P, Mg, K, Fe, total cation and clay within zone 1 were significantly low as compared to other zones, but significantly high Al, fine sand and sand. Deep $EC_a$ has significantly positive correlation to soil pH and Fe at 0.01 and significantly negative correlation to coarse sand at 0.05.

| Soil Properties | Function | $R^2$ | b0 | b1 | b2 | b3 |
|---|---|---|---|---|---|---|
| **Predictor: shallow $EC_a$** | | | | | | |
| pH | Quadratic | 0.18*** | 4.8999 | -0.0042 | $1.0 \times 10^{-4}$ | |
| EC | S | 0.07*** | -2.0959 | -5.7515 | | |
| N | Compound | 0.02* | 0.1301 | 0.9826 | | |
| CEC | S | 0.06*** | 3.0187 | -5.1707 | | |
| ESP | Exponential | 0.02* | 2.7937 | -0.0053 | | |
| **Predictor: deep $EC_a$** | | | | | | |
| OM | Inverse | 0.02* | 8.2653 | 140.3670 | | |
| C | Inverse | 0.02* | 4.7938 | 81.4321 | | |
| S | Inverse | 0.02* | -0.0907 | 16.1845 | | |
| P | Logarithm | 0.04** | 1.2987 | 2.0881 | | |
| Ca | Inverse | 0.02* | 4.5322 | -42.3070 | | |
| Mg | Power | 0.28*** | 0.1362 | 0.6004 | | |
| K | Quadratic | 0.05** | 0.1689 | 0.0028 | $-1.0 \times 10^{-5}$ | |
| Na | Cubic | 0.07** | 0.0868 | 0.0093 | $-7.0 \times 10^{-5}$ | $1.5 \times 10^{-7}$ |
| Total Cation | Quadratic | 0.13*** | 2.3696 | 0.0874 | $-4.0 \times 10^{-4}$ | |
| Al | Cubic | 0.14*** | 1.0148 | 0.0903 | $-1.0 \times 10^{-3}$ | $2.8 \times 10^{-6}$ |
| Fe | Exponential | 0.20*** | 0.2478 | 0.0054 | | |
| BS | S | 0.04** | 3.7573 | -11.1340 | | |
| Moisture Content | Cubic | 0.04* | 72.0409 | -0.6757 | $8.2 \times 10^{-3}$ | $-3.0 \times 10^{-5}$ |
| Clay | Cubic | 0.18*** | 32.8506 | 0.0487 | $2.6 \times 10^{-3}$ | $-1.0 \times 10^{-5}$ |
| Fine Sand | Quadratic | 0.14*** | 38.4191 | -0.3710 | $1.4 \times 10^{-3}$ | |
| Sand | Cubic | 0.15*** | 30.6979 | -0.0402 | $-2.5 \times 10^{-3}$ | $1.3 \times 10^{-5}$ |

Table 3. Significant Relationship of Soil Properties to $EC_a$ for the Study Area (n = 236)

### 3.8 Model for soil properties estimations

Results from the curve estimation for the independent variables of shallow and deep $EC_a$ indicate that $EC_a$ can be used to estimate multi-variables, 21 out of 24 variables. Shallow $EC_a$ has lesser soil properties compared to deep $EC_a$, where there were 16 and 21 variables, respectively. Most of the variables have high $R^2$ values, except pH, EC, N, CEC and ESP when the estimation is using deep $EC_a$ as independent variable. The relationship functions differed for some variables while others remained. The high $R^2$ values for deep $EC_a$ indicate that variables were significantly estimated using deep $EC_a$ rather than shallow $EC_a$. The best model was judged based on their $R^2$ where the estimation of soil Mg by deep $EC_a$ was the highest following by Fe, clay, pH and so on (Table 3).

### 3.9 Yield variability and soil management zones

A study was conducted to compare yield variability resulting from variability of soil ECa and other parameters for both the dry and wet seasons in the same 140 ha study area (Gholizadeh, 2011). Fig. 14 shows typical variability maps of the harvested yield compared to the variability in the bulk soil electrical conductivity, bulk soil density and soil texture. High yielding areas are associated with mid-range ECa, high clay and low sand, and low bulk density. Low yielding areas are associated with low ECa and high sand content. High yield is also associated with high pH, high EC and high OC, and vice versa. Hence water management that will allow increase in pH of the paddy soil is desirable.

## 4. Water saving practices

### 4.1 Strategies for water saving

Water saving practices, which require greater water control is associated with improving agronomic practices and the use efficiency of other inputs. Available strategies include developing improved varieties, improving agronomic management, changing the crop planting date, reducing water use for land preparation, changing rice planting practices with wet or dry seeding, reducing water use during crop growth through intermittent flooding, maintaining the soil in sub-saturated condition, alternate drying and wetting, optimum use of rainfall, supplementary irrigation of rain-fed low-land rice, water distribution strategies, water reuse or recycling and conjunctive use and alternative methods to flooding for growing irrigated rice under aerobic conditions.

High rice yield are obtained with good on-farm water management. Many researchers reported that continuous submergence with 5 to 7 cm of water is probably best for irrigated rice considering all factors. Submergence allows better weed control, higher efficiency of fertilizer use, and better insect and weed control with granular chemicals. Research has shown no difference in yield of rice grown at saturated soil condition with minimum water use but weed control is expected to be more costly.

Other researchers found optimum rice growth and production at 9 cm of ponded water depth. High values of water productivity were also found at this depth under different water regimes and fertigation levels. High water levels are required after transplanting for recovery and rooting stage and booting stage up to flowering stage. Low depths are required for tillering, panicle development and milk stage. Shallow depths promote

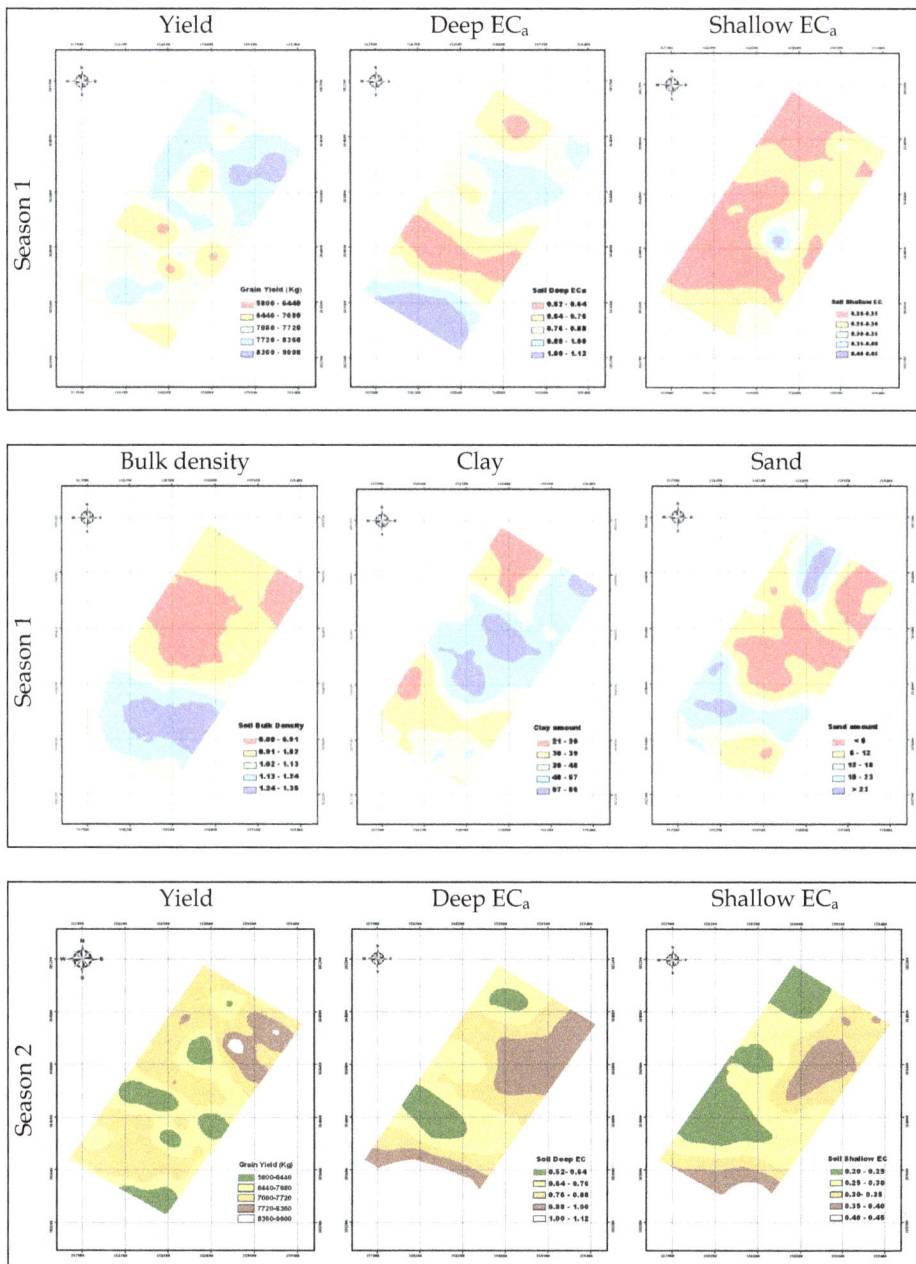

Fig. 14. (continues on next page)Variability in rice yield compared to ECa and soil physical properties in a 140 ha paddy fields for two seasons. Higher yielding areas are associated with mid-range ECad, medium ECas, low Db, high clay and low sand. Low yielding areas are associated with low ECad, low ECas, high Db, medium clay, medium sand

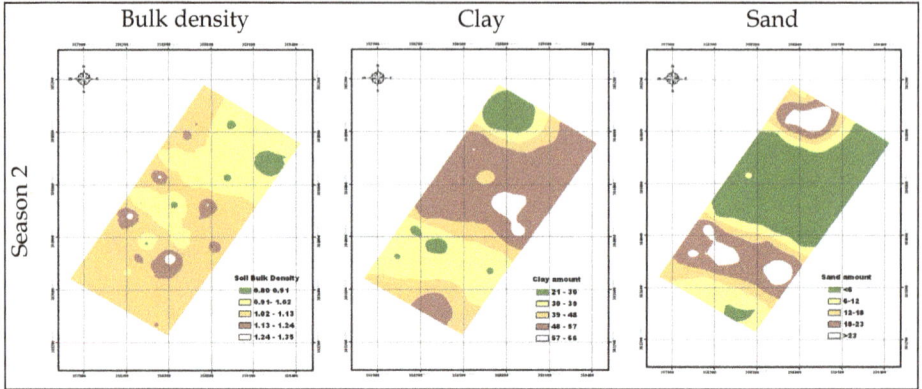

Fig. 14. (continued) Variability in rice yield compared to ECa and soil physical properties in a 140 ha paddy fields for two seasons. Higher yielding areas are associated with mid-range ECad, medium ECas, low Db, high clay and low sand. Low yielding areas are associated with low ECad, low ECas, high Db, medium clay, medium sand

vigorous tillering. Mid-season drainage is important to cut-off the supply of ammonia-N to secure desirable plant characteristics, viz. short and erect upper 3 leaves, including flag leaf, and short lower inter-node to prevent lodging, to induce favourable ear (panicle) formation conditions, and to supply soils with oxygen to ensure healthy root growth.

Mid-season drainage removes hydrogen sulphide and other harmful substances, which are produced by microbial action under reductive conditions of submergence. Water (5 cm) is needed at milk stage for translocation of nutrients stored in plant body to ear or panicle for healthy development of developing grain or spikelet.

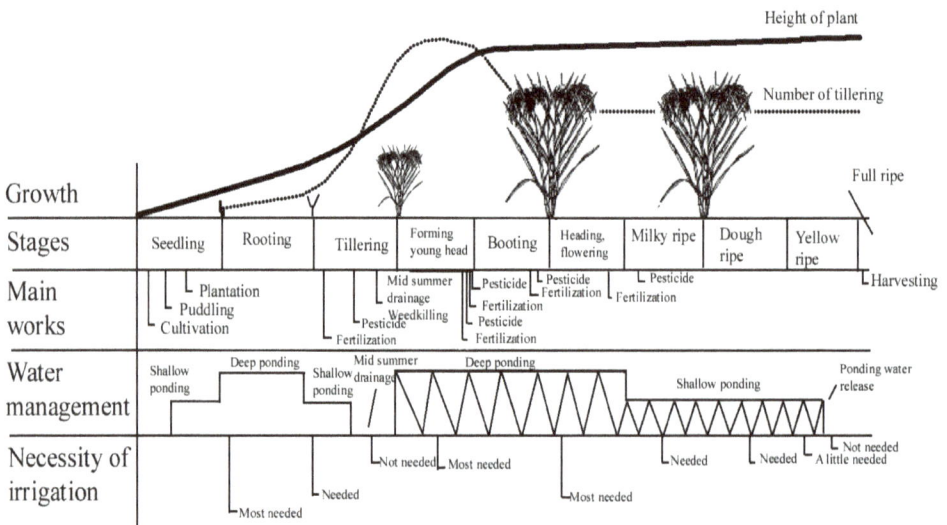

Fig. 15. Rice growth, agricultural works and water management (Maruyama and Tanji, 1997)

## 4.2 Rice growing calendar and water management

Maruyama and Tanji (1997) showed that the growth stages of rice can be divided into ten growth stages associated with water management practices in Japan as shown in Fig. 15. The paddy farmers must control the field water depth precisely according to the growth stage in order to reap the benefit of higher water productivity.

## 4.3 Water-efficient irrigation regimes to increase water productivity

Mao Zhi (2000) stated that rice is one of the most important food crops contributing over 39% of the total food grain production in China. Out of 113 million hectares of area sown under food crops 28% is covered by rice. The traditional irrigation regime for rice, termed as "continuous deep flooding irrigation" was applied in China before 1970s. Since 1980s, the industry water supply, urban and rural domestic water consumption has been increasing continuously. The shortage of water resources became an important problem and many water efficient irrigation regimes for rice have been tested, advanced, applied and spread in different regions of China.

Based on the results of experiment and the experience of spread of these new irrigation regimes, the following conclusions were drawn by the author:

- Three essential water efficient irrigation regimes (WEI) for rice as shown in Fig. 16, which include the regimes of combining shallow water depth with wetting and drying (SWD), alternate wetting and drying (AWD) and semi−dry cultivation (SDC), have been adopted in the different rice growing regions of China.
- In comparison to the traditional irrigation regime (TRI), rice yield can be increased slightly, water consumption and irrigation water use of paddy field can be decreased greatly and the water productivity of paddy field can be increased remarkably under the WEI.
- The main causes of decrease of water consumption and irrigation water use are the decrease of the percolation rate in paddy field and increase in the utilization of rainfall.
- A positive environmental impact is obtained by adopting WEI, the main cause of getting bumper yields were that the ecological environment under WEI is more favourable for the growth and development of rice than that under TRI.
- For avoiding the decrease of yield under WEI, some measures, as timely irrigation, coordinating irrigation with fertilization and weed control must be used since shortage of water resources in China is becoming more serious each year, the water efficient irrigation techniques should be further investigated and adopted on large areas.

## 4.4 Distribution variability of effective rainfall

With global warming and climate change, greater competition is expected among water users, and paddy irrigation may be sacrificed during water shortage in dry months favouring domestic and industrial users. However, rice granaries practicing multiple cropping have yet to improve on the use of "effective rainfall". Currently, the measurement of rain falling in a rice growing area is based solely on the available rain gauge network. These gauges are located at convenient locations which may not be representative of the whole rice growing area. Hence, under- or over-estimation of rainfall distribution and runoff occurs and consequently affects the management of floods during rainy seasons or base flow for irrigation during dry seasons. Therefore, better estimates of mean areal rainfall are needed as contribution of effective rainfall in the water balance during the irrigation season.

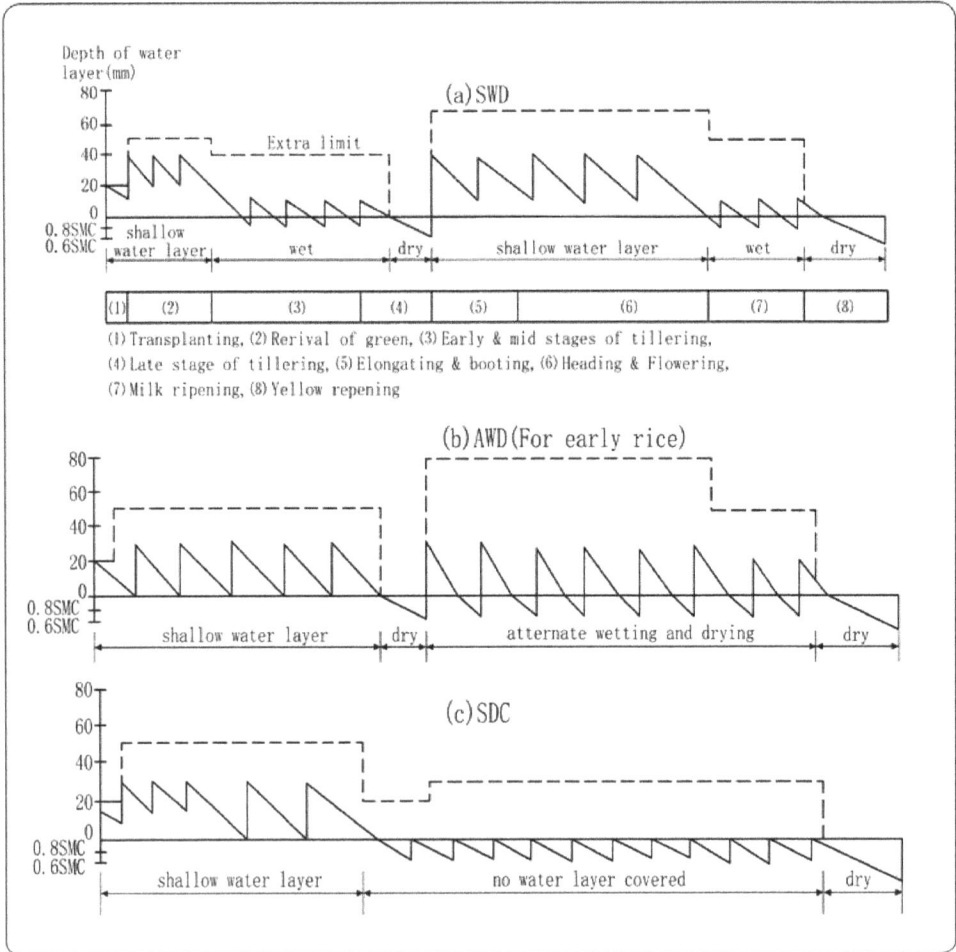

Fig. 16. Description of Different Water Efficient Regimes (Mao Zhi, 2000)

A new technique to improve rainfall distribution estimation based on weather radar-derived rainfall throughout the rice growing area was developed by UPM using GIS tools. Virtual rainfall stations are created uniformly throughout the area to improve the spatial distribution of rainfall over a rice granary or a watershed with low density rain gauge network. Virtual rainfall stations can be distributed in terms of grid centres to cover the whole study area as shown in Fig. 17. The rainfall data for these virtual rain gauges are estimated from raw radar data available from the Malaysian Meteorological Department using a newly developed Program called UPM ViRaS RaDeR ver1.0 (Amin et al., 2010). The derived rainfall data is

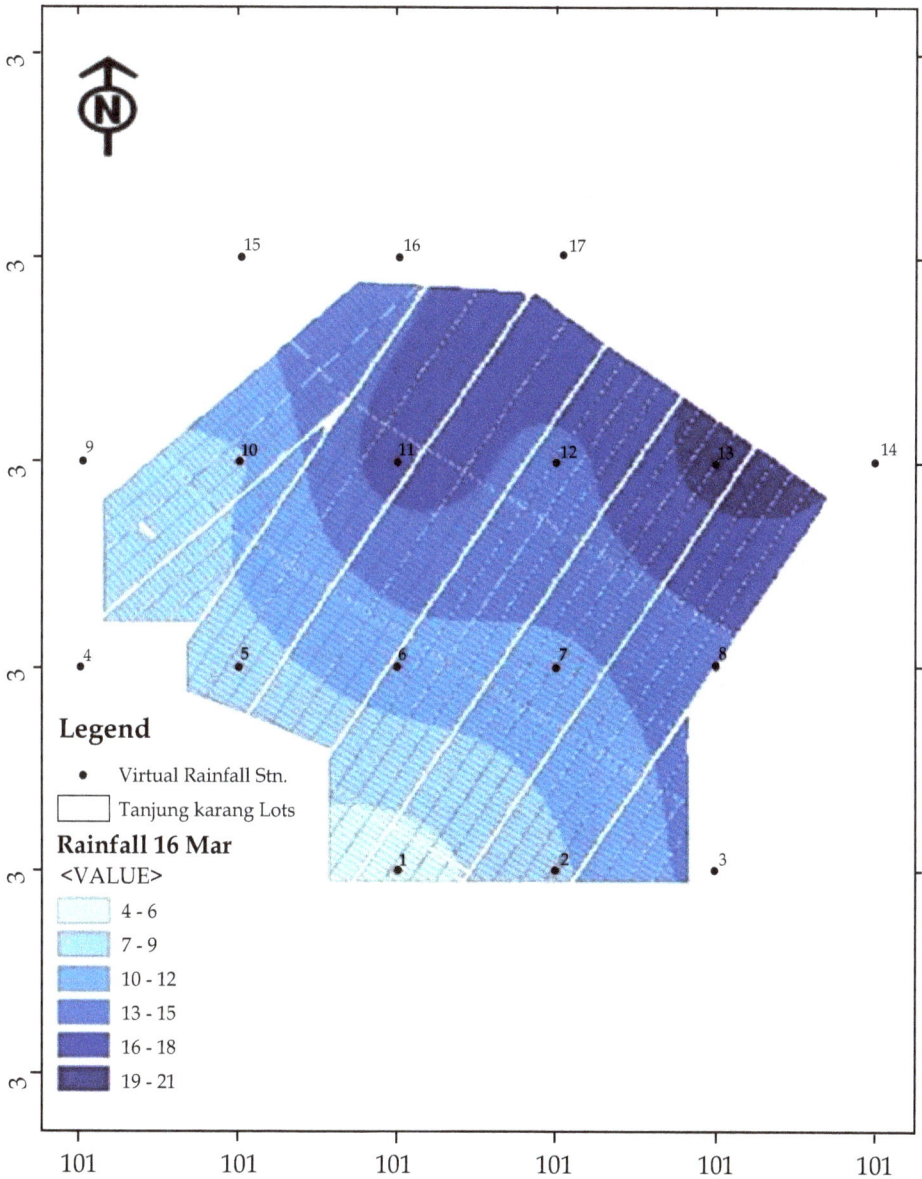

Fig. 17. (continues on next page) The radar derived rainfall data from 17 Virtual Rainfall Stations in a 2300 ha Sawah Sempadan Irrigation Compartment produced rainfall distribution pattern which otherwise would always be uniformly distributed since there is only one rain-gauge for the whole area

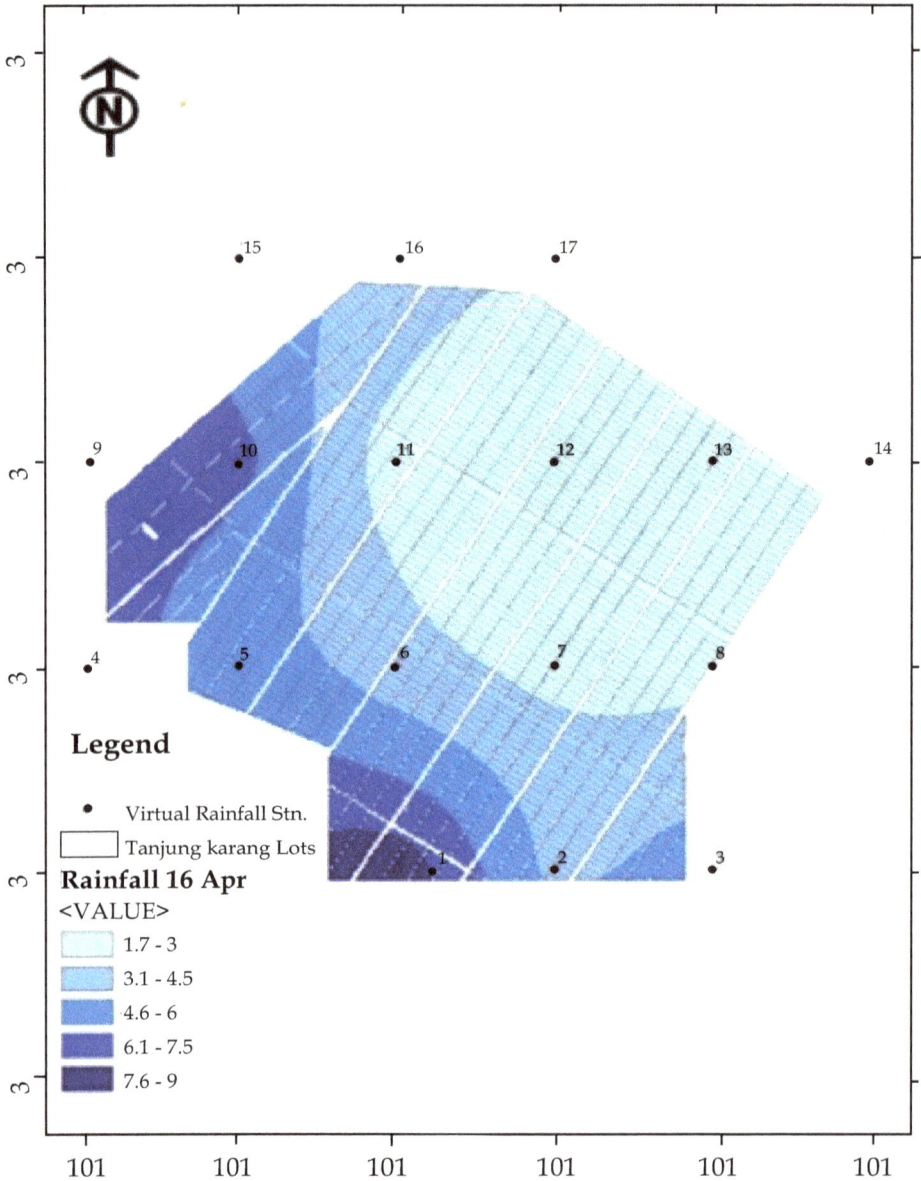

Fig. 17. (continued) The radar derived rainfall data from 17 Virtual Rainfall Stations in a 2300 ha Sawah Sempadan Irrigation Compartment produced rainfall distribution pattern which otherwise would always be uniformly distributed since there is only one rain-gauge for the whole area

then compared and calibrated with actual gauge rainfall records for the same periods to identify the calibration factor. RaDeR provides rainfall distribution pattern which otherwise would always be uniformly distributed since there is only one rain-gauge for the whole area. The calibrated radar derived rainfall data will next be used as improved rainfall input in the hydrological model for watershed runoff estimation. On the other hand, knowing the amount of rainfall that occurred in a rice granary or a farm, a suitable amount of irrigation water can be supplied precisely and better irrigation water management can be adopted. Irrigation can be stopped when enough rain water has already refilled the soil moisture reservoir or standing water depth in the paddy fields. Hence the effective rainfall will save some amount of irrigation water supply and used for other purposes.

## 5. Summary

In anticipation of future greater competition for irrigation water due to climate change and global warming, paddy water management should be more focused towards water saving and precision irrigation. This book chapter has described new indicators for evaluating the performance of different aspects of an irrigation system for rice cultivation. A GIS-based interactive assessment tool is given using a new concept to characterize the irrigation delivery performance as the season advances. The weakness of a widely used Relative Water Supply (RWS) concept is overcome by using the new indicators, viz. Rice Relative Water Supply (RRWS), Cumulative Rice Relative Water Supply (CRRWS) and Ponding Water Index (PWI). The RRWS can distinctly characterize the oversupply condition for RRWS > 1.0 and undersupply condition for RRWS < 1.0 on irrigation delivery for any given period. A value of 1.0 for RRWS indicates an irrigation delivery that perfectly matches the actual field water demand. A user-interface was developed for structuring the assessment tool within ArcGIS platform. The system can instantly give information on the uniformity of water distribution and the shortfall or excess and what decisions to adopt for the next period. The results are displayed on the computer screen together with colour-coded maps, graphs and tables in a comprehensible form. The system can be adopted as an analytical and operational tool for the irrigation managers to evaluate various water allocation scenarios and water management options.

Water savings can be obtained by practicing precision farming of rice in lowland paddy fields. However a rapid assessment of the paddy soil variability needs to be determined, for example through mapping of the bulk electrical conductivity (ECa) of the paddy fields, so that variable treatments of the management zones can be adopted to save the precious resources. ET monitoring is necessary to determine the required amount of water at each crop growth stage, and the rainfall distribution pattern in the irrigation scheme should be considered to make better use of effective rainfall with respect to the stage of crop development. The practice of precision farming (i.e. applying the right input, at the right place, at the right time, at the right amount and in the right manner using the right tools) will ensure high water and land productivity for a sustainable rice production to feed the growing world population.

## 6. Acknowledgements

The financial support for the research provided by the Government of Malaysia through MOSTI is gratefully acknowledged. Cooperation from all members of the Precision Farming

Engineering Research and Development (PREFERD) Group UPM, staff at the Tanjung Karang Irrigation Scheme and Integrated Agricultural Development Area (IADA), Kuala Selangor, Malaysia and the SMART Farming Technology Centre of Excellence UPM is highly appreciated.

## 7. References

[1] Plusquellec H., Burt C. and Wolter H.W. Modern water control in irrigation. World Bank Technical Paper no. 246 – Irrigation and Drainage series, Washington D.C. (1994).

[2] De Datta S.K. Principles and practices of rice production. Los Baños, Philippines: International Rice Research Institute, 1981, 618 p.

[3] Anbumozhi V., Yamaji T. and Tabuchi T. Rice crop growth and yield as influenced by changes in standing water depth, water regime and fertigation level. J. Agril. Water Mgmt., 1998, 37: 241-253.

[4] Abernethy C.L. and Pearce G.R. Research needs in third world irrigation, Hydraulics Research Limited, Wallingford, England, 1987.

[5] Rao P.S. Review of selected literature on indicators of irrigation performance. IIMI Research Paper, 1993.

[6] Levin G. Relative water supply: An explanatory variable for irrigation systems, Technical report no. 6, Cornell University, New York, 1982.

[7] Nihal F. Monitoring irrigation water delivery performance: The concept of Cumulative Relative Water Supply (CRWS). In: Proceedings of an International Conference on Advances in Planning, Design, and Management of Irrigation Systems as related to Sustainable Land Use, Katholic Universiteit Leuven, Belgium, 1992, pp 525-534.

[8] Shakthivadivel R., Douglas J.M., Nihal F. Cumulative relative water supply: A methodology for assessing irrigation system performance. Irrigation and Drainage Systems 7, 1993 pp. 43-67.

[9] Oad R. and Podmore T.H. Irrigation management in rice-based agriculture: Concept of relative water supply. ICID Bulletin 38(1), 1988, pp. 1-12.

[10] Weller J.A. An evaluation of the Porac irrigation system. Irrigation and Drainage Systems 5(1), 1991, pp. 1-17.

[11] Rowshon M.K., Kwok C.Y., Lee T.S. GIS-Based Scheduling and Monitoring of Irrigation Delivery for Rice Irrigation System-Part II: Monitoring. Agricultural Water Management, vol 117, 2003, pp. 117-126.

[12] Knox J.W. and Weatherhead E.K. The application of GIS to irrigation water resources management in England and Wales. Geographical Journal 165(1), 1999, 90–98.

[13] Tsihrintzis V.A., Hamid R. and Fuentes H.R. Use of geographic information systems (GIS) in water resources: a review. Water Resour. Manage, 10, 1996, 251–257.

[14] DID and JICA. The study on Modernization of Irrigation Water Management System in the Granary Area of Peninsular Malaysia, Draft Final Report, Volume –II, Annexes, March 1998.

[15] Hassan SMH. Estimation of rice evapotranspiration in paddy fields using remote sensing and field measurements. Unpublished PhD Dissertation, University Putra Malaysia, (2005), pp 212.

[16] Molden, D., Sakthivadivel, R., Perry, C. J., de Fraiture, C., and Kloezen, W.H. Comparing the performance of irrigated agricultural systems, IIMI News, Vol. 1, No. 3, Colombo, Sri Lanka, 1997.

[17] Grisso, R. 2009. Precision Farming: A Comprehensive Approach. Extension publication 442-500, Biological Systems Engineering, Virginia Tech. USA

[18] Eltaib, S.M.G., 2003. Spatial variability of soil electrical conductivity, nutrients and rice yield, and site-specific fertilizer management. PhD Thesis UPM.

[19] ESRI. 2001. Using ArcGIS, Geostatistical Analyst. ESRI, CA.

[20] Fleming, K.L., D.W. Weins, L.E. Rothe, J.E. Cipra, D.G. Westfall and D.F. Heerman. 1998. Evaluating Farmer Developed Management Zone Maps for Precision Farming. The 4th Int'l. Conf. On Prec. Agric. St. Paul, MN, p. 138.

[21] Fraisse, C. W., Sudduth, K. A. and Kitchen, N. R. 1999. Evaluation of Crop Models to Simulate Site-Specific Crop Development and Yield. In: Proc of the 4th Int Conf. on Precision Agriculture, P. C. Robert et al. (Eds.) p. 1297–1308.

[22] Jaynes, D.B., Novak, J.M., Moorman, T.B. and Cambardella, C.A., 1994. Estimating Herbicide Partition Coefficients from Electromagnetic Induction Measurements. J. Envt. Quality, 24: 36-41.

[23] Jaynes, D.B., T.S. Colvin, and J. Ambuel, 1995. Yield mapping by electromagnetic induction. p. 383–394. In P.C. Robert et al. (ed.) Proc. of site-specific management for agricultural systems, 2nd, Minneapolis, MN.

[24] Kitchen, N.R, and K.A. Sudduth, 1996. Predicting Crop Production using Electromagnetic Induction. Information Agriculture Conference Proceedings, Urbana IL.

[25] Lund, E.D., Christy, C.D., Drummond, P.E. 1998. Applying Soil Electrical Conductivity Technology to Precision Agriculture. Proceedings of the 4th Int Conference on Precision Agriculture, St. Paul MN.

[26] McBride, R.A.,A.M.Gordon, and S.C. Shrive. 1990. Estimating Forest Soil Quality from Terrain Measurements of Apparent Electrical Conductivity. Soil Sci. Soc. Am. J. 54:290–293.

[27] Sudduth K.A., C.W. Fraisse, S.T. Drummond and N.R. Kitchen. 1998. Integrating Spatial Data Collection, Modelling and Analysis for Precision Agriculture. Presented at the First International Conference on Geospatial Information in Agriculture and Forestry, Lake Buena Vista, Florida.
http://www.fse.missouri.edu/ars/projsum/erim_3.pdf.

[28] Williams, B.G., and D. Hoey. 1987. The Use of Electromagnetic Induction to Detect the Spatial Variability of the Salt and Clay Contents of Soils. Aust. J. Soil Res. 25:21-27.

[29] Maruyama, T and K.K. Tanji, 1997. Physical and Chemical processes of soil related to paddy drainage, 99-101.

[30] Mao Zhi, 2000. Water-efficient irrigation regimes of rice for sustainable increases in water productivity. In Proceedings of International Rice Research Conference, International Rice Research Institute, Laguna, Philippines.

[31] Asa Gholizadeh, 2011. Apparent Electrical Conductivity and Vis-NIR reflectance for Predicting Paddy Soil Properties and Rice Yield. Unpublished PhD Thesis, UPM Serdang Selangor Malaysia.

[32] Amin, MSM, ARM Waleed, and MY Abdullah, 2010. Improving Water Management using Virtual Rainfall Stations with Radar Derived Rainfall Data. In Proc. of Intnl. Workshop on Integrated Lowland Development and Management, Palembang City, Indonesia, 18-20 March 2010.

# Part 3

## Water Quality

# 7

# Wetlands for Water Quality Management – The Science and Technology

Vikas Rai[1], A. M. Sedeki[1], Rana D. Parshad[2],
R. K. Upadhyay[3] and Suman Bhowmick[3]
*[1]Department of Mathematics, Faculty of Science, Jazan University, Jazan, KSA;*
*[2]Applied Mathematics and Computational Science,*
*King Abdullah University of Science and Technology,*
*Thuwal 23955 – 6900, KSA;*
*[3]Department of Applied Mathematics, Indian School of Mines, Dhanbad, Jharkhand;*
*India*

## 1. Introduction

Wetlands are defined as water systems with marsh or fen and with water that is static or flowing, fresh or brackish, the depth of which at low tide is below 6m [5]. Natural wetlands are characterized by emergent aquatic vegetation such as cattails (typha), rushes (Scirpus) and reeds (Phragmites), and by submerged and floating plant species. Wetlands are categorized into three main categories: (1) fresh water coastal wetlands, (2) flood – plain wetlands and (3) constructed wetlands. These are valuable ecosystems as they support services which contribute significantly to human well - being. Some of these services are fish and fiber, water supply, water purification, flood regulation, recreational opportunities and tourism.

In this chapter, we focus our attention to potential of wetlands as water purifiers. This is important because harmful substances enter into wetland systems through animal waste from farms, toxic chemicals from factories and pesticides present in rain water. The large and diverse population of bacteria which grow on the submerged roots and stems of aquatic plants play an important role in removing $BOD_5$ from waste water. Wetland plants take up harmful substances into their roots and change the harmful substances into the less harmful ones before they are released into the water body. Harmful substances partly get buried in the wetland soil. The soil bacteria convert these into substances, which are not harmful. Soil microbes (*Bacillus subtilis* and *Pseudomonas fluorescens*) convert pesticides into simpler non-toxic compounds. This process of degradation of pesticides and subsequent conversion into non-toxic compounds is known as "biodegradation". The biodegradation is influenced by factors such as moisture, temperature, pH and organic matter content.

The chapter is organized as follows. In the next section, we present a study of a natural wetland of flood – plain type. This section summarizes a previous study of this natural wetland by the author and presents some new results. Section 3 discusses the design and construction of constructed wetlands.

## 2. Keoladeo National Park, Bharatpur, India: A case study

Keoladeo National Park (27°10'N, 77°31'E), a World Heritage Site, is situated in eastern Rajasthan. The park is 2 kilometers (km) south-east of Bharatpur and 50 km west of Agra (cf. Figure 1). Figure 1 provides a location map for the park. The Park is spread over 29 square kilometres area. One third of the Park habitat is wetland system with varying types of trees, mounds, dykes and open water with or without submerged and emergent plants. The uplands have grasslands (savannas) of tall species of grass together with scattered trees and shrubs present in varying densities. The area consists of a flat patchwork of marshes in the Gangetic plain, artificially created in the 1850s and maintained ever since by a system of canals, sluices and dykes. Water is fed into the marshes twice a year from inundations of the Gambira and Banganga rivers, which are impounded on arable land by means of an artificial dam called Ajan Bund, located in the south of the park (cf. Fig. 2). It was developed in the late 19th century by creating small dams and bunds in an area of natural depression to collect rainwater and by feeding it with an irrigation canal.

The 29 km (18 mi) reserve, locally known as Ghana, is a mosaic of dry grasslands, woodlands, woodland swamps, and wetlands. These diverse habitats serves as homes to 366 bird species, 379 floral species, 50 species of fish, 13 species of snakes, 5 species of lizards, 7 amphibian species, 7 turtle species, and a variety of other invertebrates. Keoladeo National Park is popularly known as "bird paradise". Over 370 bird species have been recorded in the park. The park's location in the Gangetic Plain makes it an unrivalled breeding site for

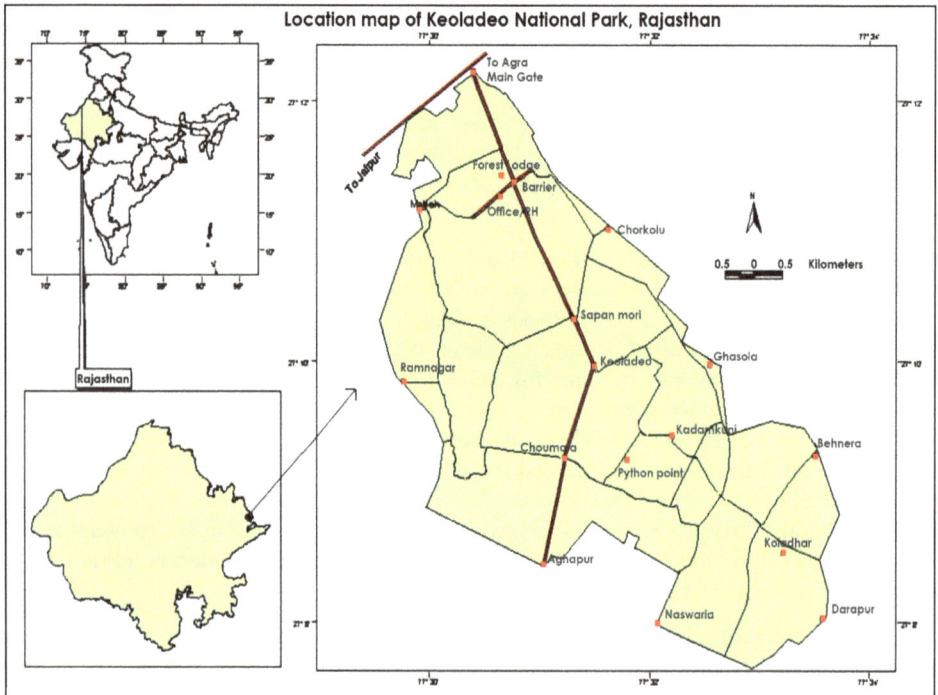

Fig. 1. Location map of the Keoladeo National park, a World Heritage site

herons, storks and cormorants, and an important wintering ground for large numbers of migrant ducks.

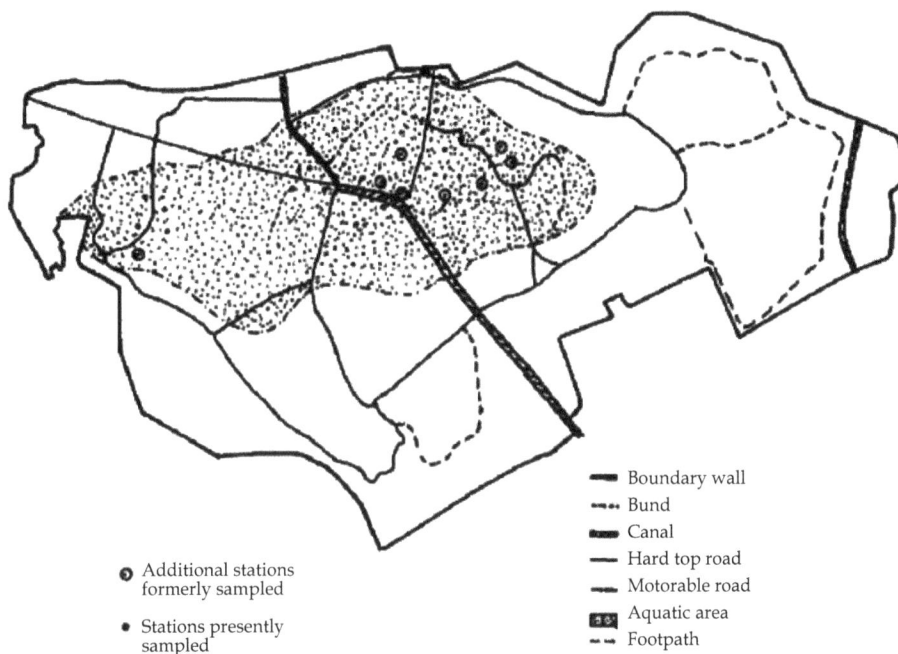

Fig. 2. Situational map of Keoladeo National Park

### Light vs nutrient supplies

It is known that the population persistence boundaries in water column depth–turbulence space are set by sinking losses and light limitation [1]. In shallow waters, the most strongly limiting process is nutrient influx to the bottom of the water column (e.g., from sediments). In deep waters, the most strongly limiting process is turbulent upward transport of nutrients to the photic zone. Consequently, the highest total biomasses are attained in turbulent waters at intermediate water column depths and in deep waters at intermediate turbulences. These patterns have been found insensitive to the assumption of *fixed versus flexible algal carbon -to – nutrient stoichiometry*. They arise irrespective of whether the water column is a surface layer above a deep water compartment or has direct contact with sediments. This helps us understand the relevant dynamical processes in the physical systems in natural as well constructed wetlands.

### Biotic part of Keoladeo National Park

KNP is a natural wetland which can be categorized as a flood – plain type. The economic value of the park is dependent on tourist activities. The tourists are mainly attracted by Siberian Crane, the migratory bird which adds aesthetic value to KNP. It provides a large habitat for migratory birds; Siberian crane being the flagship species. With reference to migratory birds, the biomass is divided into two categories: "Good" and "Bad". The excess growth of the wild grass species, *paspalum distichum* restricts the growth of bulbs, tubers and roots, on which avifauna feed on.

*Paspalum distichum* is known to deplete oxygen in the natural aquatic systems is the dominant species. The paspalum and its family acts as a bad biomass for the birds and the floating vegetation. The following species of the floating vegetation, *Nymphoides indicum, Nymphoides cristatum, Nymphaea nouchali, Nymphaea stellat,* and other useful species are categorized as "good" biomass [7, 8]. The fishes and the water fowl are the most suffered species. Although visits of the tourists bring revenue to the state, it also creates a disturbance gradient.

## 2.1 Good biomass, bad biomass and birds in the Keoladeo National Park: The biotic system

Rai [6] modeled the dynamics of the biotic system of the wetland part of KNP by the following set of ordinary differential equations.

$$
\begin{aligned}
\frac{dG}{dt} &= aG - bG^2 - cGB - d\frac{GP}{G+D}, \\
\frac{dB}{dt} &= eB - \frac{B^2}{W_1} - a_2 GB, \\
\frac{dP}{dt} &= -\theta P + \phi\frac{GP}{G+D_1}.
\end{aligned}
\tag{1}
$$

With
G:    density of the good biomass, g/cm³,
B:    density of the bad biomass, g/cm³,
P:    density of resident birds joined by migratory ones.

This system can be broken into two subsystems. (i) The competition system with "good" and "bad" biomass as component populations. (ii) The prey - predator system with good biomass and the bird population. Oscillatory dynamics are possible in the subsystems, but it is sensitive to initial conditions. The prey–predator subsystem performs oscillatory motion in significant region of the parameter space. The subsystem is essentially a Rosenzweig – MacArthur kind of system [9]. It is known to produce oscillatory dynamics in a significant region of the parameter space.

The good biomass and the resident birds occasionally joined by migratory ones constitute a subsystem. It is given by the following set of ordinary differential equations

$$
\begin{aligned}
\frac{dG}{dt} &= aG - bG^2 - d\frac{GP}{G+D}, \\
\frac{dP}{dt} &= -\theta P + \phi\frac{GP}{G+D_1}.
\end{aligned}
\tag{2}
$$

With
D:    a measure of the half-saturation constant,
a:    reproductive growth rate of the good biomass,

$b$:   measures the severity of the intra-specific competition among individuals of good biomass,

$\phi$:   conversion coefficients for the bird species (resident as well as migratory),

$t$:   time measured in days.

Critical Points for the subsystem (2) are $(0, 0)$, $\left(\dfrac{a}{b}, 0\right)$, $\left(\dfrac{\theta D_1}{\phi - \theta}, \dfrac{(a - bG)(G + D)}{d}\right)$.

For the following parameter set, the nature of these critical points $(0,0),(100,0),(20,12.8)$ turn out to be as follows.

$$a = 0.2, b = 0.002, c = 0.005, d = 0.5, D = D_1 = 20, \theta = 0.05, \phi = 0.1 \tag{3}$$

$(0, 0)$ is a saddle point. $(100, 0)$ is an unstable node. The critical point located at $(20, 12.8)$ is an unstable focus.

## 2.2 The biotic part of the wetland system in space and time

$$\frac{dG}{dt} = aG - bG^2 - cGB - d\frac{GP}{G + D} + \frac{\partial^2 G}{\partial x^2}$$

$$\frac{dB}{dt} = eB - \frac{B^2}{W_1} - a_2 GB + \frac{\partial^2 B}{\partial x^2} \tag{4}$$

$$\frac{dP}{dt} = -\theta P + \phi \frac{GP}{G + D_1} + \frac{\partial^2 P}{\partial x^2}$$

Equal diffusivity constants were assumed for good and bad biomasses. The value of the diffusion coefficient for the avian predator (birds) was taken to be $10^{-6}$. The vegetation diffusion coefficient is $10^{-7}$. Figures show contours of equal densities on a suitably chosen spatial scale.We assume that the horizontal extent of the wetland system is large compared to the depth of the wetland system. Each figure is accompanied with a scale specifically tailored to the species it represents. Dark blue represents small spatial densities, light blue represents slightly higher spatial densities, yellow and red represent higher spatial densities. The results show that "good" biomass has uniform spatial distributions in certain domains. The spatial distribution of " bad" biomass contains two distinct humps.

The "good biomass" acquires stable stationary patterns (Figure 3). The "bad" biomass performs swinging motion and selects a steady state spatial pattern given in Figure 4. Figure 5 presents stationary spatial distribution of the species clubbed under the category "bad" after the model system is allowed to run after a long period of time. Simulations in two - spatial dimensions are needed to unravel mysteries of system's spatio-temporal dynamics.

The complete system displays three kinds of oscillatory motion which represents three different health conditions of the wetland [6]. The per capita availability of water to "bad biomass" is known to be the control parameter. An earlier study by one of the authors (VR) may be helpful in this regard as vegetation plays a critical role in BOD5 removal and de-nitrification of the available nitrogen. The crucial factor for constructed wetland design for waste water treatment would be to control the density of the good and bad biomass; the

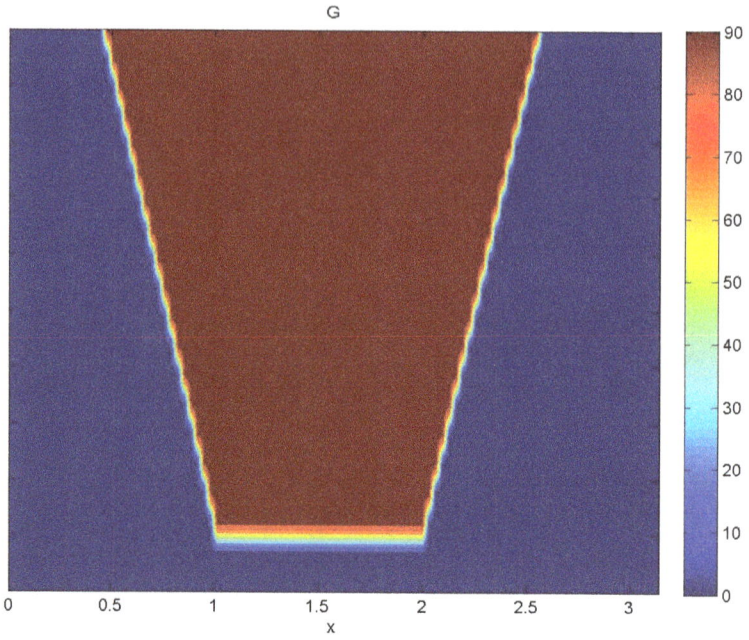

Fig. 3. Spatial distribution of the "good" biomass (G)

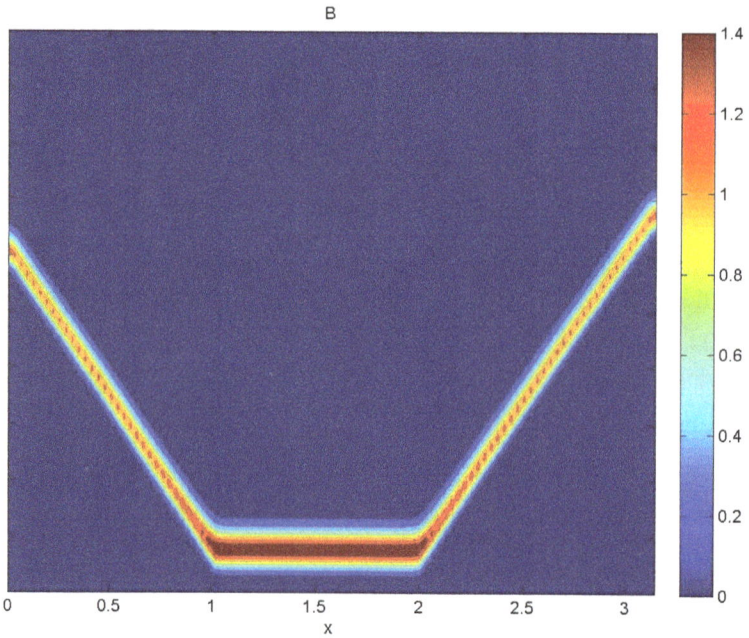

Fig. 4. The distribution of the "bad" biomass (B) - see section 2, para 2

vegetation component of the system. The chemical fertilizer upstream deteriorates the water quality (WQ) of the wetland. Construction of an artificial wetland in the vicinity of the natural one would restore the water quality standards.

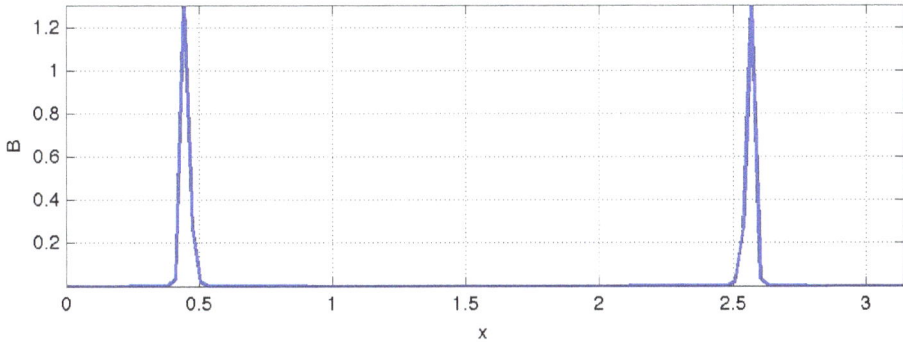

Fig. 5. The spatial distribution of the "bad" biomass after the system is run for a long period of time

## 3. Constructed wetlands for water pollution management

These man-made wetlands are used to treat aquaculture and municipal water, to regulate the water quality of shrimp ponds and manage pollution from pond effluents. The wetland treated effluents satisfy standards for aquaculture farms. Since the technology to use the constructed wetlands to treat waste water of high $BOD_5$ is limited, these are generally used to polish secondary effluents. Other applications of constructed wetlands are (a) to treat acid mine drainage, (b) to treat storm water, and (c) the enhancement of existing wetlands.

The suggestion to use wetland technology for waste water treatment is attractive for both ecological and economic reasons. Constructed wetlands are efficient in removing pathogens [2]. It performs better than conventional waste water treatment methods although the lack of knowledge of principles of pathogen removal in plants hampers optimum performance. Interactions between soil matrix, micro-organisms and plants and higher retention time of the waste water in these biologically complex systems make phyto-remediation more effective than conventional systems. Phyto-remediation involves complex interactions between plant roots and micro-organisms in the rhizo-sphere. The efficient functioning of wetland systems is hampered due to following factors:

- High redox potentials,
- Acidity of effluents, i. e., low pH and
- Microbial degradation of organic substrates (i.e. BTEX, petroleum–derived hydrocarbons, HET, phenols).

Wetland systems efficiently treat water polluted by heavy metals, chromium and magnesium.

The metal removal in these systems involves following mechanisms:

- Filtration and sedimentation of suspended particles,

- Incorporation into plant material,
- Precipitation by microbial mediated biogeochemical processes, and
- Adsorption on the precipitates.

Constructed wetlands are either free water surface systems with shallow water depth or subsurface flow systems with water flowing laterally through the land and gravel. These wetlands have been used for wastewater treatment for nearly 40 years and have become a widely accepted technology available to deal with both point and non-point sources of water pollution. They offer a land-intensive, low-energy, and low-operational-requirements alternative to conventional treatment systems, especially for small communities and remote locations. Constructed wetlands also prove to be affordable tools for wastewater reclamation, especially in arid and semi-arid areas. Although the emission of $N_2O$ and $CH_4$ from constructed wetlands is found to be relatively high, their global influence is not significant towards their contribution to global warming.

Three main components of an artificial wetland are as follows:

**(a) Construction practices**

While design should be kept as simple as possible to facilitate ease of construction and operation, the use of irregular depths and shapes can be beneficial to enhance the wildlife habitat. The site for construction should be properly chosen so as to limit damage to local landscape by minimizing excavation and surface runoff during construction and, at the same time, maximize flexibility of the system to adapt extreme conditions.

**(b) Soil**

The chosen soil must not contain a seed bank of unwanted species. The permeability of the soil should be carefully controlled as highly permeable soils may allow infiltration and possible contamination of ground water. High permeability is not conducive for development of suitable hydrological conditions for wetland vegetation. Use of impermeable barriers may be suggested in certain instances.

**(c) Selection of vegetation**

Plant species among native and locally available species should be chosen keeping in mind water quality and habitat functions. The use of weedy, invasive and non-native species should be avoided. Plants ability to adapt to various water depths, soil and light conditions should also be taken into consideration.

In the following, design and construction of two kinds of artificial wetlands which are used for water purification will be described.

## 3.1 Free water surface (FWS) wetland systems

These systems consist of basins or channels with subsurface barrier to prevent seepage, soil or another medium to support the emergent vegetation and water at a shallow depth flowing through the unit. The shallow water depth, low flow velocity and presence of plant stalks and litter regulate the water flow [3]. The soil permeability is an important parameter. The most desirable soil permeability is $10^{-6}$ to $10^{-7}$ meter per second. The uses of highly permeable soils are recommended for small waste water flows by forming narrow trenches and lining the trench walls and bottom with clay or an artificial liner.

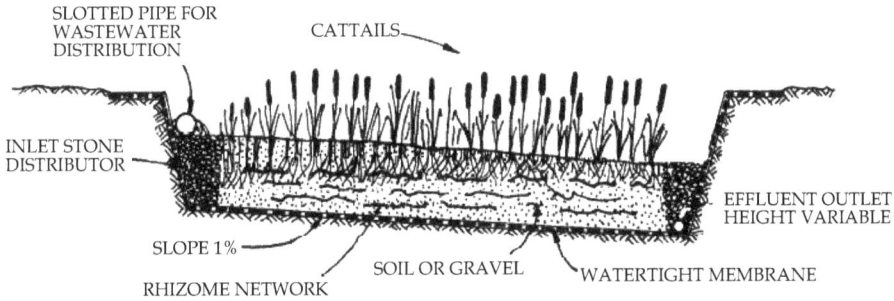

Fig. 6. A cross – sectional view of a Free Water Surface Wetland system (reprinted from the US Environmental Protection Agency design manual)

The hydrologic budget is an important part of design of constructed wetlands. The following water balance equation is generally used

$$Q_i - Q_o + P - ET = \frac{dV}{dt},$$ (5)

Where:
$Q_i$   influent waste water flow, volume/time
$Q_o$   effluent waste water flow, volume/time
$P$     precipitation, volume/time
$ET$   evapo-transpiration, volume per unit time
$V$     volume of water, and
$t$      time.

Ground water inflow and infiltration are excluded from the above equation as impermeable barriers are used. Historical climatic records can be used for estimating the precipitation and evapo-transpiration. Infiltration losses can be estimated by conducting infiltration tests [3].

Typical dimensions of a FWS are:
- Length ≈ 64 meters
- Bed width = 660 meters.
- Bed depth = 0.3 meters,
- Retention time is 5.2 days.

Divide the width into individual cells for control of hydraulic loading rate. Vegetation used in United States is Cattails, reeds, rushes, bulrushes, and sedges. Physical presence of this vegetation transports oxygen deeper than it would reach through diffusion. Submerged portions serve as home for microbial activity. The attached biota is responsible for treatment that occurs.

## 3.2 Constructed wetlands with horizontal subsurface flow (HF)

Horizontal Subsurface Flow systems (submerged horizontal flow) consist in basins containing inert material with selected granulometry with the aim to assure an adequate hydraulic conductivity (filling media mostly used are sand and gravel). These inert materials represent the support for the growth of the roots of emerging plants (cf. Fig. 7). The bottom of the basins has to be correctly waterproofed using a layer of clay, often available on site and under adequate hydro-geological conditions or using synthetic membranes (HDPE or LDPE 2 mm thick). The water flow remains always under the surface of the absorbing basin and it flows horizontally [11]. A low bottom slope (about 1%) obtained with a sand layer under the waterproof layer guarantees this.

During the passage of wastewater through the rhizo-sphere of the macro-phytes, organic matter is decomposed by microbial activity, nitrogen is denitrified. In the presence of sufficient organic content, phosphorus and heavy metals are fixed by adsorption on the filling medium. Vegetation's contribution to the depurative process is represented both by the development of an efficient microbial aerobic population in the rhizo-sphere and by the action of pumping atmospheric oxygen from the emerged part to the roots and so to the underlying soil portion, with a consequent better oxidation of the wastewater and creation of an alternation of aerobic, anoxic and anaerobic zones. This leads to the development of different specialized families of micro-organisms. It also leads to nearly complete disappearance of pathogens, which are highly sensitive to rapid changes in **dissolved oxygen content.** Submerged flow systems assure a good thermal protection of the wastewater during winter, especially when frequent periods of snow are prevented.

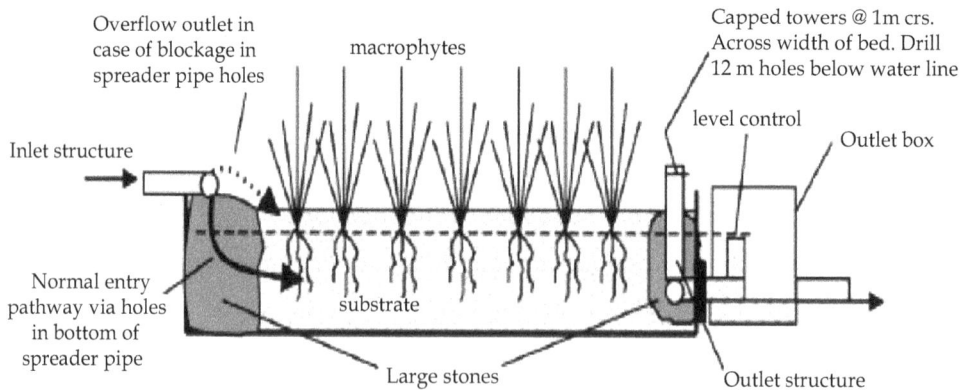

Fig. 7. Sketch of a subsurface flow wetland showing the working principles (reprinted from reference [10])

Key design parameters of horizontal subsurface flow constructed wetlands

- hydraulic loading rate (HLR),
- aspect ratio,
- size of the granular medium, and
- water depth.

Hydraulic linear loading rate is the volume of waste water that the soil surrounding a waste water infiltration system can transmit far enough away from the infiltration surface such that it no longer influences the infiltration of additional waste water. It depends on the soil characteristics. In principle, the hydraulic loading rate is equal to the particles settling velocity. A greater surface allows capture of particles with smaller settling velocities. Typical hydraulic rates in subsurface flow wetlands vary from 2 to 20 cm per day.

The aspect ratio defines the length to width ratio. This is considered to be of critical importance for the adequate flow through the wetland. Constructed wetlands are designed with an aspect ratio of less than 2 to optimize the flow and minimize the clogging of the inlet.

## 3.3 Performance evaluation

Wetland systems significantly reduce biological oxygen demand ($BOD_5$), suspended solids (SS), and nitrogen, as well as metals, trace element, and pathogens. The basic treatment mechanisms include sedimentation, chemical precipitation, adsorption, and microbial degradation of organic matter, Suspended solids and nitrogen, as well as some uptake by the vegetation.

Microbial degradation (also expressed as biological oxygen demand $BOD_5$) in a wetland can be described by a first-order degradation model

$$\frac{C_e}{C_o} = \exp(-K_T t) \tag{6}$$

Where:
$C_o$   influent $BOD_5$, mg/L
$C_e$   effluent $BOD_5$, mg/L
$K_T$   temperature-dependent first-order reaction rate constant, $d^{-1}$
$t$    hydraulic residence time, d

Hydraulic residence time can be represented as

$$t = \frac{LWd}{Q}. \tag{7}$$

Where:
$L$   length
$W$   width
$d$   depth
$Q$   average flow rate = (flow $_{in}$ + flow $_{out}$) ÷ 2

Equation (7) represents hydraulic residence time for an unrestricted flow system.

In a FWS wetland, a portion of the available volume will be occupied by the vegetation; therefore, the actual detention time is a function of the porosity (n). The porosity is defined as the remaining cross-sectional area available for flow.

$$n = \frac{V_v}{V} \tag{8}$$

With:
$V_v$   volume of voids,
$V$    total volume.

The ratio of residence time from dye studies to theoretical residence time calculated from the physical dimensions of the system should be equal to the ratio.

Combining the relationships in Equations (7) and (8) with the general model (Equation 6) yields

$$\frac{C_e}{C_o} = A\exp\left[-0.7K_T\left(A_v\right)^{1.7}\frac{LWdn}{Q}\right]$$

(9)

Where:

$A$     fraction of BOD5 not removable as settling of solids near head works of the system (as decimal fraction),

$A_v$    specific surface area for microbial activity, m²/m³

$L$     length of system (parallel to flow path), m

$W$    width of system, m

$d$     design depth of system, m

$Q$    average hydraulic loading of the system, m/d

$n$     porosity of system (as a decimal fraction).

$$K_T = K_{20}\left(1.1\right)^{(T-20)},$$

(10)

where $K_{20}$ is the rate constant at 20°C.

Other coefficients in equation (5)

$A$= 0.52

$K_{20}$ = 0.0057 d⁻¹

$A_v$= 15.7 m²/m³

$n$= 0.75

In most of the SFS wetlands, the system is designed to maintain the flow below the surface of the bed where direct atmospheric aeration is very low. The oxygen transmitted by the vegetation to the root zone is the major oxygen source. Therefore, the selection of plant species is an important factor. The required surface area for a subsurface flow system is given by

$$A_S = \frac{Q\left(\ln C_o - \ln C_e\right)}{K_T dn}$$

(11)

The cross – sectional area for the flow for a subsurface flow is calculated according to

$$A_c = \frac{Q}{k_S S},$$

(12)

Where $A_c = d\times W$, cross – sectional area for wetland bed, perpendicular to the direction of the flow, m²,

$d$     bed depth, m

$W$    bed width, m

$k_s$    hydraulic conductivity of the medium, $\dfrac{m^2}{m^3}$ d

$S$     slope of the bed, or hydraulic gradient.

The bed width is calculated by the following equation

$$W = \frac{A_c}{d}$$

Cross – sectional area and bed width are established by Darcy's law

$$Q = k_S A_S S \tag{13}$$

The value of $K_T$ is calculated using

$$K_T = K_{20}(1.1)^{(T-20)} \tag{14}$$

$K_{20} = 1.28$ d⁻¹ for typical media types.

## 4. Conclusion

Constructed wetlands are a cost-effective technology for the treatment of waste water and runoff. Operation and maintenance expenditure are low. These systems can tolerate high fluctuation in flow; with wastewaters with different constituents and concentration. Free water systems (FWS) are designed to simulate natural wetlands with water flow over the soil surface at shallow depth. FWS are better suited for large community systems in mild climates. The treatment in subsurface flow (SF) wetlands is anaerobic because the layers of media and soil remain saturated and unexposed to the atmosphere. Use of medium – sized gravel is advised as clogging by accumulation of solids is a remote possibility. Additionally, medium – sized gravel offers more number of surfaces where biological treatment can take place. Thus SF types of wetlands perform better than FWS. A properly operating constructed wetland system should produce an effluent with less than 30 mg/L BOD, less than 25 mg/L of total suspended solids and less than 10,000 cfu per 100 mL, fecal coliform bacteria.

In sum, we note that artificial wetlands are known to perform better as far as removal of nitrogen is concerned. The removal of phosphorous and metals depend critically on contact opportunities between the waste water and the soil. Performance of both kinds of constructed wetlands is poor as contact opportunities are limited in both of them. The submerged bed designs with proper soil selection are preferred when phosphorous removal is the main objective. In contrast to this, removal of suspended solids is excellent in both types of artificial wetlands. Constructed (artificial) wetlands assume special significance as natural wetlands are degrading at a rate faster than the other ecosystems. Two primary ecological agents which cause degradation of natural wetlands are

1. eutrophication and
2. introduction of invasive alien species.

The water in the wetland must be shielded from sunlight in order to control algae growth problems. Algae is known to contribute to suspended solids and cause large diurnal swings in oxygen levels in the water.

## 5. References

Jager, C. G., Diehl Sebastian, Emans, M. Physical determinants of phytoplankton production, Algal Stoichiometry and Vertical Nutrient Fluxes. The American Naturalist, vol. 175 (4), E91 – E104, 2010.

Helmholtz Association Information Booklet for Constructed Wetlands and aquatic plant systems for municipal waste water treatment.

US Environmental Protection Agency Design manual EPA/625/1 – 88/022, 1988.

EPA Guiding principles for constructed treatment wetlands, EPA 843 – B-00-003, 2000. Office of wetlands, Oceans and Watersheds, Washington DC, 4502 F.

EPA 843 – F – 01 – 002b, 2001, United States Environmental Protection Agency, Office of Water and Office of wetlands.

Rai, V. 2008. Modeling a wetland system: The case of Keoladeo National Park (KNP), India. Ecological Modeling 210, 247 – 252.

Shukla, J. B. Dubey, B. 1996. Effects of changing habitats on species: Application to KNP, India. Ecological Modeling 86, 91 – 99.

Shukla, V. P., 1998. Modeling the dynamics of wetland macro-phytes, KNP, India. Ecol. Model. 109: 99 – 112.

Rosenzweig, M L, MacArthur, R. H. Graphical representation of stability of predator – prey interactions. American Naturalist, 1963.

Davison, L., Headley, T. and Pratt, K. (2005). Aspects of design, structure, performance and operation of reed beds – eight years experience in northeastern New South Wales, Australia. Water Science and Technology: A journal of the International Association on Water Pollution Research 51, 129 – 138.

Vymazal, J. 2005. Horizontal Subsurface flow and hybrid constructed wetland systems for waste water treatment. Ecological Engineering 24, 478 – 490.

# 8

# Simulation of Stream Pollutant Transport with Hyporheic Exchange for Water Resources Management

Muthukrishnavellaisamy Kumarasamy
*School of Civil Engineering Surveying & Construction, University of KwaZulu-Natal, Durban, South Africa*

## 1. Introduction

Stream channel irregularities, meandering and hyporheic zones, is commonly seen in many riparian streams. These irregular channel geometries often influence the pollutant transport. The stagnation or dead zones are the pockets of stagnant water or water having very low velocities, which trap pollutants and release them at later time to mainstream flow at a rate that depends upon the concentration gradient of the pollutants between the two domains. The stagnation zones are formed near the concave banks of the stream and behind the irregular sand dunes formed on the bed of the stream. The stagnation zones may also be formed due to irregular stream boundaries and also due to localized channel expansions (Bencala & Walters, 1983). The hyporheic zone is a transition zone between terrestrial and aquatic ecosystems and is regarded as an ecologically important ecotone (Boulton et al., 1998; Edwards, 1998). The term hyporheic is derived from Greek language – hypo, meaning under or beneath, and rheos, meaning a stream (Smith, 2005). A number of definitions for the hyporheic zone exist (Triska et al., 1989; Valett et al., 1997; White, 1993), however, the most common connotations are: it is the zone below and adjacent to a streambed in which water from the open channel gets exchanged with interstitial water in the bed sediments; it is the zone around a stream in which fauna characteristic of the hyporheic zones are distributed and live; it is the zone in which groundwater and surface water mix (Smith, 2005). The physical process, of hyporheic exchanges, as described by many investigators (Stanford and Ward, 1988; Stanford and Ward, 1993; Triska et al., 1989; Valett et al., 1997; Brunke and Gonser, 1997, White, 1993) suggest that significant amounts of water are exchanged between the channel and saturated sediments surrounding the channel. Such exchanges have the potential to cause large changes in stream water chemistry and retard the transport of pollutants. The rates of biogeochemical processes and the types of processes governed by flow hydraulics may be fundamentally different. When the groundwater component is negligible, this is possible if there is a fine silt or clay formation underneath the hyporheic zone, the exchanges between the mainstream flow and the hyporheic zone for such condition shall be similar to the stagnation or dead zone processes, i.e., the stream water that enters the subsurface eventually re-enters the stream at some point downstream. The pollutant transport processes for such circumstances can be regarded as hyporheic exchange. Fig. 1 represents a stream with hyporheic zone.

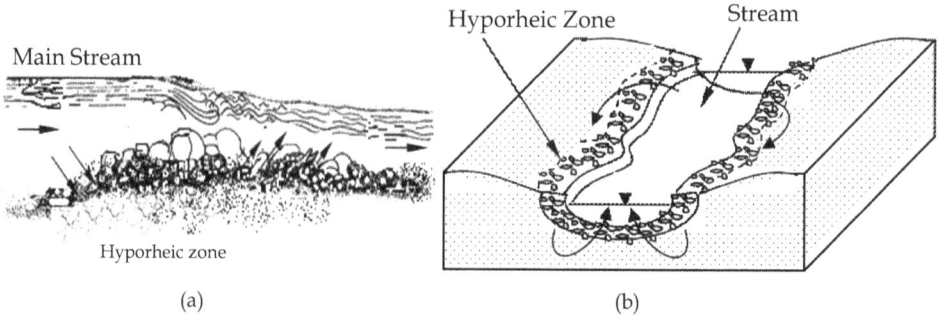

Fig. 1. Representations of a stream with underlying hyporheic zone

In the hyporheic zone, a fraction of stream water containing pollutants is temporarily detained in small eddies and stagnant water that are stationary relative to the faster moving water near the center of the stream. In addition, significant portions of flow may move through the coarse gravel of the streambed and porous areas within the stream banks (Runkel, 2000). The pollutants detained in the hyporheic zone as buffer is released back slowly to the mainstream when the concentration gradient reverses. The travel time for pollutants carried through these porous areas may be substantially longer than that for pollutants travelling within the water column. Streams with intact hyporheic zone provide more temporary storage space and residence time for water with pollutant than streams without them (Bencala & Walters, 1983, Berndtsson, 1990; Castro & Hornberger, 1991; Harvey et al., 1996; Runkel et al., 1996). The hyporheic exchange thus retards the transport of pollutants. The rate of exchange of pollutants between two domains may vary from constituent to constituent. The phenomena, that trap pollutants from the mainstream and hold into the buffer at the beginning and release them back to the mainstream water at later time, shall generate C-t profile representing characteristics of delayed transport, shifted time to peak, reduced peak concentration and also a long tail. To simulate the temporal variations of pollutant concentrations, a hybrid model which appends the retardation component with advection-dispersion pollutant transport has been conceptualized in this study. In this study, the hybrid model has been formulated for both equilibrium and non-equilibrium exchange of pollutant mass between main stream and underlying soil media. Efficacy of model was tested with the results of well known Advection Dispersion Equation and with field data collected from River Brahmani, India.

## 2. Formulation of the model considering equilibrium mass exchange

Let us consider a straight stream reach of length, $\Delta x$, having irregular porous geomorphology. This irregular porous geomorphology is comprised of uniform formations of stagnation or dead or hyporheic zones along streambed and banks. These zones represent the hyporheic exchange, and are hydraulically connected with the mainstream flow. Further, there are exchanges of pollutants with the flow through the interface between the mainstream flow and the hyporheic zones. The stream reach has a steady flow rate, $Q$ and initial concentration of pollutants $C_i$ both in the mainstream as well as in the hyporheic zone. Let a steady state concentration of pollutant, $C_R$, be applied at the inlet boundary of the stream at a time, $t_0$. It is assumed that the river reach be composed of series of equal size hybrid units. It is required to derive the model that can simulate the response of injected pollutant concentration, $C_R$ at the exit of $\Delta x$ of the mainstream due to the affects of the hyporheic exchange using a hybrid model. The hybrid model has three compartments all connected in series; first compartment represents the plug flow cell of residence time, $\alpha$; whereas second and third compartment, represent respectively two well mixed cells of unequal residence times, $T_1$ and $T_2$. The hybrid model simulates advection-dispersion transport of pollutant in regular channel for steady and uniform flow conditions when the size of the basic process unit, $\Delta x$, is equal to or more than 4 $D_L$ / u (Ghosh et al., 2008) where u...mean flow velocity in m/s, and $D_L$...longitudinal dispersion coefficient in m²/s.

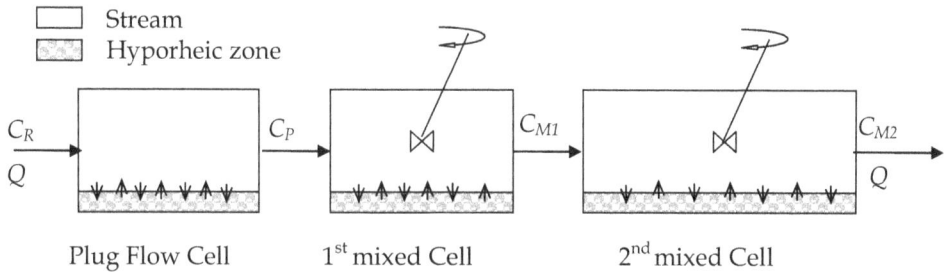

Fig. 2. Conceptual hybrid model with hyporheic zone

The arrangement of the conceptualized hybrid model and the hyporheic zone has been shown schematically in Fig. 2. It is assumed that the hyporheic zones along streambed and banks below and around the plug flow cell and two well mixed cells are in equal proportion to their ratios of: cross-sectional areas, volumes of water and also mass of solute exchanges between the hyporheic zones and the mainstream water. Let $V_0$, $V_1$, and $V_2$ be the volumes of the plug flow cell, and two well mixed cells in the mainstream, respectively. For stream flow Q, the residence time of pollutant in the plug flow and two well mixed cells of the main stream are $\alpha = V_0$ / Q, $T_1 = V_1$ / Q and $T_2 = V_2$ / Q respectively.

Let $V_0^*$, $V_1^*$, and $V_2^*$ be the volumes of the hyporheic zones below the plug flow and two well mixed cells of the mainstream water columns respectively. These volumes can be given by, $V_0^* = (\phi A_p D)$, $V_1^* = (\phi A_{M1} D)$ and $V_2^* = (\phi A_{M2} D)$; where $A_p$, $A_{M1}$, and $A_{M2}$...the interface areas (in m²) of the hyporheic zone and the mainstream flow respectively for the plug flow and the two well mixed cells, D is the depth (in m) of the hyporheic zone below and around the mainstream, and $\phi$ is the porosity of the bed materials. When this zone represents a dead or stagnant pocket with only storage of water, $\phi = 1$, and if there is no soil pores and no water storage, $\phi = 0$. If the hyporheic zone is extended all along the wetted surfaces, the $A_p$, $A_{M1}$, and $A_{M2}$, in such cases, are represented respectively by: $A_p = W_p (\alpha u)$; $A_{M1} = W_p (T_1 u)$; and $A_{M2} = W_p (T_2 u)$, in which $W_p$ is the wetted perimeter at the interface of the mainstream and the hyporheic zone and u is the mainstream flow velocity. If the ratios of the volume of water in the hyporheic zones to the three volumes in the mainstream are in proportion and constant, the ratios $V_0^* / V_0 = V_1^* / V_1 = V_2^* / V_2$ are also constant and defined as, say F. The total residence time of pollutant in the plug flow cell would thus be: $\left(V_0 + V_0^*\right)/Q = V_0\left(1 + V_0^*/V_0\right)/Q = \alpha(1+F) = \alpha R$, where R...the retardation factor, and F...proportionality constant. Similarly, the total residence times in the two well mixed cells would respectively be: $T_1R$ and $T_2R$.

It is also assumed that the retardation process of pollutants takes place in all the cells of the hybrid model due to the hyporheic exchange. In natural riparian rivers, the hyporheic exchange is a complex process which may follow non-equilibrium exchange between main stream and underlying stagnation zone. Decay of pollutant will take place both in main stream and stagnation zone, if the pollutant is of non-conservative type. It is worth trying with a conservative pollutant's equilibrium exchange between main stream channel and hyporheic zone. Hence, in this study retardation process is considered to be followed the linear equilibrium condition, which is expressed as:

$$Cs(x, t) = F\, C(x, t) \tag{1}$$

where Cs(x, t)…the concentration of pollutant which is trapped in the hyporheic zone in mg/L, F…the proportionality constant and C(x, t)…the concentration of pollutant in the main stream in mg/L.

For a steady state flow condition, performing the mass balance in a control volume within plug flow cell of hybrid model, one can get partial differential equation which governs hyporheic exchange coupled pollutant transport as given

$$\frac{\partial C(x,t)}{\partial t} + u\frac{\partial C(x,t)}{\partial x} = -\frac{\phi\, W_P D}{A}\frac{\partial C_s(x,t)}{\partial t} \tag{2}$$

where A…the cross sectional area of any control volume in the plug flow cell of the mainstream water in m$^2$.

Solving eqns (1) and (2), the concentration of pollutants at the end of the plug flow cell of residence time, $\alpha$ is given by,

$$C(x,t) = C_P(\alpha u, t) = C_R U(t - \alpha\, R) \tag{3}$$

where, $C_P(\alpha u, t)$…concentration of pollutant at the end of plug flow cell in mg/L, U(*)…the unit step function, and t…the time reckoned since injection of pollutants in min.

In the first well mixed cell, pollutants after travelling a distance of '$\alpha u$' through the plug flow cell enter and exchange to the adjoining hyporheic zone and release back to the mainstream water, before making an exit from it. The inputs to this cell are thus the outputs from the plug flow cell. The pollutants are thoroughly mixed in this cell.

Performing the mass balance in the first well mixed cell along with the hyporheic zone over a time interval t to t + $\Delta$t, one can get governing differential equation as

$$\frac{dC_{M1}}{dt} = \frac{C_R\, U(t - \alpha R)}{T_1} - \frac{C_{M1}}{T_1} - \frac{1}{V_1}\frac{dM_s}{dt} \tag{4}$$

Replacing dMs/dt by derivative of concentration, eqn (4) transforms to:

$$\frac{dC_{M1}}{dt} = \frac{C_R\, U(t - \alpha R)}{T_1} - \frac{C_{M1}}{T_1} - \frac{\phi\, W_P D}{A}\frac{dC_s}{dt} \tag{5}$$

Solving eqns (1) and (5), the concentration of pollutants at the end of the first well mixed cell of residence time, $T_1$ is given by,

$$C_{MI} = C_R\, U(t - \alpha R) \left[ 1 - e^{-\frac{(t-\alpha R)}{R\,T_I}} \right] \tag{6}$$

Eqn (6) is valid for $t \geq \alpha\, R$ which gives the time varying concentration of pollutants at the exit of the first well mixed cell coupled with the retardation due to a unit step input, $C_R$ applied at the inlet boundary of the plug flow cell.

The time varying outputs from the first well mixed cell form the inputs to the second well mixed cell. Alike first well mixed cell, pollutant before being exited from the cell exchanges to the adjoining hyporheic zone and release back to the mainstream water. Performing the mass balance over a time interval t to t + Δt, one can get a differential equation similar to the first one except residence time

$$\frac{dC_{M2}}{dt} = \frac{C_R\, U(t - \alpha R)\left[ 1 - e^{-\frac{(t-\alpha R)}{R\,T_1}} \right]}{T_2} - \frac{C_{M2}}{T_2} - \frac{1}{V_2}\frac{dM_s}{dt} \tag{7}$$

Solving Eqns (1) and (7), the concentration of pollutants at the end of the second well mixed cell of residence time, $T_2$ is given by,

$$C_{M2} = C_R U(t - \alpha\, R) \left[ 1 - \frac{T_1}{T_1 - T_2} e^{-\frac{(t-\alpha R)}{R\,T_I}} + \frac{T_2}{T_1 - T_2} e^{-\frac{(t-\alpha R)}{R\,T_2}} \right] \tag{8}$$

If $C_R = 1$, eqn (8) represents the unit step response function. Designating K(t) as the unit step response function, K(t) is given by:

$$K(t) = U(t - \alpha\, R) \left[ 1 - \frac{T_1}{T_1 - T_2} e^{-\frac{(t-\alpha R)}{R\,T_1}} + \frac{T_2}{T_1 - T_2} e^{-\frac{(t-\alpha R)}{R\,T_2}} \right] \tag{9}$$

The unit impulse response function, k (t) is the derivative of eqn (9) with respect to 't' which is given by

$$k(t) = \frac{U(t - \alpha\, R)}{R\left(T_1 - T_2\right)} \left[ e^{-\frac{(t-\alpha R)}{R\,T_1}} - e^{-\frac{(t-\alpha R)}{R\,T_2}} \right] \tag{10}$$

Eqns (9) and (10) are valid for $t \geq \alpha\, R$, and they respectively represent the unit step and the unit impulse response functions of a hybrid unit coupled with the retardation. If $R = 1$, eqns (9) and (10) respectively represent the unit step and the unit impulse response functions of the Advection-dispersion equation.

Let the stream reach downstream of a point source of pollution be composed of series of equal size hybrid units coupled with the hyporheic zone, each having linear dimension, Δx and consisting of a plug flow cell, and two unequal well mixed cells. The stream reach has identical features all along downstream; i.e., mainstream flow and hyporheic or dead zones. The exchange of pollutant takes place in all the hybrid units. Assuming that output of pollutant from preceding hybrid unit forms the input to the succeeding hybrid unit, thus the response of the nth hybrid unit, n ≥ 2 for steady flow condition can be obtained using convolution technique, as:

$$C(n\Delta x, t) = \int_0^t C\{(n-1)\Delta x, \tau\} \; k(\alpha, T_1, T_2, R, t-\tau) \, d\tau \tag{11}$$

where $C((n-1)\Delta x, \tau)$...input of nth hybrid unit in mg/L, $k(\alpha, T_1, T_2, R, \tau)$...the unit impulse response (in mg/L. min) as given by eqn (10).

The retardation factor, R can be calculated as given

$$R = \left(1 + \frac{\phi w_p D.F}{A}\right) \tag{12}$$

Eqn (12) is a function of ratio between areas of the hyporheic zone and the mainstream flow, porosity of bed materials. The retardation coefficients have been estimated with different porosity values using eqn (12) by keeping all other parameters constant as shown in Fig. 3.

Knowing parameters $\alpha$, $T_1$, and $T_2$ of the hybrid unit, and using estimated value of the retardation factor, R, one can predict concentration profiles at multiple distance downstream from source at $\{n \Delta x\}$, n = 1,2,3,…. in a stream having steady flow and homogeneous reach conditions making use of eqn (11).

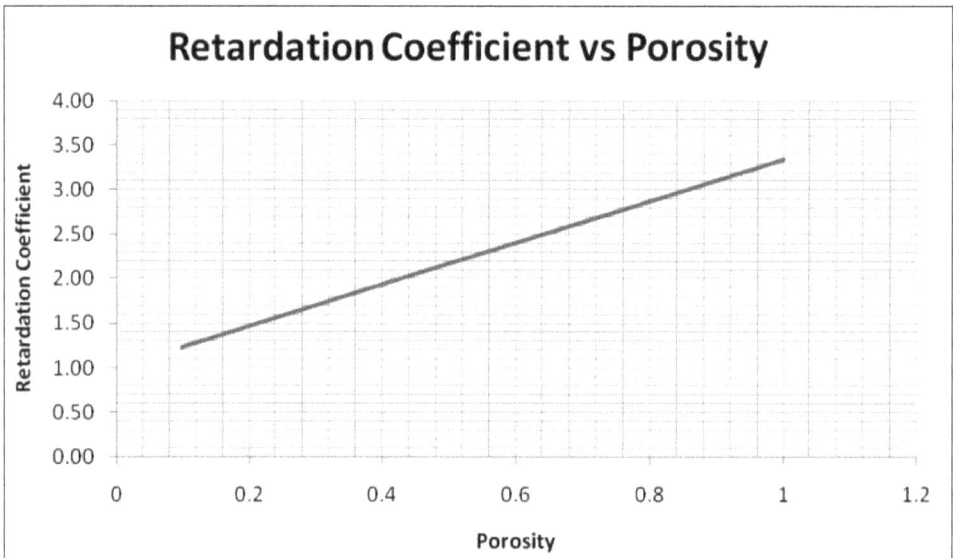

Fig. 3. Variation of retardation factor due to porosity

## 3. Formulation of the model considering non-equilibrium mass exchange

The pollutant exchange between the main stream and underlying subsoil is non-equilibrium in nature. It can be seen most of the mountainous streams where the water with pollutant re-enters the stream in an slower phase. Simulation of non-equilibrium exchange processes along with advection and dispersion is not a simple case due the complexity of the processes of exchange (Cameron and Klute, 1977). Numerous investigators (Bencala and Walters, 1983; Runkel and Broshears, 1991; Runkel and Chapra, 1993; Czernuszenko and Rowinski, 1997; Runkel, 1998; Worman et al., 2002) have studied exchange of the pollutant between main stream and porous soil media. The concentration-time profile of pollutant transport in such case is influenced significantly by the mass exchange. Cameron and Klute, 1977; Bajracharya and Barry, 1992; 1993; 1995 have illustrated that the pollutant exchange in the form of adsorption processes flatten more the concentration-time profile. Thus an exact pollutant transport simulation is important to correctly ascertain the assimilation capacity of streams. Consider a conceptualized hybrid model which incorporate non-equilibrium exchange of pollutant and which is expressed mathematically as follows

$$\frac{dC_s(x,t)}{dt} = R_D\left[C(x,t) - C_s(x,t)\right] \tag{13}$$

where, $R_D$...proportionality constant (per min), $C_s(x, t)$...concentration of pollutant adsorbed in mg/L, $C(x, t)$...concentration of pollutant in the water column in mg/L, $t$...the time in min.

For a steady state flow condition, performing the mass balance in a control volume within plug flow cell of hybrid model, one can get partial differential equation which governs hyporheic exchange coupled pollutant transport which will be same as eq. (2). Then Laplace transform has been used to solve it by combining eq. (13) and effluent concentration from plug flow zone is given by

$$C(x,t) = C_P(\alpha u,t) = C_R \exp(\gamma)\left[U(t-\alpha) + \sqrt{\eta}\int_0^t U(\tau-\alpha)e^{-R_D(\tau-\alpha)}\frac{1}{\sqrt{\tau-\alpha}}I_1\left(2\sqrt{\eta(\tau-\alpha)}\right)d\tau\right] \tag{14}$$

where $\gamma = -\dfrac{\phi W_P D_B}{A}\left(e^{R_D x/u} - 1\right)$; $\eta = \dfrac{\phi W_P D_B R_D}{A}\left(e^{R_D x/u} - 1\right)$, U(t - α)...step function which

is zero for t < α and it is 1 for t ≥ α, so eq. (14) is valid for t ≥ α; α...residence time of plug flow cell, which is x/u. As the eq. (14) is valid for t ≥ α,.it can be considered that the pollutant concentration is zero for t < α

Effluent of plug flow zone enters to the first well mixed cell, where it gets mixed before entering into the second well mixed cell. During these transports through mixed cells too, mass exchange activities follow the non-equilibrium type. Consider a unit step input, $C_R$ and perform the mass balance in the first thoroughly mixed zone which can be expressed as

$$\frac{dC_{M1}}{dt} = \frac{C_R\, U(t-\alpha)}{T_1} - \frac{C_{M1}}{T_1} - \frac{\phi\, W_P\, D}{A}\frac{dC_s}{dt} \tag{15}$$

Eq. (15) can be solved analytically and effluent from the first well mixed cell will enter second well mixed cell. Thus similar mass balance equation can be formulated. Successive convolution numerical integration can be used by combining effluent concentrations of plug flow cell and well mixed cells to get effluent concentration at the end of first hybrid cell as follows

$$K(n\Delta t) = C_2(n\,\Delta t) = \sum_{\gamma=1}^{n}\left\{C_1(\gamma\Delta t) - C_1((\gamma-1)\Delta t)\right\}\delta_{M2}\left[(n-\gamma+1),\Delta t\right] \qquad (16)$$

where $C_1(n\,\Delta t) = \sum_{\gamma=1}^{n}\left\{C_P(\gamma\Delta t) - C_P((\gamma-1)\Delta t)\right\}\delta_{M1}\left[(n-\gamma+1),\Delta t\right]$; $\delta_{M1}$ & $\delta_{M2}$...ramp kernel co-efficients.

Eq. (16) is unit step response [K(.)] of first hybrid unit and one can get unit impulse response [k(.)] by differentiating Eq. (16). The output (Eq. 16) of pollutant from preceding hybrid unit forms the input to the succeeding hybrid unit, thus the response of the nth hybrid unit, $n \geq 2$ for steady flow condition can be obtained using convolution technique which will be similar to eq. (11) and is given by

$$C(n\Delta x,t) = \int_0^t C\{(n-1)\Delta x,\tau\}\ k(\alpha,T_1,T_2,R_D,t-\tau)\,d\tau \qquad (17)$$

where C[(n-1)Δx, t)]...effluent concentration from the preceding unit, k( )...unit impulse response of a single hybrid unit.

## 4. Simulation of pollutant transport

### 4.1 Verification of model considering equilibrium exchange using synthetic data

The hybrid model simulates the advection-dispersion governed pollutant transport in a regular channel under uniform flow conditions when the size of the basic process unit, $\Delta x$, is equal to or more than 4 $D_L/u$. This means that the response of the hybrid model corresponding to that $\Delta x$ for a specific u and $D_L$, should be identical to that of the response of the Advection Dispersion Equation (ADE) model for that u and $D_L$ at the downstream distance, $x = \Delta x$.

Let the parameters,$\alpha$ $T_1$, and $T_2$ of the hybrid model for pollutant transport in a stream be known from the ADE model for a given value of u and $D_L$ satisfying Peclet number, Pe $\geq 4$. Let the value of the parameters of the mainstream flow in a stream having hyporheic zones along streambed and banks, be: $\alpha = 1.70$ min, $T_1 = 2.0$ min and $T_2 = 6.30$ min corresponding to $\Delta x = 200$ m, u = 20 m/min and $D_L = 1000$ m²/min. The retardation factor, R is assumed to be 1.25. Using the above data, the unit step responses and the unit impulse responses of the hybrid model are generated at the end of the 1st hybrid unit applying eqns (9) and (10), respectively, and they are shown in figs 4 and 5.

In these figures, the C-t profiles generated by the hybrid model corresponding to the same value of $\alpha$, $T_1$ and $T_2$ without the retardation component, i.e., for R = 1, are also shown for comparison and it can be noted that the C-t profiles represented by the hybrid model with retardation shows the characteristics of delayed pollutant transport as expected. The impulse response, presented in fig. 5, represents the following; the rising limb occurred at a later time with reduction in magnitudes, the time to peak shifted, the peak concentration reduced, and the recession limb extended for a long time to that of the distributions depicted by the hybrid model with retardation.

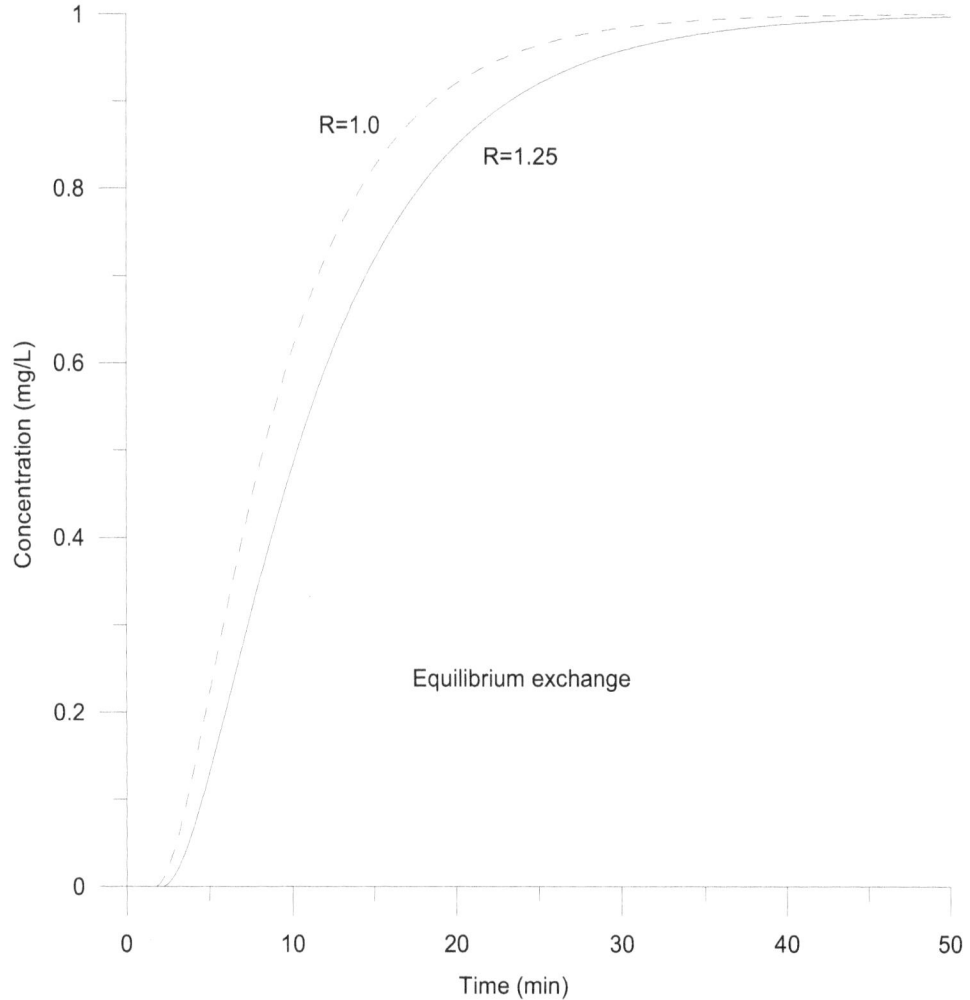

Fig. 4. Unit step responses of the hybrid model at the end of one hybrid unit of size $\Delta x = 200$ m for $\alpha = 1.7$ min, $T_1 = 2.0$ min, $T_2 = 6.3$ min

The above data are adopted again and the C-t profiles of the hybrid model for a unit impulse input have been generated for n = 3, 6, and 11 using eqn (11) and shown in fig. 6 and it can be noted from the figure that as the pollutants move from the near field to the far field the C-t distributions of the hybrid model get more and more attenuated, elongated and delayed in terms of occurrence of the rising limbs, the times to peak, and the peak concentrations. This means that the total residence time of the hybrid unit is increased by a factor of R for one unit as compared with the hybrid unit without mass exchange. As pollutant move down stream by n units, the residence time increased by a factor of "n times R". These characteristics of the C-t profiles for a steam with retardation component are in the expected lines. This may be due to more permanent loss of pollutant within the hyporheic zone or the water along with pollutant may take very long time to re-enter stream. The model is also capable of simulating non-equilibrium exchange which has been demonstrated in section 4.2 below with synthetic data.

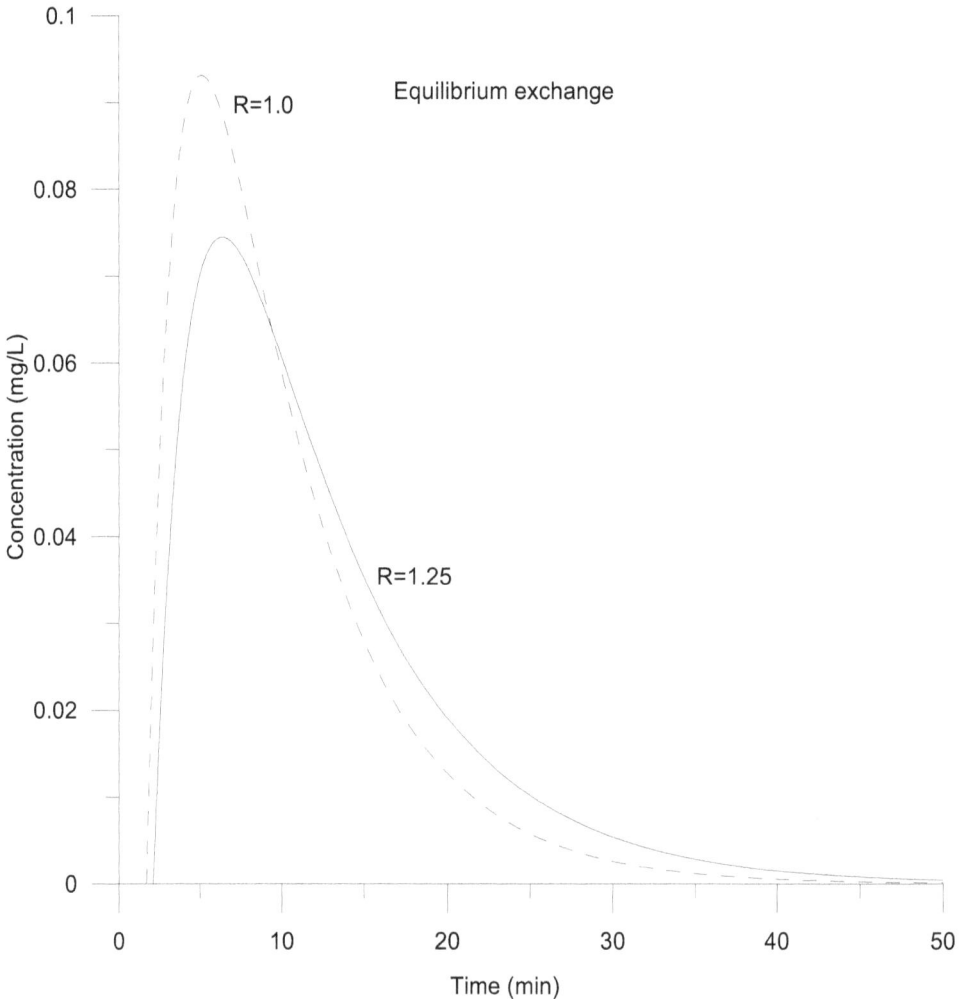

Fig. 5. Unit impulse responses of the hybrid model at the end of one hybrid unit of size $\Delta x$ = 200 m for $\alpha$ = 1.7 min, $T_1$ = 2.0 min, $T_2$ = 6.3 min

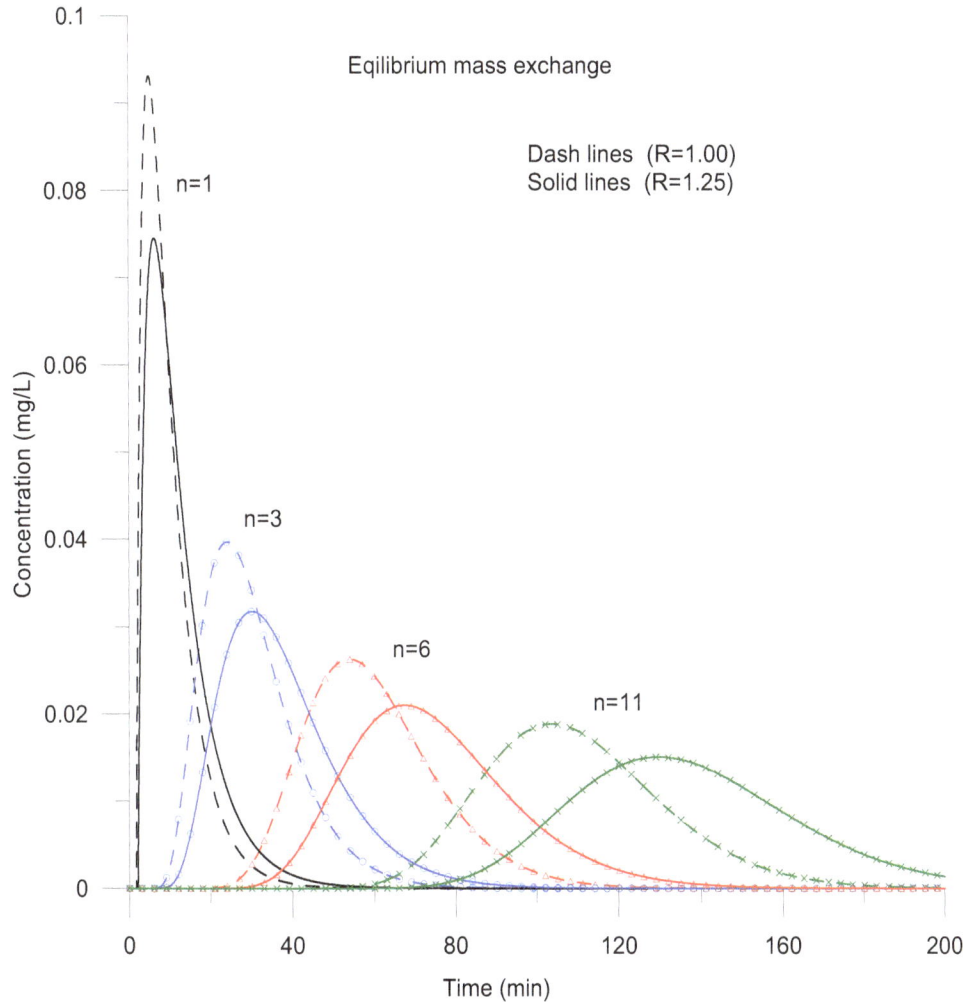

Fig. 6. Unit impulse responses of the hybrid at the end of first (n = 1), third (n = 3), sixth (n = 6) and eleventh (n = 11) hybrid units.

## 4.2 Verification of model considering non-equilibrium exchange using synthetic data

Let the parameters, $\alpha$ $T_1$, and $T_2$ of the hybrid model are being 1.7 min, 2.0 min and 6.3 min respectively. Non-equilibrium mass exchange rate $R_D$ is assumed to be 0.25 per min which is a time reciprocal parameter. The C-t profiles of the hybrid model for a unit impulse input have been generated having parameters ($\alpha$ $T_1$, $T_2$ and $R_D$) for n = 3, 6, and 11 using eq. (17) which is presented in fig. 7. In this study, the synthetic data was used for non-equilibrium exchange because of absence of field data to calculate mass exchange rate ($R_D$). However, the section 4.3 explains the simulation of pollutant transport with equilibrium exchange using field data collected from a river.

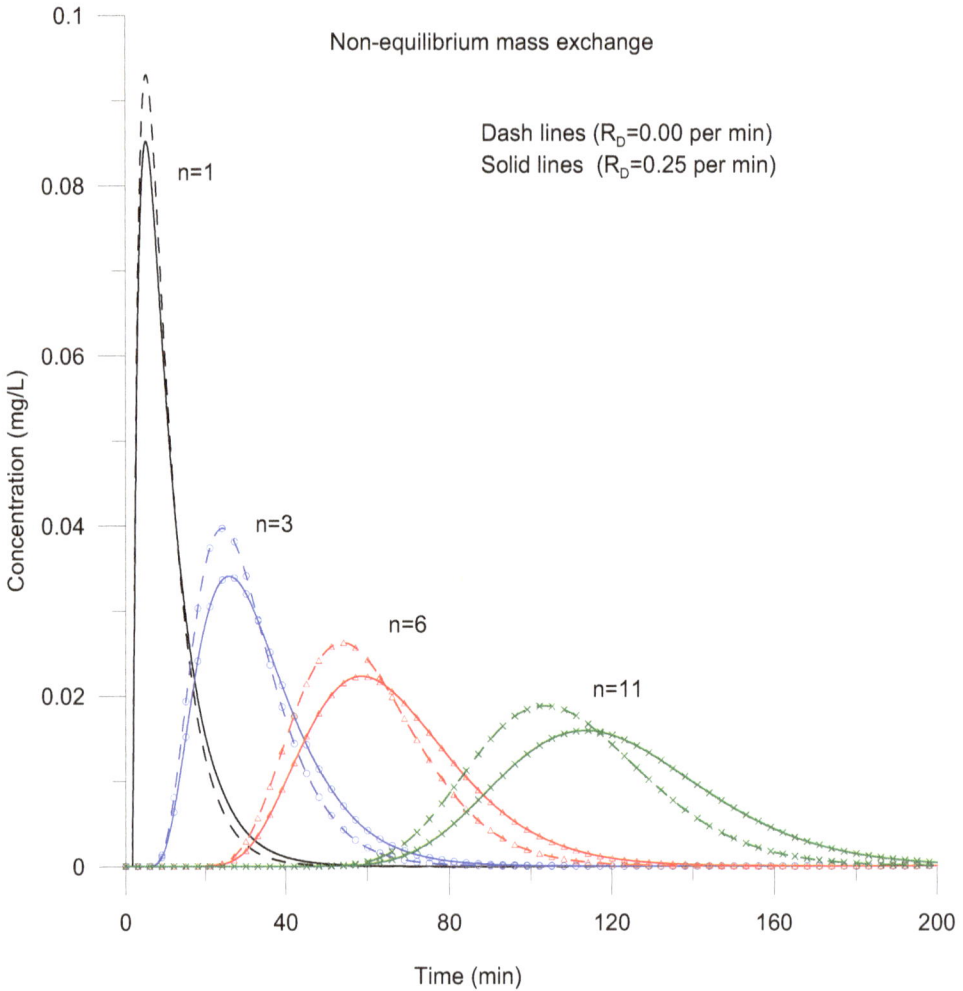

Fig. 7. Unit impulse responses of the hybrid at the end of first (n = 1), third (n = 3), sixth (n = 6) and eleventh (n = 11) hybrid units

## 4.3 Model verification using field data

In order to verify the performance of the model, field data from the river Brahmani, India has been collected. A river reach from Rengali dam to Talcher is affected seriously by the waste water discharged by river Tikira, tributary of the main river. The Talcher Township is located 26km downstream of Tikira confluence. Fig. 8 shows the study area with locations of sampling points. Data collections from the field have been tabulated in Table 1.

If observed $C$-$t$ profile is available, as an inverse problem model parameters can be estimated using optimization. In absence of observed $C$-$t$ profile, model parameters can be obtained by relating with longitudinal dispersion co-efficient, $D_L$, satisfying the condition of

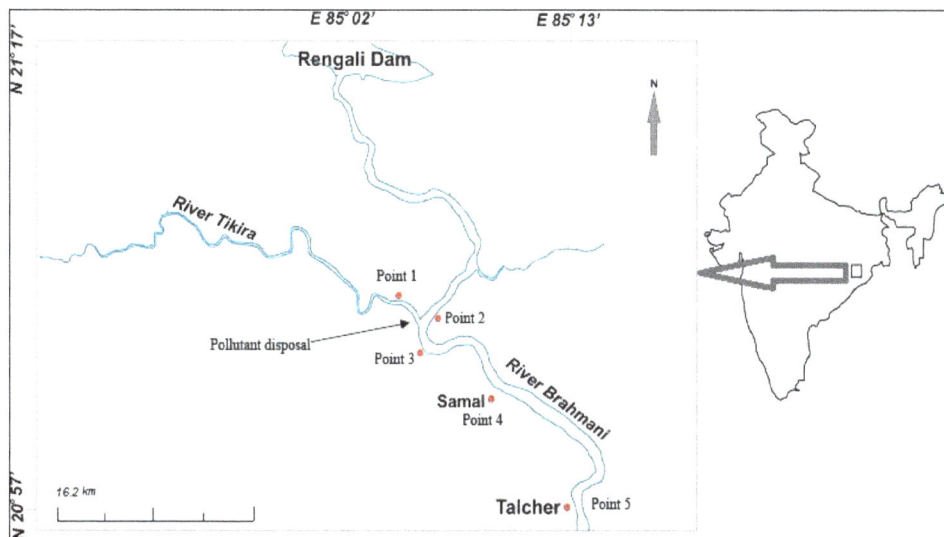

(Adopted from Muthukrishnavellaisamy K, 2007)

Fig. 8. Map showing study river reach and sampling points

| Channel geometry, flow characteristics and dispersion co-efficient | | | | | | |
|---|---|---|---|---|---|---|
| Location | $Q$ (m³/s) | $U$ (m/s) | $A$ (m²) | $H$ (m) | $W$ (m) | $D_L$(m²/s) |
| Before Tikira (Point 2) | 195.98 | 0.83 | 218.12 | 3.21 | 67.91 | 430.74 |
| After Tikira (Point 3) | 239.72 | 0.92 | 257.99 | 3.43 | 75.10 | 490.05 |
| Talcher | 238.62 | 0.9 | 257.00 | 3.42 | 74.93 | 488.61 |
| Model parameters having U = 0.91 m/s and D$_L$ = 489.3 m²/s *(these are average values for river reach between Tikira & Talcher)* | | | | | | |
| Cell size ($\Delta$x), m | $P_e$ | | $\alpha$, min | $T_1$, min | | $T_2$, min |
| 3200 | 5.9 | | 15.5 | 19.4 | | 25.7 |
| Water Quality data | | | | | | |
| Location | Measured concentration | | Location | | | Measured concentration |
| Point 1 (Q = 13.3 m³/s) | 242.3 mg/L | | Point 2 (Q = 195.98 m³/s) | | | 25 mg/L |
| Point 3 | 38 mg/ L | | Point 4 | | | 36 mg/L |
| Point 5 | 32.2 mg/L | | | | | |

(Source: Muthukrishnavellaisamy K, 2007)

Table 1. Flow and water quality data collected from the river Brahmani

Peclet number, $P_e = \Delta x \, u \, / \, D_L$ which should be greater than or equals to 4 and less than 8 (Ghosh et al., 2004; Muthukrishnavellaisamy K, 2007; Ghosh et al., 2008) as follows:

$$\alpha = \frac{0.04 \, \Delta x^2}{D_L} \tag{18}$$

$$T_1 = \frac{0.05 \, \Delta x^2}{D_L} \tag{19}$$

$$T_2 = \frac{\Delta x}{u} - \frac{0.09 \, \Delta x^2}{D_L} \tag{20}$$

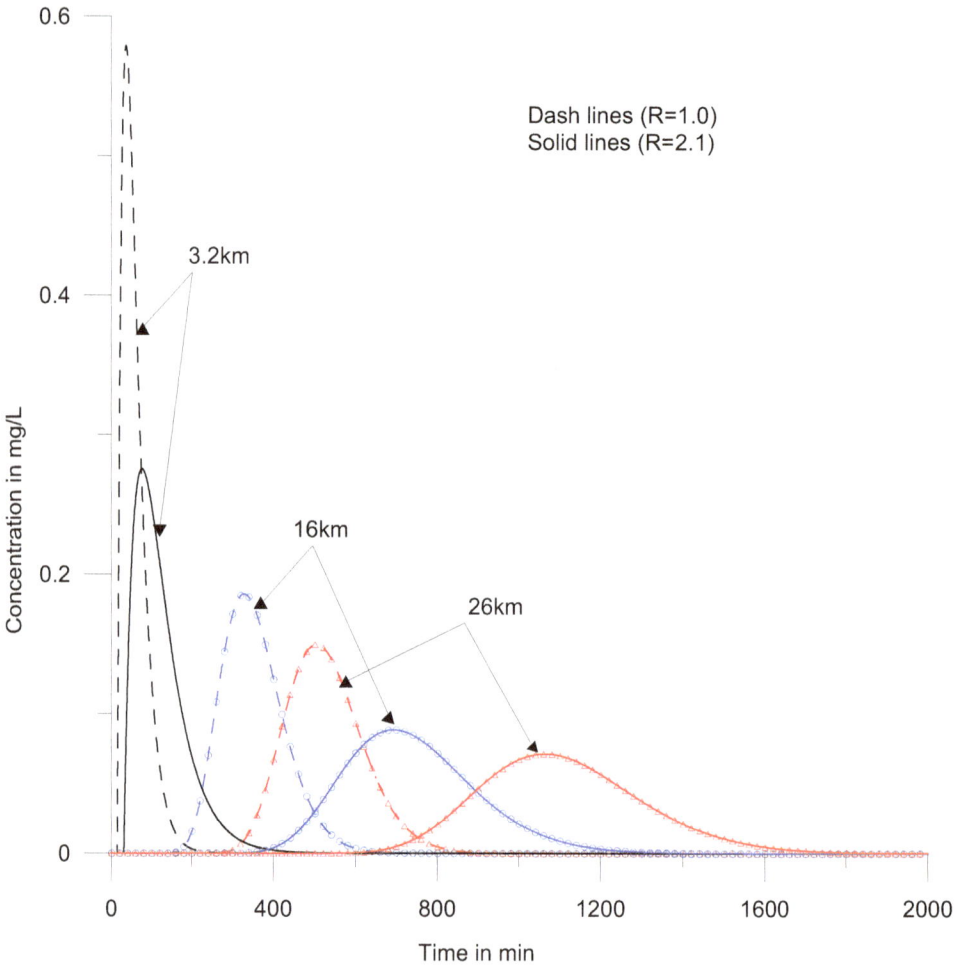

Fig. 9. Pulse responses of the hybrid at the end of 3.2 km, 16 km and 26 km for the data collected from river Brahmani

One can choose any peclet number between 4 and 8 in order to match the response of hybrid model with the ADE model. Having u = 0.9 m/s and $D_L$ = 490 m²/s, the hybrid unit size ($\Delta x$) has been chosen as 3200 m by satisfying the condition of peclet number and the parameters of the hybrid model ($\alpha$, $T_1$ and $T_2$) are approximately estimated as: $\alpha$ = 15 min, $T_1$ = 19 min and $T_2$ = 25 min. The reach length of 26 km (from Tikira confluence to Talcher) is covered with 8 hybrid units. By successive convolution, the pollutant concentrations at Talcher (26 km downstream of pollutant source) are predicted for step input. Having the above data collected from river Brahmani, using eqn 11, pulse response at 3.2km (1st hybrid unit), 16km (5th hybrid unit) and 26km (8th hybrid unit) have been simulated with different retardation factors (R = 1.0 & 2.1) and presented in fig. 9. The maximum pollutant concentration at various locations downstream of pollutant disposal point can be derived by numerical integration of pulse responses obtained using hybrid model for those downstream locations. The maximum pollutant concentration at Talcher is about 34.5 mg/L, which is very close to the measured value of 32.2 mg/L. It clearly demonstrates the influence of retardation process of pollutant transport. In order for complete verification of model, numerous data are needed towards calculating mass exchange rate constant (R). However, this chapter theoretically compares the model with limited field data.

## 5. Conclusions

A Hybrid model coupled with hyporheic exchange has been derived by incorporating a time delay factor termed as "retardation factor" with each of the three compartments in the hybrid model to simulate retardation governed pollutant transport in riparian streams or rivers. A linear equilibrium condition between the concentration of pollutants in the hyporheic zone and the mainstream water has been considered. The stagnation or dead or hyporheic zone retards the transport of downstream pollutants. The hybrid model is a four-parameter model representing three time parameters and one constant factor. Theoretical study on non-equilibrium exchange of pollutant has also been done to demonstrate the model.

The unit step response and the unit impulse response functions of the hybrid model have been simulated with synthetic data and limited field data. The characteristics of the concentration-time profiles generated by the hybrid model are comparable to the physical processes of pollutant transport governed by the advection-dispersion-retardation both in equilibrium and non-equilibrium exchanges in a natural stream. This present model can be used to obtain theoretically exact solutions and can be compared with results of ADE model considering with and without retardation of pollutant transport in a stream along with advection and dispersion processes.

Data regarding the influence of the hyporheic zone to pollutant trap in streams are rare due to the absence of simple techniques to get necessary parameters and complexity of the phenomenon. The pollutant exchange between the main channel and the hyporheic zone is very variable and estimation of exchange rate is mostly inaccurate due to channel irregularities and other complexities. In depth analysis and understanding about the hyporheic exchange will over-come the problem in collecting relevant data from natural streams.

It can be concluded that the presented hybrid model for pollutant transport in streams affected by hyporheic exchange is a useful tool in predicting water quality status streams.

Simulation of non-equilibrium exchange of non-conservative pollutant between main channel and stagnation zone is vital. As we have considered conservative pollutant in this chapter, the maximum concentrations of pollutant at different locations were same due to mass conservation, but the residence times were different. Thus, this study also gives a retrospect for the extension of the model considering non-equilibrium condition of decaying pollutant exchange for natural streams.

## 6. Nomenclature

$A_P$, $A_{M1}$, $A_{M2}$ – interface areas of the hyporheic zone and the mainstream flow under plug flow, first and second well mixed cells respectively
$C(x, t)$ – Concentration of pollutant in the main stream
$C_P$, $C_{M1}$, $C_{M2}$ – Pollutant concentrations at the end of plug flow, first and second well mixed cells respectively
$C_R$ – Input pollutant concentration of a hybrid unit
$C_s(x, t)$ – Concentration of Pollutant trapped in the hyporheic zone
$D$ – Depth of effective soil layer/hyproheic zone
$D_L$ – Longitudinal dispersion co-efficient
$F$ – Proportionality constant
$K(t)$ – Unit step response of a hybrid unit
$k(t)$ – Unit impulse response of a hybrid unit
$M_s$ – Mass of the pollutant trapped in hyporheic zone
$n$ – Number of cells
$Q$ – Stream discharge
$R$ – Retardation factor
$R_D$ – Mass exchange rate constant
$t$ – Time
$T_1$ – Residence time of first well mixed cell
$T_2$ – Residence time of second well mixed cell
$u$ – Stream flow velocity
$U( )$ – Unit step function
$V_0$, $V_1$, $V_2$ – Volumes of mainstream plug flow, first and second well mixed cells
$V_0^*$, $V_1^*$, $V_2^*$ - Volumes of hyporheic zones under plug flow, first and second well mixed cells
$W_P$ – Wetted perimeter at the interface of hyporheic zone and the main stream
$x$ – Distance
$\alpha$ – Residence time of plug flow cell
$\delta_{M1}$, $\delta_{M2}$ – Ramp kernel co-efficient of first and second well mixed cells respectively
$\Delta t$ – Small time interval
$\Delta x$ – Size of control volumes within plug flow cell.
$\phi$ – Porosity

## 7. References

Bajracharya, K., and Barry, D. A. (1992). "Mixing cell models for nonlinear non-equilibrium single species adsorption and transport." *Water Res.*, 29 (5), 1405-1413
Bajracharya, K., and Barry, D. A. (1993). "Mixing cell models for nonlinear equilibrium single species adsorption and transport." J. of Contaminant Hydrology, 12, 227-243

Bajracharya, K., and Barry, D. A. (1995). "Analysis of one dimensional multispecies transport experiments in laboratory soil columns." *Envir. International*, Vol. 21, No. 5, 687-691

Bencala, K. E. & Walters, R. A. (1983). Simulation of solute transport in a mountain pool-and-riffle stream - a transient storage model, *Water Resources Research*, 19(3), pp. 718-724

Berndtsson, R. (1990). Transport and sedimentation of pollutants in a river reach: a chemical mass balance approach, *Water Resources Research*, 26(7), pp.1549-1558

Boulton, A. J.; Findlay, S.; Marmonier, P.; Stanley, E. H. & Valett, H. M. (1998). The functional significance of the hyporheic zone in streams and rivers, *Annual Rev. Ecol. System*, 29, pp. 59–81

Brunke, M. & Gonser, T. (1997). The ecological significance of exchange processes between rivers and groundwater, *Freshwater Biology*, 37, pp. 1-33

Cameron, D. R., and Klute, A. (1977). "Convective dispersive solute transport with combined equilibrium and kinetic adsorption model." *Water Resour. Res.*, Vol 13 (1), 183-188

Castro, N. M. & Hornberger, G.M. (1991). Surface-subsurface water interactions in an alluviated mountain stream channel, *Water Resources Research*, 27, pp. 1613-1621

Czernuszenko, W., and Rowinski, P. M. (1997). "Properties of the dead zone model of longitudinal dispersion in rivers." *J. of Hydraul. Res.*, 35 (4), 491-504

Edwards, R. T. (1998). The hyporheic zone, In: *River Ecology and Management*, Naiman RJ, Bilby RE (Eds.), pp. 399–429, Springer: Berlin

Ghosh, N.C., G.C. Mishra, and C.S.P. Ojha. (2004) A Hybrid-cells-in-series model for solute transport in a river. *Jour. Env. Engg. Div., Am. Soc. Civil Engr.* 130 (10), 1198-1209

Ghosh, N. C.; Mishra, G. C. & Muthukrishnavellaisamy, Kumarasamy. (2008). Hybrid-cells-in-series model for solute transport in streams and relation of its parameters with bulk flow characteristics, *J. of Hydraul. Eng., ASCE*, pp. 497-503

Harvey, J. W.; Wagner, B. J. & Bencala, K. E. (1996). Evaluating the reliability of the stream tracer approach to characterize stream subsurface water exchange, *Water Resources Research*, 32, pp. 2441–2451

Muthukrishnavellaisamy. K (2007) A study on pollutant transport in a stream, PhD Thesis, Indian Institute of Technology Roorkee, Roorkee, India

Runkel, R. L. (1998). One Dimensional Transport with Inflow and Storage (OTIS): A Solute Transport Model for Streams and Rivers. USGS Water Resour. Invest. Report No 98-4018., Denver, Colorado

Runkel, R. L. (2000). Using OTIS to model solute transport in streams and rivers, *U.S. Geological Survey Fact Sheet FS-138-99*, pp 4, 2000.

Runkel, R. L.; Bencala, K. E.; Broshears, R. E. & Chapra, S. C. (1996). Reactive solute transport in streams, 1. Development of an equilibrium-based model, *Water Resources Research*, 32(2), pp. 409-418

Runkel, R.L., and Broshears, R.E. (1991). One dimensional transport with inflow and storage (OTIS): a solute transport model for small streams, Tech. Rep. 91-01, Center for Advanced Decision Support for Water and Environmental System, University of Colorado, Boulder

Runkel, R.L., and Chapra, S.C. (1993). "An efficient numerical solution of the transient storage equations for solute transport in small streams." *Water Resour. Res.*, 29 (1), 211-215

Smith, J. W. N. (2005). Groundwater–surface water interactions in the hyporheic zone, *Environment Agency Science report SC030155/SR1*, pp. 70, Environment Agency, Bristol, UK

Stanford, J. A. & Ward, J. V. (1988). The hyporheic habitat of river ecosystem, *Nature*, 335, pp. 64-66

Stanford, J. A. & Ward, J. V. (1993). An ecosystem perspective of alluvial rivers: connectivity and the hyporheic corridor, *J. of the North American Benthological Society*, 12(1), pp. 48-60

Triska, F. J.; Kennedy, V. C.; Avanzino, R.J.; Zellweger, G.W. & Bencala, K.E. (1989). Retention and transport of nutrients in a third-order stream in Northwestern California: hyporheic processes, *Ecology*, 70, pp. 1893-1905

Valett, H. M.; Dahm, C. N.; Campana, M. E.; Morrice, J. A.; Baker, M. A. & Fellows, C. S. (1997). Hydrologic influences on groundwater-surface water ecotones: heterogeneity in nutrient composition and retention, *J. of the North American Benthological Society*, 16, pp. 16:239-247

White, D.S. (1993). Perspectives on defining and delineating hyporheic zones, *J. of the North American Benthological Society*, 12(1), pp. 61-69

Worman, A., Packman, A.I., Johansson, H., and Jonsson, K. (2002). "Effect of flow-induced exchange in hyporheic zones on longitudinal transport of solutes in streams and rivers." *Water Resour. Res.*, 38 (1), 1001

# Part 4

# Politics, Regulation and Guidelines

# 9

# From Traditional to Modern Water Management Systems; Reflection on the Evolution of a 'Water Ethic' in Semi-Arid Morocco

Sandrine Simon
*Open University*
*United Kingdom*

"Society is like a pot: it can't carry water when it is broken" (African proverb)

## 1. Introduction

Which strategic water policy options are semi arid, developing, Muslim, countries going to take in order to face the dilemmas that typically characterize the dual – and potentially conflicting – aspiration to modernize the economy *whilst* respecting traditional socio-political practices and ways of life? This chapter focuses on the case of Morocco, described as one of the most liberal countries of the Muslim Arab world - and yet as a country that is keen to balance traditions and modernity -, in view of articulating a reflection on the conflicting interests that can clash when critical environmental and economic choices have to be made to position a developing country into the 21st century's globalised world.

The chapter focuses on water because of the crucial importance of that resource in a semi-arid country and because the ways in which it has been managed throughout centuries illustrate the changes in socio-political structures in the society. The focus on water in a semi arid country is symbolic of how precious natural resources are in the development of economies and societies. Morocco provides a fascinating terrain to explore ingenuous traditional water management structures and processes both in urban and in rural environments. Thus, for instance, traditional water management systems represent one of the architectural and urban pillars of the medina of the UNESCO World heritage - and cultural and spiritual capital of Morocco - Fes, whilst *khetarras* in the rural South (for instance), provide a remarkable example of a well-thought, long-lasting system of water collection and distribution. This country also developed, in the last decades, massive modern water policies focused on the construction of dams and water transfers. Economic principles constituted one of the main drives in the *politique des barrages* of the previous king (Hassan II), with a strong focus on agricultural production and exports targets. Morocco has however somehow questioned its development path in this beginning of the 21st century, with the arrival of its new king and a sense that the development of the country could be re-thought and targeted differently. A new Charter of the environment was created, massive investments were geared towards renewable energies and, more importantly, governance systems were questioned. Centralized versus more local – and potentially more traditional – approaches of resource management were discussed. In this context, could the expression " 'modern water management' *versus* 'traditional water management systems'" become "revisiting traditional water management systems in order to re-think and question the notion of modernity in the context of water management"? If so, it would mean that a new type of water ethics is progressively emerging in a context where both the notions of economic development and centralized environmental governance are being questioned.

The objective of this chapter is to demonstrate that this could be the case in a country like Morocco and to explore what this would imply for the years to come. The chapter starts by presenting traditional and modern water management techniques. The evolution of the political dimensions of water management is then explored, allowing the reader to appreciate the extent to which technical choices are of a political nature. The chapter then concludes on the emergence of a new water ethics, with a particular focus on new understandings of development and environmental governance.

## 2. Traditional versus modern techniques in water management in Morocco

Morocco is a semi-arid country where both traditionally and through modern techniques, management systems have had to be found to store, distribute, allocate fairly, treat ... clean and dirty water resources. Whilst Islam has equipped this North African country with ingenious traditional water management systems, growing populations and hence domestic needs, growing demands coming from economic development, as well as climate change extreme weather events (both droughts and floods) have altered the way in which people, technology and governance systems have been allocated to that scarce natural resource, *l'maâ*. This section concentrates on traditional and then modern management approaches, after having first presented the country's physical environment, its constraints and its potentials.

### 2.1 Water issues in a semi arid country

Morocco is located in the North West tip of Africa, with a small Mediterranean coast and a very long Atlantic coast, important mountain ranges – the Anti, High and Middle Atlas as well as the Rif mountains –, agricultural plains West of the mountains 'crescent', and deserts, East of it. It is therefore subject to the influence of highly diverse climatic conditions. Rainfall is distributed unevenly: it can vary from more than 1,800 millimetres per year (mm/year) in the northern part of the country to less than 200 mm/yr in the southern parts. More than 50% of the rainfall is located in 15% of the country's surface. The average precipitation - 340 mm per year - therefore has to be apprehended in a context of great climatic diversity. On the whole, it is fair to say that the country is essentially semi-arid, if not arid, with 79% of the country located in an arid and Saharian zone, 14% in a semi-arid zone and 7% only in sub-humid and humid locations. Besides, it has been considered, since 2001, as being in condition of 'water stress' – that is, benefiting from less than 1000 cubic meters of water per inhabitant per year.

Various phenomena are aggravating the situation. First of all, due to **high population growth** (2% growth rate), water availability per inhabitant would have dropped from 1200 cubic metres per inhabitant per year (m3/inh/y) in 1990, to 950 m3/inh/y in 2000. In 2030, that figure should drop to 500 m3 (Et Tobi, 2003, p.6). The increase in water demand is therefore daunting. Second, **climate change** has resulted in a series of droughts (1982-1983, 1994-1995, 1999-2000) and localised floods (1995, 1996, 2001, 2002, 2010, 2011) and will make the average surface and underground water flow decrease by 15% between 2000 and 2020, following studies carried out in 2001 (Agoumi, 2005, p.36,37). As this author stresses, climate models predict a warming up of the North African region of 2 to 4 degrees Celsius throughout the 21st century, accompanied by a reduction in rainfall of 4% between 2000 and 2020. In Morocco, research centers estimate that the increase in temperature between 2000 and 2020 will probably be in the range of 0.6 to 1.1 degrees C., considerably affecting the

| Internal Renewable Water Resources (IRWR),1977-2001, in cubic km | | Morocco | Middle East & North Africa |
|---|---|---|---|
| Surface water produced internally | | 22 | 374 |
| Groundwater Recharge | | 10 | 149 |
| Overlap (shared by groundwater | and surface water) | 3 | 60 |
| Total Internal Renewable Water Resources | (surface water + groundwater - overlap) | 29 | 518 |
| Per capita IRWR, 2001 (cubic meters) | | 936 | 1223 |
| **Natural Renewable Water Resources** | | | |
| | Total, 1977-2001 (cubic km) | 29 | X |
| | Per capita, 2002 (cubic meters per person) | 936 | X |
| Annual River Flows: | | | |
| | From other countries (cubic km) | 0 | X |
| | To other countries (cubic km) | 0 | X |
| **Water Withdrawals** | | | |
| Year of Withdrawal Data | | 1998 | |
| Total withdrawals (cubic km) | | 11 | X |
| Withdrawals per capita (cubic m) | | 399 | X |
| Withdrawals as a percentage of Actual | Renewable Water Resources | 42.6 % | X |
| Withdrawals by Sector (as a percent of total) | | | |
| | Agriculture | 88 % | X |
| | Industry | 2 % | X |
| | Domestic | 10 % | X |

Source: Water Resources. COUNTRY PROFILE – Morocco. World Resource Institute 2006.
http://earthtrends.wri.org/text/water-resources/country-profile-126.html

Fig. 1. Provides general data on water in Morocco, the Middle East and North Africa

water cycle. The effects of climate change on groundwater recharge in Morocco are particularly well documented in Van Dijck et al (2006). Third, some **existing modes of water management** could be considered as worsening the situation. Slimani (2010, p. 60) stresses that Morocco is particularly behind with regards to water treatment management: whilst Morocco produces more than 750 million cubic water annually, only 100 million cubic meters are treated and 10 million are re-used. According to the Ministry of Trade and Industry, the cost of environmental damage is calculated to be around 8% of Morocco's annual GDP – 2.5 billion dollars (ref. http://www.fm6e.org/fr/notre-fondation.html). This includes serious problems of water pollution: organic pollution and pollution by heavy metals, salination of water and siltation, increased in case of climate change. Currently, 88% of water is used for agricultural irrigation in Morocco (compared to an already very high 80% water allocation to agriculture observed in the MENA region), and many consider that water is being managed in their country in an irrational manner, with over-exploitation of groundwater resources and use of clean, expensive tap water in inappropriate ways (ACME, May 2011). In order to deal with the current Moroccan water crisis, authors such as Agoumi (2005) have identified main areas in which efforts should be focused such as integrated energy-water policy, de-pollution and water savings, optimization of water demand, better monitoring of water supply and demands. Others have suggested to also jointly manage water and forest resources in an integrated way (Et Tobi, 2003). Responses to crises have adapted to needs and constraints and we will soon see that, whilst traditional techniques were valuing small scale coherent management systems that protect the social and political fabric of communities, more modern techniques have opted for intensive economic development based on the prioritization of agricultural exports and less concerned with the social and political impacts of technological choices opted for. I will come back to this very important point throughout the chapter, but I suggest to take it for now as a point of reflection – a point that is captured in the following quote by Allan (in Turton and Henwood, 2002, p.30): "The most important solution for water deficit economies is *socio-economic development*. With socio-economic development comes the adaptive capacity to deal with the challenge of water scarcity. Water scarcity has two orders. First-order water scarcity is the scarcity of water. Second-order scarcity is that of the capacity to adjust to the scarcity".

## 2.2 Traditional water management approaches

Islamic civilization has acquired a great reputation in terms of its ability to develop ingenious approaches to water management and to agricultural practices that are well adapted to particularly harsh climatic conditions. Some authors have explored in detail the reciprocal influences that North Africa and Andalousia (in the South of Spain) had on each other, in particular during the period from 700 to 1100 where a genuine Islamic agricultural revolution took place. Authors specialised in North Africa, such as Pérennès, highlight the fact that numerous and diverse water management systems can be found in that area of the world which developed, in its past, subtle water distribution systems that he describes as a 'social management system of scarce resources'. The prestigious water management heritage of North African countries goes back to pre-Islamic times when numerous irrigation techniques emerged from the creation of sedentary urban and rural civilisations of the desert, mostly Berbers. The Arab and hence Islamic expansion towards the West spread various water management and agricultural techniques. The introduction of new plants coming from India and the Sassanide Empire (spread from the Khorassan to Mesopotamia, 226-651), such as rice, lemon trees, cotton, spinach – to quote only a few -, acclimatised and then largely spread by Arab dynasties during the VIIIth and XIIth centuries, lead to the introduction of new irrigation techniques themselves leading to the intensification of agricultural processes. It is in the VIIIth century that the Arabs introduce the *noria* (the water wheel activated by an animal, a donkey or a mule, generally), for instance, as well as the *qanats* (small scale dams).

A few characteristics of traditional water systems in the Muslim, North African, world must be highlighted and are particularly important in the light of this chapter's argument. First, these techniques were extremely well adapted to the natural environment and consequently varied in their types. They dealt with urban as well as rural environments, and used groundwater as well as surface water. Second, water management was paying attention not only to the diversity of the physical environment but also to the variety of users and water management conflicts were integrated in the management system of the rare resource. Finally, it is worth noting that ingenious traditional water management systems in North Africa, and Morocco in particular, were never born out of great, big-scale projects such as those observed in the times of Pharaonic Egypt. As Pérennès explains, "the rise of irrigation techniques in the Muslim world did not emerge from a strong, despotic and centralised State"(1993: page 77). Quoted in Pérennès, Paul Pascon also explained that "traditional water management systems had numerous functions other than solely that of managing water resources. In semi-arid areas where water can be disputed, they captured the complex way in which societies functioned" (in Pérennès, p. 19). In the context of this chapter, the point is the following: the move from traditional to modern water management techniques is controversial in that it is seen as one of the causes of current difficulties the country is encountering in the management of water rare resources – at least by certain authors. It is not only a physical, technical problem but, above all, a political societal problem in that "technical choices are first of all social choices: the choice of the 'great water management systems' based on the construction of large dams, for instance, is justified by its centralising objective and its capacity to create, for rural communities, dependency situations generated by the need to manage these large infrastructures and equipment" (Pérennès, p.19). The dismantlement of certain aspects of the traditional Moroccan society through the introduction of modern, foreign, natural resource management systems is one important example of how development can be counter-productive. If "Water tells the story of societies"- (Pérennès, 1993, p.21) then, what comes next is part of the stories to be told about old, modern and future Morocco.

### 2.2.1 Water canals and water clocks in rural Morocco

How do traditional water management systems adapt to arid environments? What are examples of the techniques that characterize the great Islamic hydraulic heritage? Here, we have chosen to describe a few techniques because they contributed particularly well to two aspects of traditional management. The first one relates to the adaptation to arid environments. The second one captures unusual water conflict management characteristics.

One first example of traditional water management system that is worth presenting is the system of *khetarras*, or "subterranean aqueducts engineered to collect groundwater and channel it to surface canals which direct it to fields and community wells" (Lightfoot, 1996, p.261). Authors agree to say that the most spectacular networks of *khetarras* that still exist can be seen in Morocco: a network of 600 *khetarras* could still be seen in the 1980s in the Haouz plain of Marrakesh, whilst around 400 others exist in the area of the Tafilalet and the Souss. *Khetarras* provided the only reliable irrigation water for North Tafilalt until the early 1970s, when new water management systems were introduced by the government. In the Haouz, the network – when in full use - was 900km long and contributed to the brilliant growth of the city of Marrakesh. The network was originally introduced through the Iranian technique of the *qanat* by the engineer Abdallah ben Yunus at the end of the 11th century, improved in order to adapt to the physical specificities of the plain and transformed into a very specific technique of the *khetarras*, built and managed by the *khatatiriya*. The *khetarras* that are still active nowadays are maintained by all community inhabitants and sometimes friends from neighbouring villages against some money or favors. In the 1990s, in order to encourage the exploitation of every possible water source, the local government of certain areas (e.g. Errachidia in the South East of Morocco) was providing small grants for the maintenance of *khetarras*. Within a *khetarra*, the water flow is equivalent to 10 litres per second, on average – water never stops flowing. *Khetarra* irrigation is a sustainable water recovery method. Because it relies entirely on passive tapping of the water table it does not upset the natural water balance, whereas the withdrawal of water by pumping can lead to aquifer depletion (Lightfoot, 1996, p.262).

Traditionally, another system of, this time, surface water distribution, is well known in Morocco and still in use: it is the system of *seguias,* main system of collect, distribution and transfer of water. In the Haouz plain, 150000 hectares of land are irrigated by a system of 140 km of *seguias* and 1000km of smaller canals derived from the seguias (the *mesref*). The *seguias* are organized in the shape of fish bones, with the *seguia* itself, deviating water from the river, to the *mesrefs*, distributing water much further from the oued (river) to the fields to be irrigated. The loss of water through infiltration can be very high (up to 50%), but it ensures that there is still water to be captured downstream – since little dams upstream capture the majority of available water. In order to avoid conflicts over irrigation, an alternative system of *seguias* irrigation to the left and to the right of the river is put in place. This distribution

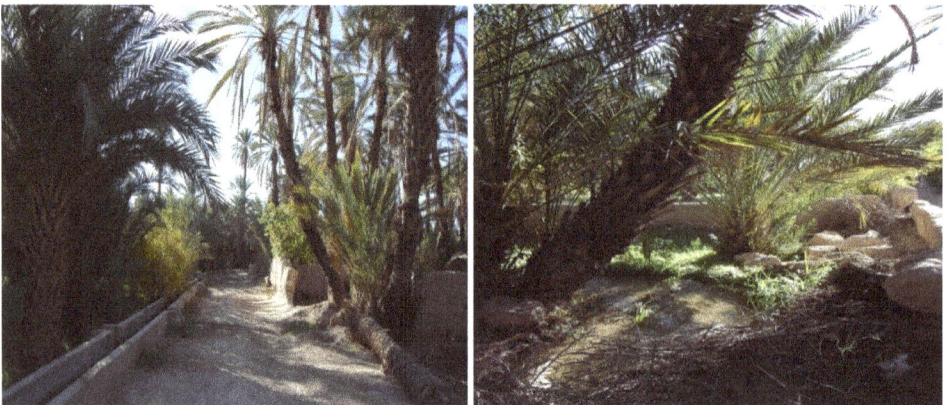

(a)                                                                      (b)

Fig. 2. a) Cement *seguia*; b) Natural *seguia* in Tata, Southern Morocco and allocation of water requires a set of agreements between tribes living along a river. These agreements led to the creation of two types of *seguias*, observed, in particular, in the valley of the Draa in the south of Morocco. The *'melk seguias'* (56 out of 89) allow people to privately own part of the water. Depending on the water pressure in the *seguia*, and the number of farmers allowed to benefit from it, people take it in turn to benefit from a certain 'time of irrigation': the *noubas* (water days) are subdivided in these specific amounts of time. Other *seguias* have a collective status – they are the *'allam seguias'* (27 out of 89 in the Draa valley). Any transaction concerning the land then also includes the water with which it is intimately linked. The order in which land is being irrigated is entirely of topographical nature. The *allam* system exists in communities that are particularly coherent and united. Despite this, in both systems, an *amazzal* ensures that there is fair distribution of water and manages conflicts in case these rise, as the next section describes in more detail. The objective of this section was to show that the *seguia* system is well adapted to both the physical geography and also respects a human network.

A second dimension of traditional water management systems in Morocco is the way in which these manage *conflicts over water*. This dimension has been the subject of a thorough study by Wolf (2000) and has been explored by various other authors (Pérennès, etc.). What caught their attention was the fact that Berbers' methods of conflict resolution on water were based on:

- the *allocation of time, not quantity* - villages, family, individuals benefit from set days for irrigation of their crops. In some instances of water markets (e.g. in the Draa valley, in the South of Morocco), the commodity bought is time for water usage, not water quantity, which circumvents the need for storage and Islamic codes. Wolf reported in one of his interviews that 'Berbers felt the idea of buying and selling water was both repugnant and contrary to the tenets of Islam) (see Figure 3 water clock below);
- the *prioritizing of use* – this is a method used to deal with a fluctuating supply, emphasizing the fact that it is necessary to prioritize the use to which water is put. In Islam, the priority is drinking for humans, followed by drinking for animals. Next is irrigation (which flows through canal systems, such as *seguias* or *khetarras*, presented earlier). Next is water for mills, and the last priority is irrigation water brought to the land through modern means (pumps, etc.)
- the *protection of downstream rights* – upstream users could be tempted to over-use water, but this is prevented both by the allocation of water by time and also by regional laws that forbid the use of modern materials (cement, in particular) for canal intakes. The traditional methods of piled rocks, although potentially qualifies of 'inherently inefficient', guarantees that a substantial quantity of water can still reach downstream users.
- *process techniques* of conflict resolutions – these tend towards the formulation of mutually acceptable solutions through facilitators (marabouts, or *'a'alam'* – also designed as the *'amazzal'* in other publications (Pérennès) – who represent the users in dispute and are in charge of negotiating a solution. These facilitators are also in charge of choosing the irrigation timing schedule, as well as a ceremony of forgiveness (the *'sulkha'*), once the dispute has been resolved. The *a'alam* rotates from a family lineage to another.

Water conflict management techniques were also developed in towns, such as Fes, where water circulated through a 70km long network of canals and was regulated by a corporation of specialists called the *qanawiyyum*.

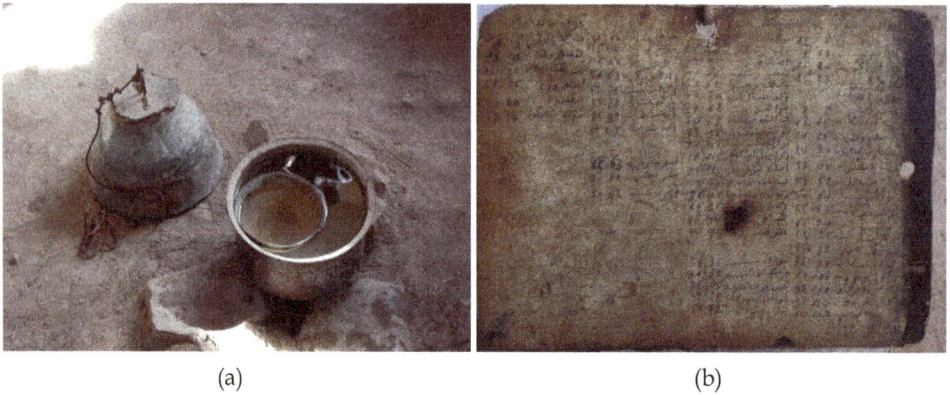

(a)                                                              (b)

Fig. 3. a) Water clock in Southern Morocco – it takes 45 minutes to fill the pot, time used as an irrigation unit; b) List of benefiting water users

The ability of the Berber communities to traditionally deal with water conflicts in this way is based on their very strong **social capital**, a term extensively explored originally by Robert Putnam and revisited in the context of Morocco by sociologist Fatima Mernissi who defines it as a wealth that improves efficiency when people respect each other and prioritise common public interest before individual wants. She explains that a people who, like the Berbers, have *tiqa* (trust), *ta'awun* (the capacity to co-operate), *tadamum* (solidarity), and *hanan* (unconditional kindness), have a very strong social capital. Thus, she defends the principles of tribal democracy (in which collective and specific rights are being defended) against 'occidental democratic principles' that protect individual and universal rights. "Only nations who protect traditions of co-operation and of solidarity, encouraging people to invest in common projects", she concludes, "will triumph in centuries to come" (1997, p.19)

### 2.2.2 Urban traditional management: The example of Fes

The Islamic prestigious water management heritage also refers to urban water management and the integration of water within Islamic architectural concepts. Muslim communities such as Moroccan ones, have traditionally dealt with water issues and shortages also through their selection of sites for settlements, urban integrated design and through the constant reminder of the spiritual value of water in their mosks. As Michell explains, "In both the hot and dry and the hot and humid areas of the Islamic world, architecture has been a means of controlling the environment by the creation of domestic micro-climates, of which the courtyard house is the most common example. In Islamic popular architecture, the insulation properties of many natural materials have been exploited and a range of ventilation systems developed (…) Water is an essential component to, and an illustration of, the nature of Islamic architectural decoration. Its use for decoration, as well as for coolness, is best seen in house and palace architecture rather than in religious buildings, where the paramount function of water is for ritual purposes" (Michell, 1995, p.201, 173). It is no surprise if the spiritual and cultural capital of Morocco, Fes, was created in an area that is rich in springs and where various sites have been, and are still used, for health purposes (the *stations thermales* of Moulay Yacoub and Sidi Harazem, for instance, extensively described by Doctor Edmond Secret in his 'Sept Printemps de Fes' for their health benefits).

Architecture is indeed more than a history of form and style. It illustrates cultural and environmental factors, as well as the way of life of the people from whom it is built. Elements of traditional architecture are actually being re-used through 'green architecture' nowadays - architecture that seeks to "construct a human habitat in harmony with nature" (Wines, 2000, p.8). It does so under the popular appellation of 'Sustainable architecture'. In ancient cities, such type of architecture usually meant relying on construction technology development based on regionally accessible materials which satisfy the demands of climate, topography, agriculture, as the main means for survival.

(a)                                                                                (b)

Fig. 4. Water in architecture in UNESCO urban World Heritage, Fes a) 14th century water clock; b) Nejarrine fountain

Fes provides an interesting urban example of how ancient Islamic civilizations have developed urban strategies to distribute and manage water resources strategically. Serrhini, Director of the ADER Fes, explains (Serrhini, 2006) how, following Moulay Idriss' selection of a water rich site to create it, the city expanded in the hands of a water conscious and ingenious Berber dynasty, the Almoravids. From then until the XIXth century, three types of complementary water infrastructures were developed and became exemplary.

The first one was the network of clean river waters, used for house cleaning, filling of basins, irrigation of gardens (with the help of *norias* – see Figure 5b), artisanal usages and to fuel mills…. – but not drinking. The water came mainly from the Oued Fes, divided into three smaller rivers directly North of the old town, where the water was distributed through a *répartiteur urbain* (Figure 5a), depending on the priority and volume of the water usage to be made (use for public baths, the *hammams*, or domestic usages, or uses for the tanneries).

The infrastructure was based on a system of underground tunnels and surface canals where water flew simply following gravity, down through the old town, the medina. There was also the network of spring water, the drinkable water, linking some twenty springs around the old town into a pottery canals network called the *maâda*, feeding seventy fountains for public water use, as well as private houses. Finally the water sanitation network was organized underground through the *Sloukia* which took the water outside the city towards the Oued Sebou. At the time when that structure was built, the network was sufficient for the population it served and the wastes led outside were sufficiently rapidly biodegraded. The various infrastructures and the architectural attention that was paid to them showed how much water was valued. Economically, a fair allocation of water resources was ensured thanks to a careful management of the volume being directed towards various types of activities. Spiritually, water was beautifully present in fountain and ablution rooms of mosks. Environmentally, water was used carefully and recycled as much as possible (e.g. the water from fountains was re-orientated for it to be re-used in gardens) before it was got rid of.

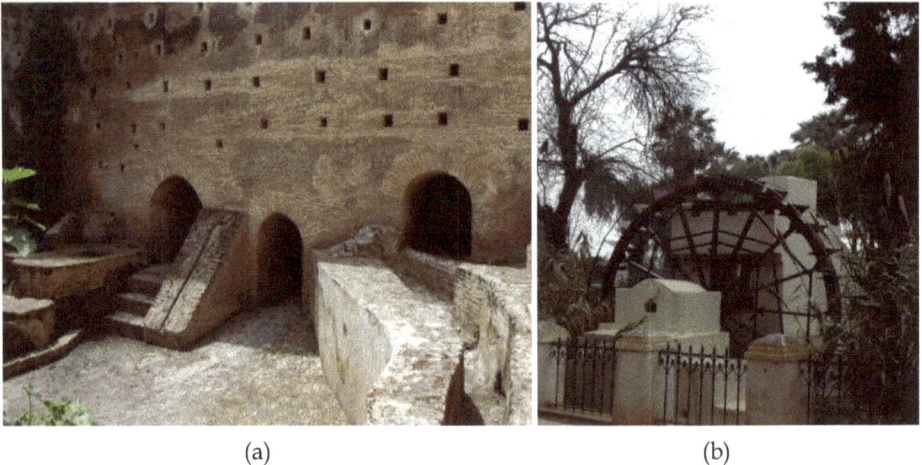

(a)                                                                          (b)

Fig. 5. a) Water distributor, Fes b) Noria from which the water flowed to the distributor

The urban structure, with public fountains accessible to all, was respectful of the right to all to access and consume water freely. Moreover, the tradition of the public bath, the *hammam*, allowed a more efficient use of natural resources – both water and fuel to heat it. The *hammam* has been described by Sibley as "a sustainable urban facility which not only promotes cleanliness and health of the urban dwellers but also social interaction: it serves as a meeting place for both male and female communities" (Sibley, 2006, p.1). She also explains that the religious requirements for washing in Islam played an important role in the way *hammams* developed. *Hammams* are generally well embedded in the urban fabric of the city, located along the underground water channels and built on sloping sites to facilitate drainage. The furnace of the *hammams* is often integrated to a bakery to make economic use of the firewood or by-products from other industries such as olive pits used to heat up the water. "The amount of water each client receives is limited to four to six traditional buckets – anything above which has to be paid for" (Sibley, 2006, p.3).

Things have changed. The size of the medina population to be provided with both water and sanitation, in particular, has considerably increased. Family size has reduced and the French presence during the Protectorate (1912-1956) led to the creation of the 'ville nouvelle' where many people ended up moving to. Private bathrooms were built in modern houses and the tradition of the *hammam* is slowly declining – although 30 plus *hammams* are still in use in the medina. Parallel problems are nevertheless growing, notably the source of fuel to be used - or not - for public baths in a densely populated part of town. Economic activities and modes of production (for instance substances now used to treat animals' skins in the tanneries) have changed in such a way that relying on a traditional system of water distribution and usage is not adapted anymore. This is an example where tradition and modernity cannot go hand in hand anymore, despite the wishes of King Mohamed VI. But the various initiatives aimed at restauring these old hydraulic structures and to learn from them demonstrate that they could still be useful one day and that they are still highly valued.

This is true not only nationally but also internationally: the Arab Fonds for Economic and Social Development is contributing to restauration works and the old medina of Fes is one of UNESCO's World Heritage Sites, mainly due to its integrated urban organization. Beside, and thanks to the systemic nature of each component of the urban fabric of a medina such as Fes, learning lessons related to restauring one component (for instance how to draw lessons in sustainability when improving the running of *hammams* in terms of construction, heating system, water uses, social dimensions, etc.) will lead, in a domino effect, to re-learning about the sustainable dimensions of traditional urban Islamic structures, as a whole.

## 2.3 Modern technical approaches

The modernisation of water management in Morocco was very much geared towards 'fueling' agricultural production, itself targeted as the main component of the country's modernisation and development.

In line with development objectives, themselves emerged out of the post-independence era in the 1960s and focused on modernisation objectives for the country, Morocco chose to put together a strong *politique des barrages* (dam building) in place, aimed at strengthening its irrigation potential throughout its territory. The irrigation strategy was characterised by the need to separate the country into various irrigation zones. The Doukkala plain, in particular, received special attention because of the citrus production it generates for exports. The technical irrigation equipment that that area benefited from increased from 4700 hectares in 1960 to 14000 hectares in 1967. The whole question focused on how to modernise agriculture and move away from traditional models of production and irrigation. An important issue became the size of the production unit, which had not only technical but also political implications. French 'colonisers' under the Protectorate had large-size agricultural pieces of land, located in fertile areas of Morocco and focused on export crops. The economic Plan of 1960-1964, which encouraged intensification of agricultural practices, somehow stagnated – fought against by 75% of rural families who still only had less than 2 hectares to work on and were still subject to traditional systems that were against intensification practices. The re-organisation of land ownership in Morocco was very slow, after independence, and it is only after 1973 that private colonial pieces of land were re-distributed. But, as Pérennes explained (1993, p.165), out of 100 hectares of colonial land, 35 went in the hands of rich

Moroccan land owners, 35 started being managed by the State and only 30 were distributed amongst little agricultural production units within the framework of the agricultural reform. Irrigation equipments and infrastructures benefited the richer and inequalities remained.

In order to modernise irrigated lands and agricultural practices, the State then decided to focus on new crops (with new crop contracts guaranteeing fixed prices to farmers as well as compensation in case of loss of crops). That system of 'integrated crops' included cotton, beetroot and industrial tomatoes. The State also funded part of the irrigation equipment under an agreement in which farmers had to progressively reimburse parts of the advanced sums and to pay a fee for water usage. In terms of technical choices, traditional irrigation systems based on gravity and *seguia* networks were extended into a fuller network (the *trame d'irrigation*). But the superimposition of a complex physical irrigation network based on traditional systems and a centralised mode of management did not lead to satisfactory results and many objectives of the *Réforme Agraire* could not be reached. Nevertheless, King Hassan II, in his famous Erfoud speech in 1974, and whilst international prices were in turmoil, announced a new technical option: *la politique des barrages*, in which he announced his target of irrigating 1 million hectares by the year 2000 – which was reached. Overall, budgetary choices went in that direction: in the early 1970s, 41% of the budget was aimed at the irrigation and agricultural sector - the building of dams, the purchasing of modern equipment, would overcome climatic constraints and water shortages.

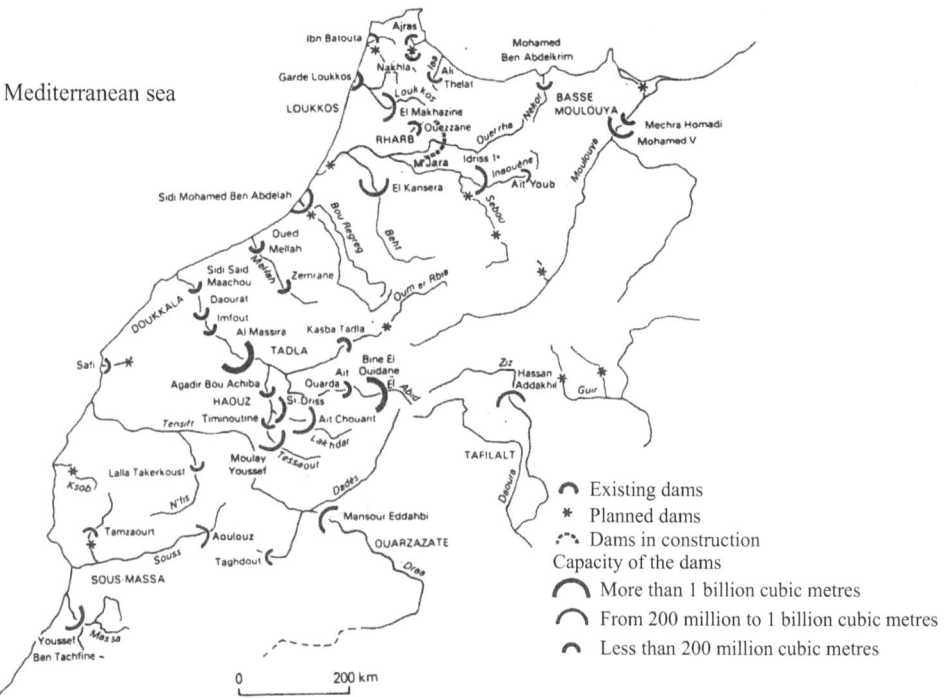

Source: Pérennès (1993, p.173)

Fig. 6. Carte des barrages au Maroc Pérennes

A series of events (the mid 70s collapse in the prices of phosphate - resource which, at the beginning of that decade, had ensured high economic growth in the country through its exports -; in the mid 70s again, the increase in military expenditures; and then severe droughts in the 1980s) broke that momentum and led to sectoral adjustments that included a questioning of the 'grande irrigation' option. The PAGI (Programme of improvement of large scale irrigation) of 1985 changed the management system of these agricultural zones (which often became privatised).

The big question of 'what should be the place of the State and that of the farmers in agricultural development?' was once again asked and once again as technically as politically relevant. The 1980s led to programmes of structural adjustments and dependencies on the IMF services that had heavy consequences in the irrigation sector and economic orientations. Morocco respected IMF conditions, orientations and advises – more and better intensification of agriculture, less water wastes, choice of revenue yielding crops, etc. But the management structure – re-organisation of agricultural land; financial reform with subventions on the one hand and taxes on irrigated land on the other; re-allocation of management roles – was difficult to integrate and included new components such as the payment of water by farmers. Between 1980 and 1984, irrigation water pricing was such that certain irrigation practices were discouraged (aspersion). The method of 'economic sanctions' was therefore introduced but not politically backed up by regional authorities who considered water as a socially sensitive issue. The State – and the World Bank - then decided to help farmers with various credit systems that would allow them to become more entrepreneurial. But whilst average and big size exploitations benefited from that system, smaller ones remained marginalised because of the land ownership structure: the lack of ownership titles on collective lands and micro-funding systems made their access to credit difficult. Globally, big irrigation policies, in particular in the Doukkala, were successful. However, interestingly, farmers who had the choice, preferred to carry on cultivating on plots whose size was smaller than the encouraged 'official production sizes', and chose to carry on cultivating vegetables and cereals - crops perceived as giving them more independence and food security.

Technically speaking, at the beginning of the 21st century, as Abdelkader Benomar - director of research and planning at the Moroccan Secretary of State in charge of water and environmental issues – reported to Lamia Mahfoud (2011), Morocco is prepared to invest in massive initiatives in order to implement a strategy that will help in anticipating water shortages. The current trends, with regards to the management of water supplies, focus on the construction of more dams (60 large dams, also used for hydropower – a source of energy that saves on average 70,000 tons of oil per year (Doukkali, 2005, p.73), and 1000 little dams before 2030), the transfer of water (800 Million cubic metres planned to be transferred) from the North of the country to the South, the re-use of treated water, the de-salination of sea water. And when it comes to managing water demand, the Moroccan strategy focuses on improving water use efficiency in industries and in touristic units, to re-adjust water tariffs, to integrate water saving practices in the building industry, and to improve water usage practices in irrigated agriculture.

Many initiatives are undertaken in public-private partnerships. Beside, a new system of subventions established in 2010 aims at intensifying agricultural practices through massive subsidies by the state on agriculture machinery.

Technical efforts geared towards irrigated agriculture therefore remain high on the agenda. As Bennis and Tazi-Sadeq explain (1998), in Morocco there are still two types of irrigation: Large-scale hydraulics (GH), involving vast areas fed by high-capacity dams and providing year-round water supply (presently about 500,000 out of a potential 830,000 hectares); and small-scale hydraulics (PMH), involving small areas of several hundred hectares fed by water sources that are not highly regulated (e.g., pumps, water diversion, co-lineal reservoirs, spring water catchments, and flood waters). "The goal of the state is to reduce the amount irrigated by seasonal waters to 170,000 hectares, and increase the amount irrigated by year-round water to 510,000 hectares (60 percent). This measure should contribute in a major way to nutritional security, job creation, and the effort to slow rural exodus throughout the country. The goal will be reached through rehabilitation and modernization of equipment in the areas concerned, using traditional irrigation systems based on customary rules of water distribution" (1998, p.8). The way in which Morocco is trying to achieve these targets is therefore to create agricultural water users associations that will embrace these technical changes. This new approach, described by Bennis and Tazi-Sadeq as a 'very modern and complex concept of hydro-agricultural equipment', despite attempting to bring socially more friendly solutions to water supply and demand crises, has been questioned by these authors: "Will the population accept high annual costs for participation in investments that were decided without their consent, water fees based on consumption rates, and unit prices that exceed those that are customary to the region?".

One of the crucial questions of the *Plan Vert* in Morocco therefore remains centered on finding appropriate ways in which to engage people's participation in the making of its country's sustainable development. This is a political as well as an economic and above all ethical question, which will be discussed in the last section of this chapter.

## 2.4 The impacts of modern approaches on traditional water management systems

In terms of water and agricultural management techniques, historians have talked of a collapse of the Islamic civilization on cultural and technical fronts. Swearingen (in Pérennès, 1993, p. 17) explained that phenomenon by demonstrating that in Morocco, in particular, colonization had broken the coherence of rural societies in order to impose to them new and contradictory interests; those of the 'colons', those of the State, and those of metropolitan lobbies (such as those encouraging the construction of major infrastructures such as great dams in the colonized country). Important questions are currently being asked by certain people (and should be asked by a wider circle of, in particular, policy makers) concerning the appropriateness of certain modern water policies in the light of both ecological and climate but also social, cultural, and political changes. The example of the abandonment of the use of *khetarras* is merely one example amongst other significant changes, but it usefully illustrates the combination of factors involved in such changes.

In 1996, Lightfoot extensively studied that issue and explained how the *khetarra* system of the Moroccan Tafilalt were in the process of being abandoned as surface and groundwater supplies and of being replaced with diesel and electric pumping devices, as well as large dams. His studies highlight the regional impact of the entire *khetarra* system and emphasizes both the problem of water recharge as well as the social implications derived from the imposition of new techniques. A 300 km network of such canals was excavated in the Tafilalt region. When great dams were constructed upstream from the Tafilalt on the Oued (river) Ziz, and concrete-lined government canals as well as unregulated use of diesel pumped wells were introduced, dessication started occurring. The availability and distribution of water changed dramatically. As Lightfoot explains "No longer is the oasis fed by the occasional flood or heavy spring runoff from the Ziz, and because the Ziz now infrequently flows at the Tafilalt, and only in concrete lined canals, groundwater recharge has been greatly inhibited while growing quantities of groundwater are being pumped out to make up for the dam-induced deficit of surface water. (...) The government canals provide a measured, cheap, reliable amount of water, but government resource officials and Tafilalt farmers concede that Ziz water is now insufficient – providing half their needs – and good only for supplementing the water coming from other sources" (Lightfoot, 1996, p.266). If, from the point of view of water availability, the introduction of new techniques in the Tafilalt is questionable, it also is from a legal perspective. The introduction of a new distribution system (the release of water from the reservoir) has meant that the 'water timers' no longer regulate water allocation. Moreover, in 1996, there was no authority to actually regulate the various diesel water pumps installed in the region (750 private ones). Modern techniques have therefore, at least in this case, proved to be potentially capable of providing greater quantities of water but not in a way that allows for groundwater recharge in the long run - not in a *sustainable* way. Culturally, the abandonment of the traditional irrigation techniques have altered the land use patterns of the oasis – less and less palm trees produced dates, traditionally traded from the oasis, as a result of sustained dessication and poorer groundwater reserves. Moreover, there have been important social impacts, such as the loss of local control over water resources."*Khetarras* were qsour-operated and collectively maintained, and intricate relationships had evolved to manage them and distribute their benefits according to each shareholder's inputs of land, labor, tools, and money. Diesel-pumped wells are often privately owned and, as a result, the traditional ties that bind village society are breaking down (p.268). With the overexploitation of water and the large scale depression of the groundwater levels in the area due to the construction of dams, new *khetarras* parallel to the old galleries would have to be excavated in parallel to the old galleries which would prove to be labor intensive and expensive – prohibitively so. Similarly, in urban areas, such as Fes, the traditional water management systems – based on the water distributor, in particular – was stopped due to the fact that the flow of the Oued Fes (which fed that distribution system) was considerably decreased following the construction of a large dam North of Fes, in the 1980s.

The choice of water management techniques therefore has important political impacts and the political dimension of water management in Morocco is explored in the next section of this chapter.

## 3. The politics of water management in Morocco

It is not rare to hear people talk of water as the new gold, or to associate the idea of having water as having power. Water security has grown as a major concern for the 21st century. In brief, who manages water inevitably has to deal with political issues. As Turton stressed, "Because water is scarce, and because it is essential for life, health and welfare, it has become a contested terrain and therefore a political issue" (Turton, 2002, p.9). In the literature, one talks of **hydropolitics** as "the authoritative allocation of values in society with respect to water". This definition implies the issue of **scale** (ranging from the individual, to the household, village, city, social, provincial, national and international level with a number of undefined levels in between) and the **range of issues** that are covered (water conflicts and their mitigation, states and non-state actors, water service delivery, water for food, the social value of water, the political value of water, the psychological value of water, water demand management (WDM), water as a target of aggression, water as an instrument of peace, water and gender, …). Including in the range of issues is the core place occupied by water in a specific type of development, as mediator between people and nature. The politics of scarce natural resource management offers a particularly interesting terrain of research in the context of both a *Muslim country* and a rise in *sustainable concerns*, as well as when attempting to **re-think the notion of economic development**. It is on these three aspects that I will be focusing next.

### 3.1 The evolution of water politics in Morocco: Towards 21st century 'new departures'?

Morocco is a Muslim country. Its King is the religious chief of 'his' believers, and the conduct of economic, social, political affairs all have to be in agreement with Islamic principles. In the context of environmental management, and with regards to water, in particular, this could and should present an advantage since water is so central to Islam. As Caroline Pestieau, then president of the International Development Research Centre, emphasised in her preface to Faruqui et al.'s book on Islam and water management, "since Islam is the religion of about one-fifth of the world's population and the official faith of a number of countries, in many of which water is the key scarce factor for development, understanding its actual or potential role is important" (in Faruqui et al, 2001, p.vii). Reminding ourselves of the Islamic principles related to water management is important in order to understand the present structure and functioning of water institutions in Morocco (Table 1).

This is because "the laws and rules governing the functioning of land and water uses in the country have actually emerged from the historical superimposition of three bodies of laws and rules: the *Orf* (customary sets of rules and admitted practices), the *Chraa* (religious interpretation of the Islamic law and rules) and the modern legislation introduced by the French protectorate and later reinforced by independent Morocco, since 1956" (Doukkali, 2005, p.75). When Islam was introduced to Morocco, the religious jurist (*Ulema*) accepted the very heterogeneous customary practices adapted to different physical milieu in the country. The only impact that the *Chraa* then had was to give some moral references that remained very theoretical and that didn't have real impact on the management of water resources. Lakes, groundwater and rivers, uncovered by customary laws, were at the time defined as public goods under the control of the sultan. At the beginning of the French Protectorate, in 1914, all surface waters were put under public domain following two arguments, as explained by Doukkali (2005). a) All waters were traditionally owned by central authority in Morocco. b) The concept of public domain was more in accordance with the true precepts of Islam. A state control over water was thus imposed by a protectorate very keen to manage the resources as it wanted. The conditions of water usage through irrigation and other purposes were described in a 1925 new legislation on water which strangled any private initiative on water development in the country.

The Protectorate also issued a law regarding water user associations to formalise and initiate the creation of an irrigation network. Three major rights systems were then in place (modern registered rights over water, customary rights registered and customary rights unregistered).

| Social and spiritual dimensions of water | Water is considered a blessing from God that gives and sustains life and purifies humankind and the earth. Water is especially important for muslims for its use in *wudu* (ablution before praying) and *ghusl* (bathing). Equity: A Muslim cannot hoard excess water- he is obliged to allow others to benefit from it. The priority of water use rights is: first the right of humans to quench their thirst, second, the right of cattle and household animals and third the right of irrigation. In Islam, human-environment interactions are guided by the position of humans as *khulafa*, stewards of the earth. |
|---|---|
| Non economic instruments for water management | The Qu'ran states clearly that the supply of water is fixed and that it should not be wasted. Given the importance of cleanliness in Islam, and that many MENA countries have minimal waste water treatment, it is common to hear muslims declare that *waste water reuse* is undesirable or even haram (unlawful according to Islam). Family planning is allowable in Islam but should not be encouraged solely for material reasons. In many countries continued high population growth is severely stressing existing water resources and the environment. Family planning could help prevent further reductions in water availability of water per capita. |
| Economic instruments for water management | Economic measures over water usage are controversial in Muslim nations because of the **Islam precept that water cannot be bought or sold**. In Islam, water is considered the gift of God, so no individual literally owns it. According to Islam, a fair tariff will lead to equity across society. In endorsing fair markets, the prophet refused to fix prices for goods in the market, including water, except in special circumstances. Most Muslim scholars agree that a just price for water is that determined by the market, providing that the market is free from unfair practices. Economic practices also deal with **intersectoral transfers of water**, which will change radically in the next decades. Are intersectoral water markets allowable in Islam? As a population evolves from rural and agrarian to urban and industrial, reallocation is not only permissible but is required to preserve equity, and the primacy of the right to quench thirst. Intersectoral transfers are considered inevitable. |
| Integrated water management (IWM) at different levels | **IWM** should address all water resource management issues in relation to each other and to the water sector as a whole, with the goal of promoting equity, efficiency and sustainability. The water sector has many horizontal and vertical linkages and IWM needs to address micro and macro level decisions. **Community based water management**: The input of the community on matters that concerns it, including water management, is mandatory in Islam. This consultation is required of all of those who are entitled a voice – women too. **National level**: haddith command not to harm oneself and others, and the environment – hence encouraging an integrated approach that protects humans and nature. **International levels**: *Shura* (consultation on matters of mutual interest) and *fassad* (avoiding harm and damage) should be applied internationally. |

Table 1. How does Islam regards various approaches to water management - adapted from Faruqui et al. (2001)

The dominant concerns for resource protection inherent in the 1914 and 1925 laws gave place to the emerging economic requirements of the independent country after 1956, and the Moroccan 'policy of dams', centred on irrigated agriculture, emerged. The State got involved in large scale development works and, "for the purposes of balancing growth requirements and poverty concerns, agricultural development became an important component of the economic strategy since the mid-1960s and through the 1970s and the 1980s" (Doukkali, 2005, p.78). The increase in State intervention in the water sector was then based on the 1969 Agricultural Investment Code. Since the 1980s, the emerging macro-economic necessities and resource related constraints have prompted diverse types of water institutional reforms in Morocco.

### 3.1.1 The great institutional water reforms

In the 1980s, droughts provided an impressive impetus for changes in favour of a rapid expansion of private and groundwater based irrigation systems which, although they expanded or stabilised farm production, did it in the absence of effective regulatory arrangements and resulted in heavy aquifer depletion and in serious decline in the flow of several springs and watercourses that supported medium and small scale irrigation perimeters. The macro economic crisis of 1983 and subsequent economic liberalisation led to readjust agricultural institutions to allow them to cope with a market-oriented agricultural sector. The State decided to make the publicly managed large scale irrigation system more flexible and responsive to local needs. This involved the promotion and involvement of farmer organisations in water allocation and management through a revised legislation on Water Users Associations in 1990. WUAs were able to adjust positively but only in the context of small scale irrigation perimeters. In large scale irrigation perimeters, WUAs were still dependent on the State for funding and functioning. The revision of water pricing proved to depend on the involvement and empowerment given to the WUAs who refused to adhere to a 'full cost recovery' system they did not understand since they had had no say about the nature and make up of costs, nor in irrigation decisions.

Then, by the 1990s, there was little scope for further development of water resources through large scale schemes and the need to integrate surface and groundwater management became apparent. "The state was compelled to shift focus from water development and irrigated agriculture to the most difficult and challenging frontiers of water reallocation and integrated water resources management from the perspective of the whole economy" (Ait Kadi, 1998, in Doukkali, 2005, p. 83). The Water Law of 1995 was passed. It aimed at integrating and coordinating the allocation and management of all water resources and users under a single but decentralised institutional arrangement centred on river basin agencies (RBAs). De Miras and Le Tellier (2005) described the so called 'Loi 10-95' as "establishing the legal framework for the politics of water for the next decades and includes a number of legal instruments aimed at addressing the problems of decreasing reserves of water, increasing water demand, increasing water prices and the deterioration of water quality and the environment" (p. 222). As Doukkali explained (2005, p. 73), the 1995 Water law had also assigned top priority to the security of the drinkable water supply: by 1990, most urban households had been provided secure access to water supplies – whilst only 14% of rural households had secure water supplies. The RBAs still work with water sector partners (such as WUAs) and have authority to manage surface water storage and allocation, groundwater pumping and water pollution and quality. Their main responsibility is to prepare river basin management plans that are respectful of IWRM principles and well integrated in the twenty year National Water Master Plan, as indicated by the national government through the heavy inputs it provides in such enterprises. As Doukkali explains, the RBAs have considerable managerial and regulatory responsibilities beside their role in developing and supplying water. They can monitor and regulate water use and quality as well as plan and organise flood control and water related emergencies within their respective basins. That third institutional phase, that of the 1990s, therefore changed once again the configuration of water actors on the national scene. It also changed it on the international scene, since it is at that point that the government of Morocco decided to significantly reorganise its water administration and to promote the private sector in water resource development and management. Thus, the concessions for water distribution in four large cities (Tangier, Casablanca, Rabat and Tetouan) were granted to private water companies, and the private sector involvement was also extended to the irrigation sector in 2002, encouraged by World Bank enthousiasts, through two projects – the construction of a transmission pipeline (Guerdane project) and a distribution network (the Gharb project).

In the urban context, political choices related to water management options are characterised by three main phases, as Haouès-Joune explains. The final phase of the Protectorate (1946-1953) was one during which the urbanism strategy was to ensure that the poorest communities could have access to urban services. In practice, that meant that efforts were put into providing services through infrastructures whose costs would be optimised. From the end of the 1970s, efforts were then put into identifying water tariffs that would be economically viable whilst still allowing the poorest communities to have access to water and water services. This was done in a context where the management and growth of accommodation was not well coordinated with the management of the water sector. Until the 1990s, water distribution was done both through 'régies' and through public services. To simplify, until the 1990s, the *institutional organisational* system, lead by the State, was as follows in the water sector (Allain-El Mansouri, 2005, p.166). Consultation bodies, such as the Superior Water and Climate Council, created in 1981, were aimed at coordinating the various actors in the water sector. *Administrative bodies* were originally represented by the Interior Ministry (through 'régies') and the Land, Environment and Water management Ministry. The creation of the Secretary of State for Water was aimed at illustrating the political will to prioritise water issues in the context of moves towards sustainable development strategies. At a local level, the directors of the 'agences de bassins' were in charge of implementing water policies imposed from above. *Public establishments* such as the ONEP (National Office in charge of drinkable water and régies) were in charge of the planning, production and provision of water in urban and rural centres and are autonomous financially. Finally, *local interventions* were undertaken by local collectivities and users associations.

Major changes took place in the 1990s with regards to political choices in the water sector. The risks of real water shortage had then increased in the country, together with the difficulties in meeting a growing demand for water because of the heavy expenditures associated with the equipment of new large scale infrastructures such as big dams, water transfers, water pollution treatment, and more extensive groundwater exploitation. The 1990s were marked by a desperate need to start saving water (water leakages were estimated to reach 40% of the resources provided), to improve water treatment and recycling, and to diminish the water infrastructure bills. Choices to be made were perceived as being technical as well as socio-political. In practice, Morocco chose to manage the resource in a semi private, semi public way. This resulted in the State delegating water management and treatment as well as electricity provision to private companies. Thus, the Spanish consortium REDAL1 was put in charge of water distribution in Rabat, for instance. The *concession* is defined as a convention through which a public body gives the mission of exploiting a public service following certain conditions that are described in detail in a list of duties and requirements, and against a payment that, most of the time, comes from tax payers' money, to a 'concessionnaire' of its choice. This has been described as *gestion déléguée*. The resulting 'marketisation' of water in Morocco has been considered unsatisfactory, on two fronts. With current poverty levels, especially in urban environments, identifying appropriate water pricing has been a difficult problem to solve and the macro-economic situation a difficult terrain in which to integrate a natural resource market. The *gestion déléguée* therefore encountered a few hic ups, with REDAL1 being stopped after 3 years (instead of 30) and replaced by the great Veolia Environment. On a second front, politically speaking, this water management system has been far from operating in a participatory manner.

This is how the 1990s approaches led to the creation of the new Moroccan Environmental Charter.

### 3.1.2 The 'new environmental charter'

In these early decades of the twenty first century, Morocco seems to be focusing on a new political environmental strategy where its visibility in terms of environmental initiatives is of prime importance. In 2000, the King Hassan II Great World Water Prize, an international award, was jointly established by the Government of Morocco and the World Water Council, "in memory of his Majesty King Hassan II of Morocco's distinguished leadership and encouragement of cooperation and sound management of water resources". The Prize is to be awarded to an institution, organization, individual or group of individuals in honor of outstanding achievements in any aspects of water resources such as scientific, economic, technical, environmental, social, institutional, cultural or political.

The award is presented every three years in conjunction with the World Water Forum, during a special ceremony. The award winner receives a prize sum of US$ 100,000, a trophy and a certificate. The theme for the Prize is "Cooperation and solidarity in the fields of management and development in water resources". The 4th edition of the King Hassan II Great World Water Prize will be held in Marseille in March 2012 during the 6th World Water Forum. Beside such grandiose initiatives, and maybe more importantly, a new Environmental Charter (together with the Fondation Mohamed VI pour la protection de l'environnement (http://www.fm6e.org/fr/notre-fondation.html) was announced in April 2010 and is aimed to provide a solid framework for all environmental laws, a 'de-facto constitution for environmental policy'. Amongst many of its targets are objectives to reduce external energy dependency and to ensure that half of energy usage comes from renewable energies (solar energy, use of methane from landfills, ...). The Charter also puts special attention to water management and, in particular, aims at increasing waste water recycling to more than 96%. As Slimani explains (2010, p.59, 60), Morocco is commiting to a program costing 7 billion euros to improve the water treatment network and recycle used waters. All in all, the charter will thus provide proof that Morocco will sign up for a progressive policy to reconcile the imperatives of socio economic development with the preservation of the

From Traditional to Modern Water Management Systems; Reflection on the Evolution of a 'Water Ethic' in Semi-Arid Morocco

199

environment and sustainable development'. Such a radical step, even if not fully accompanied by practical measures (yet), illustrates the wish of Mohamed the VIth and its government to approach development and economic issues in a more integrated and more independent way.

The new Environmental Charter thus emerged at a time when governance issues were being questioned and debated, and water management issues therefore found themselves linked to a whole new 'Moroccan environmental ethics'.

## 3.2 'New waves' in Moroccan water ethics

In previous sections, we explored how both the technical and political dimensions of water management in Morocco evolved throughout time. We saw that technical choices also had political dimensions. Here, we are going to examine the ethical implications that water politics can have.

| Human dignity | There is no life without water and those to whom it is denied are denied life |
| Participation | All individuals, especially the poor, must be involved in water planning and management with gender and poverty issues recognized in fostering this process |
| Solidarity | Upstream and downstream interdependence within a watershed continually poses challenges for water management resulting in the need for an integrated water management approach |
| Human equality | All persons ought to be provided with the basic necessities of life on an equitable basis |
| Common Good | Water is a common good, and without proper water management human potential and dignity diminishes |
| Stewardship | Protection and careful use of water resources is needed for intergenerational and intra-generational equity and promotes the sustainable use of life-enabling ecosystems |
| Transparency and universal access to information | If data is not accessible in a form that can be understood, an opportunity will arise for an interested party to disadvantage others |
| Integrated Water Management (IWRM) | A means to ensure equitable, economically sound and environmentally sustainable management of water resources |
| Empowerment | The requirement to facilitate participation in planning and management means much more than to allow an opportunity for consultation. Best ethical practice will enable stakeholders to influence management |
| Inclusiveness | Water management policies must address the interests of all who live in a water catchment area. Minority interests must be protected as well as those of the poor and other disadvantaged sectors |

Source: COMEST, 2004

Table 2. Principles of water ethical practices

First, what does **water ethics** mean? UNESCO previously examined that question through working group meetings organized under the auspices of the World Commission on the Ethics of Science and Technology (COMEST) and the International Hydrology Programme (IHP) in 1998 and published a report on Best Ethical Practice in Water Use (COMEST, 2004) which identified the fundamental principles presented in Table 2.

Morocco has been embracing, at least in political discourses, terms such as **ethics** and **integrated management** in the context of environmental and developmental strategies. Mohamed the VIth has been keen to show the world that his country was aligned to considerable international reflection on environmental ethics throughout the world and initiatives such as the adoption of the Universal Declaration of Bioethics and Human Rights (UDBHR) by all member countries of UNESCO in 2005.

| | |
|---|---|
| Social and spiritual dimensions of water | Many people don't have access to water and still rely on expensive informal sources of water. The situation is inequitable and the primary water right under Islam is being compromised. |
| Non economic instruments for water management | Policy-makers are beginning to appreciate the value of some haddiths with regards to water conservation and environmental education has been taking place in some mosques in the Middle East. The WHO launched health education programs through mosques in Afghanistan.<br><br>Considering huge water constraints in the MENA area, waste water reuse in irrigation has been explored in view of ensuring that it was safe, not harmful to human health. |
| Economic instruments for water management | Supplying water almost free under today's conditions of polluted ansd scarce water supplies has resulted in severe inequities – the poor often pay immorally high prices for water in informal markets, or receive water of poor quality. Under changing conditions, Muslim leaders must adapt their water policies to meet timeless objectives such as social justice. Recovering costs for providing water is allowable in Islam – but what is a fair tariff?<br><br>As a consensus in the rest of the world, private public partnerships are best recommended. If regulated markets are to be used then they must put in place legal, institutional and regulatory mechanisms to ensure that the markets operate fairly and efficiently. This includes developing better participatory processes. |
| Integrated water management (IWM) at different levels | **Community level:** In many Muslim countries, there is a very centralized decision-making system. Beside, decision-makers, often men, haven't invited wide participation. Changes have to accelerate and to happen at grassroots levels.<br><br>**National level:** IWM that include principles such as equitable tarrifs, environmental protection and food security need to integrate social policies sustained by grassroots inputs and discussed at national levels.<br><br>**International levels:** there are currently many international water sharing disputes where states are not following the principles of shura or fassad. Legal agreements need to be reached. |

Source: adapted from Faruqui et al. (2001).

Table 3. Issues and recommendations on how to integrate Islamic principles in a new 21st century water ethics

The debates and international agreements on environmental and water ethics also encouraged a contemporary re-visit of Islamic principles used in the context of water management. Although it is clear that Islam generally advocates a fair distribution of water resources and a prioritization of usage, in practice there are currently a few issues that need adjusting for water management practices in a Muslim country like Morocco to be realigned with principles of water ethics (Table 3).

In addition to the alignment with international principles of water ethics and to Morocco's efforts to recreate links between modern water management and Islamic water management principles, the question of which type of 'economic development' to strategically embrace in order to help the country's development whilst generating people's participation to creating a sustainable economy is the central theme of this chapter. Politically, Morocco chose, after its independence, to base its economic development largely on natural resources – including agricultural – exports. Technology was one of its main tools in doing so. However, the creation of the new Environmental Charter as well as a new set of human and social reforms introduced by the King Mohamed the VIth at the beginning of the 21st century put the country in new 'tracks', in terms of a) how 'economic development' is being apprehended with regards to environmental protection and b) which actors could and/or should be involved in making 'sustainable development' happen. It is on these two aspects that I want to finish my reflection on the evolution of a water ethics in Morocco.

### 3.2.1 Ecological economics and human ecology: The role of water in alternative modes of 'development'

Earlier, we examined the political dimensions of water issues. Another crucial link exists between 'water issues' and 'economic development': water is needed in agricultural as well as industrial activities.

It is crucial to any type of production and to human life. It is also a much more complex, a more systemic type of natural resource than other natural resources that are used in economic activities in that, like air, it is indispensible to human life – without water, a human being will die in only 3 days. For this reason, human civilizations have valued water for all it brings to life: spiritual richness, a habitat for certain species, a support for navigation and for the generation of electricity, a crucial component of ecological cycles, a natural resource that can be directly consumed or that can contribute to the production of food, etc... Whilst 'economic development', in a mainstream neo-classical sense, will focus on the productive nature of water resources, alternative understandings of the term 'development', such as the ones introduced by disciplines such as ecological economics and human ecology (both preoccupied with the interactions between human economic systems and ecological systems functioning) will help in widening our understanding of 'development' and might help countries such as Morocco in dis-engaging themselves with old, quasi-colonial, styles of economic development, in order to enter the 21st century more innovatively and independently.

As Slimani explains (2010, p. 60), "Now that industrialized countries seem reluctant to fully engage, as the disappointing outcomes of the Copenhagen Summit on Climate Change in December 2009 have shown, Morocco's strategy constitutes a strong signal to developing countries. Instead of being an additional constraint, the environmental imperative could well be a new developmental tool and a stepping stone towards a stronger, at least more sustainable, type of growth". Slimani's point is also reinforced by Tazi-Sadeq, Moroccan researcher specialized on water issue, who adds that "The diminishing supply and the increasing and ever more diversified demand in water entail a change of paradigm. This change has political and ethical implications having to do with efficient management – on the usage side - and fair distribution of water resources. It consists in placing water policies on the side of the demand and the human person at the centre of the debate" (Tazi Sadeq, 2005, p.13-15).

These reflections are in line with definitions of 'development' that include both ecological and human dimensions. COMEST, in particular, explain that "*development* can only take place if the people who are both its beneficiaries and its instrument also are its justification and its main objectives. Development must be integrated and harmonized. In other words, it must favour the complete development of human beings in spiritual, moral and material ways, hence ensuring people's dignity in society, in agreement with the Declaration of Human Rights" (COMEST, 2004, p.10).

If a country like Morocco is envisaging thinking about 'development' in more ecological and human ways, it is partly because it came to recognize the need for alternative models of development. Things are not fully working, big technical pushes, reforms, water pricing methods, have not been fully embraced and, worse, have led to uprisings that had been unseen in the past (we will come back to this in the next section). The so called Human Development approach arose in part as a result of growing criticism to the leading development approach of the 1980s, which presumed a close link between national economic growth and the expansion of individual human choices. The need for an alternative development model was then seen as being due to many factors, including:

- Growing evidence that did not support the then prevailing belief in the "trickle down" power of market forces to spread economic benefits and end poverty;
- The human costs of Structural Adjustment Programs became more apparent;
- Social ills (crime, weakening of social fabric, HIV/AIDS, pollution, etc.) were still spreading even in cases of strong and consistent economic growth;
- A wave of democratization in the early 90's raised hopes for people-centred models.

In Morocco, the human and ecological impacts of economic growth must also be stressed. As Leila Slimani (2010) explains, Morocco wants to use the protection of the environment as a central tool for development policies. For the last ten years, Morocco has experienced economic growth in all economic sectors: industrial, agricultural, tourism, urban development, infrastructures...These evolutions have had negative impacts on the environment. The Ministry of Trade and Industry estimated that environmental degradation costs 13 billion dirhams each year: 3,7% of its GNP (1.6 billion US dollars). Not only these costs are going to have direct consequences on the pace of developmental activities but they also impact lifestyle and the habitat of citizens. But the last point listed above (the democratization phenomenon) also resonated particularly loud and is motivating people to start thinking about development differently.

This is true both for internal reasons (Mohamed the VIth made a point of initiating social reforms in his country from the moment he replaced his – much more authoritarian – father) and international reasons (Foreign investors favor democratic regimes). Thus, on top of wanting to set an example to other countries and describing the Charter for the

Environment as leading the Arab and African nations in becoming more energy dependent via their renewable energy sources, hence deciding to use the charter as a 'blueprint' for other countries to follow as a collective, homogenised set of initiatives to fight climate change, Mohamed the VIth developed the concept of 'proximity' by inviting his citizens to participate in the writing of the Charter through an online consultative process, between January and February 2010 (Slimani, L. 2010, p. 59).

However, an *online, one month*-long consultative process might not have been enough for people to feel they could genuinely participate and be heard. And so, as the next section shows, there is still a long way to go in order to refine the new Moroccan vision of water ethics and governance, and to make it work.

### 3.2.2 Alternative environmental governance or 'Watering' the 'Printemps Maghrébin'

The Commission Mondiale d'Ethique des connaissances Scientifiques et des Technologies, COMEST (the World Commission on Ethics, Scientific knowledge and technology) was still considering the debate on governance (in particular water governance) as relatively new, in 2004. It explained that "In general, governance is defined by the ways in which traditions and institutions allow to balance power in the running of a country. Water governance", it stressed, "deals with levels of governance where reality takes over theory. Good governance means that a genuine dialogue takes place. It allows people to define or re-define good shared principles, rights and responsibilities in view of improving the co-ordination of all involved parties, and stimulating development" (COMEST, 2004, p.8). In Morocco, a lot of shortcomings existing in the legal system as well as problems related to the lack of official recognition of certain rights, will have to be addressed if new modes of environmental governance are to really exist. Problems related to the 1995 water law were, for instance, illustrated by Boukhima (2009) who explained that the unrealistic financial conditions set by the law (payment of high fees to get the permission to drill a well, notably) had led to all sorts of illegal, de-regulated and ecologically destructive digging of wells by Syrian enterprises in the area of Souss-Massa-Darâa where the annual water deficit had already reached 233 million cubic meters. Similarly, economic and financial options taken by the Moroccan government in favour of the 'gestion déléguée' (private-public partnership) has been highly criticised and has been the object of numerous demonstrations. The right to accessing water is being jeopardised by current practices in favour of privatization and water pricing, and Non Governmental Organisations such as the ACME have been expressing their dissatisfaction and communicating the views of the Moroccan population, especially its wish to make water management more communal, since 'water is a common good that should not be privatized in any way, as well as the need to include the right to access water in a new constitution.

Moroccan researcher Tazi-Sadeq spent relentless efforts defending the human right to access water and sanitation services, and has done so in an official context, from a UNESCO office in Rabat. As she put it, "The right to water emerges as a concept around which changes and reconciliations can crystallise. It is necessary to reconcile economy and ecology over water. But this vital resource calls for other reconciliations. It makes it necessary to remedy different inequalities, to create an international legal and institutional framework followed at the level of states – first guarantors of effective access to water – and establish links between local and global action. Each of these changes represents an argument in favour of the promotion and proclamation of the right to water. The right to safe water would make it possible to ensure access to water without discrimination, in a sustainable and enduring manner and at a cost that is socially and economically acceptable; to avoid its becoming a threat to the environment, to aquatic systems, to health, to peace and security; to determine responsibilities; to put in place an effective governance and define its operation modes at the international, national and local levels; to mobilise necessary resources, coordinate partnerships and organise cooperation and solidarity" (Tazi Sadeq, 2005, p.13-15).

Other Moroccan stakeholders, such as numerous NGOs and, in particular, ACME-Maroc (Association for the world contract on water), are functioning in more participatory ways and communicating equally important messages, if more practically demanding, when it comes to political and institutional changes. Thus, the ACME for instance demanded that a public enquiry should be conducted - by the Parliament and the legal profession – to determine in which conditions, and in exchange of what, the decision to delegate the management and distribution of water, sanitation and electricity to private enterprises had been taken. It also demanded the re-opening of enquiries from anti-corruption instances because it suspected that the creation of delegated water management contracts had been corrupted and illegal in their applications. As a very active and militant association (NGO), it somehow characterises what many other NGOs are doing in Morocco – creating an alternative system of governance and expression by the people, calling for more justice and participative processes, demanding more recognition. The ACME approached issues of

---

Considering that:
a.  the constitution of a democratic State must take citizen's fundamental rights into account
b.  the right to life is the most fundamental of human rights
c.  the right to life depends on access to water
d.  water is part of nature, essential to life
e.  water must be considered as a common good, shared by the national community
f.  Morocco has adopted in 1995 a Water Law, considering water as a common good, and because this Law must be promoted to a higher level in our legal system (Dahir n° 1-95-154 du 18 rabii I 1416 -16th of august 1995)
g.  the adoption of the UN resolution 64/292 on the 28th of July 2010 that declares that the right to access water and sanitation is a human right – resolution which Morocco officially signed
h.  the adoption of the UN resolution A/HRC/15/L.14 of the Council of Human rights on the 24th of September 2010, re-asserting the right of humans to have access to water and sanitation
i.  certain States have already included the right to water and sanitation in their constitutions – for instance Bolivia, Venezuela, Uruguay, Nigeria..
j.  ACME-Morocco demands that the right to access water and sanitation should be included in the Moroccan constitution as well as the notion of water as 'common good', property of the whole national community and protected by it and for it, with a priority given to meeting the water domestic needs.

Box 1. ACME's demand to include water rights in the Moroccan constitution

water management from an educational angle, considering that environmental awareness and communication with communities will be needed if these are to take part in the implementation of sustainable development principles. In doing so, it showed its alignment with international initiatives such as the International decade of education for sustainable development 2005-2014. It also organized projects with women in rural communities. More recently, it also officially requested the inclusion of the right to access water and sanitation in a new constitution (Box 1).

The introduction of new 'voices' in the water decision-making process is both encouraged (through participatory principles concepts advocated in the new environmental charter) and feared by authorities used to hold the reins and relatively unfamiliar with democratic and human-scale development practices they are wishing to bring back into place. Morocco is currently experiencing, through its prolonged 'spring', a change in governance which, for the first time, also includes environmental considerations and re-link people to their land (and their water). This is a true 'revolution' in an 'ecological economics' sense of the term which, although it is only the beginning, could provide fascinating alternative modes of water governance – provided that the authorities dare listen to the various successful initiatives currently being undertaken to prove that modern and traditional can be happily reconciled in order to re-understand which practices work well for the Moroccan citizen and for the country. Such initiatives include efforts by the architect Aziza Chaouni (Aga Khan prize of architecture), who is working on the re-introduction of sustainable water management principles in the rehabilitation of the medina of Fes and people working on integrated forest and water management, or again efforts by numerous environmental NGOs to educate rural and urban populations in view of re-energizing their wish to value the natural resources they depend on and they used to know how to protect.

## 4. Conclusion

Through a reflection on the evolution towards a new water ethics in Morocco, this chapter has attempted to explore the practical ecological, technical, and political implications of trying to put into practice concepts such as 'Integrated, sustainable, water management' for a developing, Muslim country. I started by describing the physical constraints this North African country has to deal with, its aridity and the irregularity in precipitation that make finding appropriate and locally adapted water policies a – difficult – necessity. I then gave a few examples of how traditional water management systems used to (and still do, in some regions) deal with water shortages and potential water conflicts, both in rural and in urban environments. I then explained how the French Protectorate, followed by the independence of the country, provided a new uneasy framework (of land tenure and water prioritization) that seemed to both go 'against the tide' (in terms of social structures and geographical specificities) but also open the door to 'modernisation' and economic development, a realm that the newly independent country was keen to embrace. Institutional reforms, new water laws and the creation of new stakeholders (water users associations, etc.) constituted a set of initiatives that intertwined with time and resulted in an emerging and growing need and wish to function within a new developmental and more environmentally friendly paradigm. This new paradigm includes notions such as participation, democratisation, decentralisation, integrated water management (Box 2), environmental charters and laws, ... Through its new 'green approach', the king of Morocco is both keen to be internationally perceived as respectful of 'green initiatives' undertaken worldwide – Morocco took part in the World Environmental Summit in Rio in 1992 – and to show his people that he is hearing its requests and needs (numerous demonstrations followed the famous 20th of February 2011).

---

**Principle 1:** Fresh water is a finite and vulnerable resource, essential to sustain life, development and the environment.
**Principle 2:** Water management should be based on a participatory approach, involving users and policy-makers at all levels.
**Principle 3:** Women play a central part in the provision, management and safeguarding of water. In order to ensure full and effective participation of women at all levels of decision making, account should be taken of approaches that public agencies use to assign social, economic and cultural functions to men and women.
**Principle 4:** Water is a public good and has a social and economic value in all its competing uses.
Principle 5: Integrated water resources management is based on the equitable and efficient management and sustainable use of water. The real challenge with IWM is to find ways of integrating various policy tools in a socially, politically, economically and ethically acceptable way.

---

Box 2. Dublin principles presented at Rio and from which the notion of IWM is based

But the 'Printemps Maghrébin', in Morocco, will certainly experience a few seasons. For if the notion of development is being currently challenged, economic pressures are still high and often influence the choice of water technologies and policies that are not yet appropriately participatory nor ecologically sustainable. In order for water management in Morocco to become more humanly and ecologically sustainable, a stronger respect for and re-visit of traditional practices as well as a thorough exploration of the following definition of sustainable development will be needed. As Allan explains, (2002, in Turton and Henwood (eds), p.25) "Sustainable water policies are not achieved through the adoption of sound environmental principles alone. Nor are they achieved by efficient water use based on principles of economic efficiency. Sustainable water use is achieved in the political arena. National hydropolitics is a mediating discourse. The voices of society, the economy and the environment impose their often conflicting priorities and demands on the national water resource". Similarly, a stronger confidence in the cultural potential of the country's environmental practices could help in re-defining the type of 'development' that Morocco is keen to pursue. As UNESCO reports on 'creative cultural diversity in the world' put it, "development efforts often fail because the importance of the human factor – that complex web of relationships, beliefs, values and motivations which lie at the very heart of a culture – is being underestimated in many development projects. (…) Development cannot be seen as a single, uniform, linear path, for this would eliminate cultural diversity and experimentation, and dangerously limit humankind's creative capacities in the face of a treasured past and an unpredictable future" (Perez de Cuellar, 1996, p.7).

*To be developed is not to have more, but to be more* **Ghandi**

## 5. References

ACME   http:// www.acme-eau.org/ACMEMaroc/index.php
African proverbs from
        http://www.special-dictionary.com/proverbs/keywords/water/23.htm

Agoumi, A. (2005). La vulnérabilité hydrique du Maroc face aux changements climatiques. La nécessité de stratégies d'adaptation. Objectifterre Special Changements Climatiques (Novembre 2005), pp.36-37, Agence intergouvernementale de la francophonie, Paris, France

Ait Kadi, M. (1998). Water sector development through effective policies, institutions and investment: a country perspective, Paper for workshop 3, international Conference on Water and Sustainable Development, Paris, March 1998

Allain-El Mansuri, B. (2005). La délégation au secteur privé de la gestion de l'eau potable au Maroc : le cas de Rabat-Salé. P. 163-189. In De Miras, C. (ed) *Intégration à la ville et services urbains au Maroc*. IRD, Paris.

Allan, T. (2002). Water resources in semi-arid regions: Real deficits and economically invisible and politically silent solutions, In *Hydropolitics in the developing world*, A. Turton and R. Henwood (Eds), pp. 23-36, African Water Issues Research unit, ISBN 0-620-29519-8, Pretoria, South Africa

Bennis, A: and H: Tazi-Sadeq (1998). Population and Irrigation Water Management: General Data and Case Studies, in de Sherbinin, A. and V. Dompka (eds) Water and population dynamics. Case studies and policy implications. IUCN, AAAS. http://www.aaas.org/international/ehn/waterpop/morroc.htm

Boukhima, A. (2009). Irrigation, les pilleurs d'eau, In *Focus, EcoPlus* (December), p.26, 27

COMEST (Commission mondiale d'Ethique des Connaissances Scientifiques et des Technologies) (2004). L'eau et la gouvernance : Meilleures pratiques éthiques. UNESCO, Paris, France. Available from www.unesco.org/shs/ethics

Crouch, D.P. and Johnson, J.G. (2001). *Traditions in Architecture*, Oxford University Press, ISBN -10: 9780195088915 Oxford, United Kingdom

De Miras, C. and Le Tellier, J. (2005). Le modèle Marocain d'accés à l'eau potable et à l'assainissement. Casablanca et Tanger-Tétouan; La gestion déléguée, entre régulation sociale et marchandisation, In de Miras, C. (Ed) Intégration à la ville et services urbains au Maroc. Institut de Recherche pour le Développement, Paris, France pp.219-254

Doukkali, M.R. (2005). Water institutional reforms in Morocco. Water Policy, Vol.7, No.1, pp.71-88, ISSN 1366-7017

Et Tobi, M. (2003). Histoire d'eau. Pour une stratégie de gestion de secteurs naturellement liés: les ressources hydriques et les ressources forestières, in *L'Opinion* (janvier), p.6,7.

Faruqui, N.I.; Biswas, A.K. & Bino, M.J. (2001). *Water Management in Islam*, International Development Research Centre, ISBN 0-88936-924-0, Ottawa, Canada

Global Water Partnership http://www.gwp.org/The-Challenge/What-is-IWRM/Dublin-Rio-Principles/

Haouès-Jouve, S. (2005). L'accés au grand nombre aux services urbains essentiels. L'experience casablancaise dans le long terme, entre ruptures et continuités. In de Miras, C. (Ed) Intégration à la ville et services urbains au Maroc. Institut de Recherche pour le Développement, Paris, France pp. 191-217

Jellali, M. and Geannah, M. (1998). La gestion décentralisée de l'eau au Maroc. Situation actuelle et perspectives. *Proceedings of the International Conference 'Eau et Développement durable', Paris, 19-21 March 1998*

Lightfoot, D.R. (1996). Moroccan ketarras: traditional irrigation and progressive desiccation. *Geoforum*, Vol.27, No. 2, (May 1996), pp.261-273, ISSN 0016-7185

Mahfoud, L. (2011). Stratégie de l'eau: anticiper la pénurie. Pertinences. Maroc Vert An III, Numero 8 (May), pp. 52-54, ISSN 1114-8772

Mernissi, F. (1997). *Les Ait Débrouille du Haut Atlas*, Le Fénnec, Casablanca, Maroc.

Michell, G. (1995). *Architecture of the Islamic World. Its history and social meaning*, ISBN 9780500278475 Thames and Hudson, London, United Kingdom.

Pascon, P. (1978). De l'eau du ciel à l'eau d'Etat . Psychosociologie de l'irrigation. HTE , no. 28, sept, p.3-10.

Pérennès, J. (1993). *L'eau et les homes au Maghreb. Contribution à une politique de l'eau en Méditerranée*, Karthala, ISBN 2-86537-357-6, Paris, France

Perez de Cuellar, J. (1996). *Our creative diversity, Report of the World Commission on Culture and Development*. UNESCO, Paris, France.. Des airs, des eaux, des lieux : Fes, capitale thermale. APS, Tours, France.

Secret, E. (1990). Les sept printemps de Fes

Serrhini, F. (2003). L'eau à Fès, symbole d'une civilisation urbaine. Revue H.T.E., No.26 (Juin 2003), Fes, Maroc

Sibley, M. (2006). The historic hammams of Damascus and Fez : lessons of sustainability and future developments, 23rd Conference on Passive and Low Energy Architecture, Geneva, Switzerland, 6-8 September 2006

Slimani, L. (2010) Environnement une place au soleil, *Jeune Afrique*, No2570, pp. 58-60

Swearingen, W. D. (1987). Moroccan mirages : agrarian dreams and deceptions, 1912-1980, Princeton, Princeton University Press, USA.

Tazi Sadeq, H. (2005). For an effective right to water. The Green Cross OPTIMIST, Winter 2005, p.13-15, Available from http://www.greencrossinternationale.net

Turton, A. (2002). Hydropolitics: The concept and its limitations, In *Hydropolitics in the developing world*, A. Turton and R. Henwood (Eds), pp. 13-22, African Water Issues Research unit, ISBN 0-620-29519-8, Pretoria, South Africa

UNESCO (2005). *United Nations Decade of Education for Sustainable Development (2005-2014): International Implementation Scheme*. Technical paper ED/DESD/2005/PI/01. UNESCO Education Sector, Paris.

Van Dijck, S.J.E. et al. (2006). Desertification in Northern Morocco due to the effects of climate change on groundwater recharge, In *Desertification in the Mediterranean region. A Security Issue*, W.G. Kepner, J.L. Rubio, D.A. Mouat and F. Pedrazzani (Eds), chapter 26. ISBN 9781402037603, Springer, Netherlands.

Wines, J. (2000). *Green architecture, The Art of Architecture in the Age of Ecology* Taschen, London

Wolf, A. (2000). Indigenous approaches to water conflict resolution and implications for international waters. *International Negotiations, a Journal of Theory and Practice*, Vol.5, No.2, (December 2000), pp.357-373 Available at http://www.transboundarywaters.orst.edu/publications/indigenous/

# 10

# Public Private Partnerships in the Privatization of Water Service Delivery in Kenya

Okeyo Joseph Obosi
*Department of Political Science & Public Administration*
*University of Nairobi,*
*Kenya*

## 1. Introduction

The paper is divided into four sections. Section I presents background information on the water supply situation including the hydrological situation. Section II presents the privatization of the water supply and the institutional management of the privatisation process globally. International experiences on privatization and water sector reforms including the public-private partnership as a strategy is also discussed. Section III discusses the Water Management, reforms, and governance including monitoring mechanisms in Kenya. Section IV provides a conclusion as it summarises the salient features and challenges of the processes discussed in the chapter.

### 1.1 Hydrological situation

Kenya is mainly an agricultural country with an expanding economy whose basic element for development is water. Water is required for agricultural, commercial, and domestic use Mogaka et al, (2003). However, the climate in Kenya varies by region and season to an extent that whereas some parts of the country would be experiencing floods, others will hardly receive a drop in a year. This makes accessibility to clean water unreliable in Kenya even to the areas where rainfall is abundant.

Kenya's surface water resources are distributed within five drainage basins, namely, the Lake Basin, Rift Valley, Tana, Athi, and Ewaso Ngiro as shown in the table below.

| Basin | Surface Water | % | Groundwater | % | Total | % |
|---|---|---|---|---|---|---|
| Lake Basin | 11,993 | 59.2 | 539 | 18.7 | 12532 | 54.1 |
| Rift Valley | 211 | 1 | 586 | 20.3 | 797 | 3.4 |
| Athi River | 582 | 2.9 | 405 | 14 | 987 | 4.3 |
| Tana River | 6,789 | 33.5 | 685 | 23.8 | 7474 | 32.3 |
| Ewaso Ng'iro | 674 | 3.3 | 663 | 23 | 1337 | 5.8 |
| Totals | 20249 | 100 | 2878 | 100 | 23127 | 100 |
| | Safe yield in '000 cubic metres per day | | | | | |

Source: Mogaka et al., (2006). Climate variability and Water Resources Degradation in Kenya: p.9.

Table 1. Showing safe yield from water resources by Major Drainage Basins in Kenya

Each of the seven the Water service Boards (WSBs) and their corresponding Water Service Providers (WSPs) in Kenya fall within respective drainage Basins for management and development of water resources and services purposes. For Example Kisumu, Homa Bay, Kisii, Kericho towns served by Lake Victoria South Water Service Board, fall within the Lake Basin. Nakuru and Eldoret towns fall within the Rift Valley Water Service Board and Rift Valley Drainage Basin. Nyeri town fall within Tana drainage Basin and Tana Water Service Board.

The precipitation and subsequent run off across parts of the country is exceptionally variable and unpredictable, hence endemic drought in the country. This is also responsible for pronounced differences in average annual rainfall, evapo-transpiration, and groundwater, hence high variability within the same season, between different seasons, and over several years. The country has a mean annual rainfall of about 500 mm, which varies from between250mm in the Arid and Semi-Arid Areas (ASALs) to 2,000 mm in the high mountain eco-systems. About 66% of the country receives less than 500 mm of rainfall as shown in the table below.

| Mean Annual Rainfall (mm) | Land Area (sq.km) | % of Land Total |
|---|---|---|
| >1000 | 64070 | 11.2 |
| 800-1000 | 32960 | 5.8 |
| 700-800 | 24260 | 4.3 |
| 500-700 | 73140 | 12.8 |
| 300-500 | 270410 | 47.4 |
| <300 | 105730 | 18.4 |
| Total | 570570 | 100 |

Source: Mogaka et al(2006). Climate variability and Water Resources Degradation in Kenya: p. 28.

Table 2. Showing distribution of Rainfall in Kenya

## 1.2 Water supply system in Kenya

According to Onjala, 2007, accessibility to water in Kenya has been compounded by three legacies: First, the natural legacy where Kenya has extremely limited per capita (696 m³/person per year) endowment of fresh water resources and high hydrological variability, both temporal and spatial. The amount of water available for utilization in any one year (among other factors) is dependent on the rate of run-off, the aridity of the catchment area and the methods of interception at various points of hydrological cycle. The second legacy is management one characterized by a rapidly growing demand for water from various sectoral uses and, on the other hand, a diminution of natural storage capacity(wetlands, catchment and aquifer recharge areas) and lack of development for artificial storage capacity(dams and reservoirs) to meet the demand as shown in the table below.

Finally, the country has a colonial legacy where national boundaries were drawn irrespective of geographic and social realities. Consequently, Kenya shares over half of its rivers, lakes, and aquifers with neighbouring nations.

| Demand by Category | 1990 | 2000 | 2010 |
|---|---|---|---|
| Domestic Water | | | |
| Urban | 573 | 1169 | 1906 |
| Rural | 532 | 749 | 1162 |
| Industrial | 219 | 378 | 494 |
| Irrigation | 3965 | 7810 | 11655 |
| Livestock | 326 | 427 | 621 |
| Inland fisheries | 44 | 61 | 78 |
| Wildlife | 21 | 21 | 21 |
| Total/day | 5680 | 10615 | 15937 |
| Total cc/year(millions) | 2073 | 3874 | 5817 |

Source: Mogaka et al. (2006). Climate variability and Water Resources Degradation in Kenya: p.16

Table 3. Showing estimated water demand, 1990-2010 (thousands of cubic metres /day)

In the rural areas, a large number of homesteads are still far from water points, especially those in the low potential areas where rivers are mainly seasonal. On the other hand, ground water resources are either limited or underdeveloped. Although ample water resources may exist, the patterns of use and accessibility may be a serious problem (Onjala, 2007). The level of coverage goes down as low as 20% during the dry seasons while seasonal water sources often dry up, making distances to water long and often exceeding 5 kilometres (Kenya. 1996). The shift in time taken to source water is occasioned by two reasons:

First is the low service coverage and inability of the local water authorities to sustain supplies of piped water to all segments of town hence consumers switching to alternative sources which are often contaminated locations. Secondly, in some of the towns, there is increased concentration of piped water uses to inferior alternative sources such as bore-holes, wells, springs and nearby rivers even during wet seasons due to unreliability. Of great concern is Kenya's vulnerability to hydrological variability. According to Onjala,(2007), Kenya lacks the buffering capacity to deal with the shocks of either too little or too much rainfall to the extent that most parts of Kenya would experience insufficient access to water a few months after it has experienced floods.

Consequently Kenya's per capita total water storage for all uses at 60 cubic metres compared to South Africa's 746, Thailand's 1287, China's 2486 (Mogaka et al 2003) is too low to meet the increasing demand and provide a buffer against floods and droughts. It is therefore, a priority to enhance water accessibility through increased storage. According to a world Bank Survey in 2001, there were, in Kenya, approximately 742,000 water connections in about 680 piped systems,350 community run water schemes, 1800 water supplies out of which about 1000 were public operated schemes, 1782 small dams, 669 water pans, 9000 boreholes. The water accessibility is more rural than an urban problem. It is not surprising that most of these alternative sources of water are majorly in the rural areas where piped water is negligible. They are manifested more in small scale water service providers.

Whereas some are in somewhat agency relationship with main water service providers, some operate independently of the existing national water regulatory framework. As a way of enhancing regulatory controls, the government of Kenya has instituted water sector reforms which saw WRMA and WSRB responsible for both governance and provision of water services, respectively.

The Government of Kenya had from independence (1963) to the year 2003, undertaken the responsibility to supply Water to its citizens. This was done through the Ministry of Water development. The government did this in two ways: First, through Local Authorities, especially in urban centres. Here, the government would sell/deliver water in bulk to the local Authorities who would in turn sell water to its customers. Each Local Authority in Kenya had distinct Department of Water and Sewerage. Secondly, where the government felt there was no viable water service provider (Local Authority), it would do so directly through National Water and Conservation & Pipeline Corporation (NWC&PC). This was as provided by an act in Kenya's law chapter 372. However, from the year 2003, the government implemented the new Water Act of 2002 (Water Sector Reform Secretariat, 2005). The new Act was enacted to repeal the erstwhile Water Act chapter 372 of the laws of Kenya that had been in operation since 1962 (Republic of Kenya, 1972). This was done in order to usher in reforms in the water sector.

The reforms are entrenched in the new Act in two main aspects; the management of water resources and the management of water services. The latter is considered under Part IV of the Act covering water supply and sewerage (O. A. K'Akumu, 2005).The new Act established various institutions that were to manage both water resources and water services provision.

Alternative/Independent water suppliers supplement the supplies of water to urban dwellers that get unsatisfactory or no service from the conventional piped water network. A number of alternative water suppliers have emerged to address these shortfalls. There are two types of alternative private sector participation in the water sector that have emerged in the form of water kiosks and private water vendors.

Water kiosks are a form of public-private partnership whereby the government provides water to the kiosk where it is re-sold to the local customers.

The 'Private' component can be a private company but also a group of citizens united in a Community-based organization (CBO) and either or not supported by one or more NGOs(O.A. K'Akumu 2007). Private water vendors – also known as the "other" private sector (Solo 1999) – are "informal" and/or small-scale operators who provide water (and sanitation) services in mostly low- and middle-income neighborhoods. They operate outside the government influence and may even be illegal. These types of WSPs encompass a wide range of suppliers including drinking water companies that supply water in disposable bottles, which are sold in supermarkets, shops, kiosks and even by hawkers. There are also those who supply drinking water to offices using bigger and returnable containers. Water tankers supply homes during periods of serious water shortages although at higher rates, and finally, well-owners and cart-pushers. Cart-pushers provide tap water for those who are not served by taps, while well-owners provide cheaper water than the official supplier, catering for those unable to afford official rates, while also providing water for all during shortages.

## 2. The privatisation of the water supply and the institutional management of the process

The section presents various issues on water privatisation in particular and water sector reforms in general as expressed in international literature. It presents cases of small scale water service providers as an alternative water service provision mechanism whose success could either be impeded or enhanced by the methods of the privatization adopted by a particular country.

### 2.1 The independent small scale water service providers

The Independent Small Scale Water Service providers include: Small companies, cooperatives, Water Kiosks, Cart Vendors, Water Tankers, Individual bore hole owners, Community Water projects, and NGO funded projects. They are independent to the extent that some are self-employed entrepreneurs or local artisans.The independent water service providers are therefore expected to register with WRMA through a lengthy process before its application to WSRB through a WSB is granted. It has to meet the respective conditions set by the two institutions, respectively, for abstraction and use of water to provide water services. Most work without formal recognition from local authorities, and are neither sub-contracted by the large water distribution companies nor in any agreement with the public sector providers. According to ADB, 2002, there are three main types of SCWSP, namely; Partners of water utilities; Vendors and Resellers; and pioneers of piped water networks include water kiosks and local standpipes. This category buy water from the large water companies and resell to the end users at a profit. Vendors and Resellers include mobile carters, truckers and household resellers. They provide water to where the water utilities are unable to serve. The category of pioneers of piped water networks provide piped water, often ground water to communities which have not accessed the piped water from the utility companies.

In Kenya, WSRB regulations clause 5.3 stipulates that the Licensee shall undertake to ensure that all small scale water service providers operating within an area of a WSP are duly registered with a Licensee and are supervised by the main Water Service Provider through a Small Scale Service Provider Agreement in order to provide a safe, efficient and affordable service to the consumers. The main Water Service Provider shall charge a reasonable administrative fee for the supervisory roles rendered on behalf of the Licensee. The Licensee shall undertake to pursue a clustering strategy on all its publicly funded boreholes in its area of operation in order to create small scale water service providers with appropriate supervisory arrangement with due regard to creation of service providers capable of financial sustainability, efficiency and growth.

In Eldoret town (Rift Valley Water Service Board) of Kenya, the water kiosks have been crucial for the supply of water to low income residents to the extent that whenever there was a disruption, residents of 6 out of 15 low income residential areas, namely Munyaka, Langas, Kamukunji, Huruma, Bondeni and Pioneer are greatly affected. The same is true of Nyeri town (Tana Water Service Board), Nakuru Town (Rift Valley Water Service Board) and Nairobi city (Athi Water Service Board) in Kenya. According to Asingo (2007), residents from low income areas, find the charges for Eldoret Water Company and Sanitation (ELDOWAS) so expensive that they restrict it to drinking and cooking only while relying on water from bore-holes for other needs like washing. In a study of India, Mackenzie Ray (2010), established that water vendors play an intermediary role in parts of the cities which are underserved such as Rajkot, Ahmadabad, and Chennai whereby they are either re-selling water from municipal water standpipes or obtaining water from groundwater sources and transporting it by tanker to the slum areas where residents purchase it. The role played by small scale water providers cannot be underestimated despite accusations of exploiting the poor. Solo (1999) argues that small scale water and sanitation providers play a big role in extending access of key services, especially in Latin America. Kjellen (2000) argues that given the inadequate state of water infrastructure in Dar es Salaam, the small scale water providers complement the water distributive system to the City's distributive system and do not provide poorer quality of water than the City does to its official customers. Similar observations have been made by Njiru (2006) on the role of small scale water providers in sub-Saharan Africa. About 20-45% of residents in Ho chi Minh City, Cebu and Manila (Philippines), and Jakarta depend on water supplies from SCWSPs.

An ADB funded survey of six Asian cities: Cebu, Kathmandu, Jakarta, Ho Chi Minh City, Manila, Shanghai, and Dhaka, revealed that because of the failure of the conventional water utilities to serve many low income households, a large number of the population rely on alternative water supplies which are run by either community groups or local entrepreneurs (ADB, 2002). With the existing tariff and management structures, the large water companies which are usually favoured by the government, are unable to supply a large population with water, hence a large population from low income areas turn to either illegal connections or other alternative water suppliers. The study established that in the cities surveyed, a large population remain unconnected from the municipal/city connections because of the following reasons:

- The connection fees are so high and lump sum payments upfront usually exclude the poor.
- The amount of water supply is usually insufficient, and the vulnerable poor are the first to be left out.
- The cost of extending water to low income areas is regarded uneconomical.
- Most of the low income dwellers do not own their respective lands legally hence impossible to be connected to the water systems.

Where the services are extended to the low income areas, the large water companies do not know how to do it since: First, the services including technical requirements are not tailored to the demand of the low income households thus most poor are kept out of connection. Secondly, the payment systems precludes the poor with irregular income and finally, employees of large companies do not communicate well with the poor, hence the risk of being overcharged or penalized in case of improper billing.

## 2.2 Water provision services sector reforms and interventions in Kenya

The Water Act established Water Services Regulatory Board (WSRB), and seven Regional Water Services Boards (WSBs), namely, Coast, Nairobi, Rift Valley, Central, Northern, Lake

Victoria South, and Lake Victoria North. Each of the seven WSBs were as per the Act incorporated as Public enterprises, and were each expected to apply for Water Service Provision from the WSRB. Once granted, the licence for water provision, the licence is expected to be leased to the PLC so incorporated which will act as a Water Service Provider (WSP) . Even NWC&PC was turned into an "interim" WSP.

The Act states that "the water services board shall by force of this section be constituted a corporation" (Republic of Kenya, 2002: 982). WSBs remained as asset owners and financiers. Section 53(2) states that a WSB is mandated to "purchase, lease or otherwise acquire on such terms as the Minister may approve, premises, plant, equipment and facilities; and purchase, lease or otherwise acquire land, on such terms as the Minister may approve" (Republic of Kenya, 2002: 983). A WSP on the other hand is generally responsible for operations or control of water assets, although the degree of responsibility may be varied according to the agency agreement between the two bodies.

The Act also provides for the participation of Independent Water Service Providers (WSPs) outside the local authorities' registered public limited companies. These include water service facilities owned or operated by NGOs, CBOs, community self-help groups and other local water undertakers. These are directly registered by WSRB but are supervised by respective WSBs. Section 113 of the Act provides the WSBs legal rights to:

1. Assume overall administrative and legal responsibility for provision of water services that was previously directly under the Central government, that is, the Department of Water Development except the direct operation of facilities that the Act reserves to the WSPs;
2. Assume ownership of water services facilities owned or used by the Central Government (Department of Water Development and its parastatal – NWC&PC);
3. Access water service facilities owned or operated by local government service providers; and
4. Influence the use of water service facilities owned or operated by NGOs, CBOs, community self-help groups and other local water undertakers. (Ministry of Water and Irrigation, 2004).

Kenya has several development partners in the water sector including Swedish International Development Agency (SIDA), Danish International Development Agency (DANIDA), World Bank, German Development Agency (KFW/GTZ), French Agency for Development (AFD), United Nations Children's Fund (UNICEF), Japan International Cooperation Agency (JICA), Department for International Development (DFID), African Development Bank (ADB), Finnish Development Agency (FINNIDA), and the European Union (EU), among others. Currently, International Development agency (IDA) and French Agency for Development (AFD) support commercialisation of water utilities serving mainly urban centres (Nairobi and Mombasa) while the German cooperation (KFW) is focusing on commercialisation of water utilities in medium-sized urban centres. Japan is interested in supporting smaller urban centres and rural areas, Denmark, Finland and Belgium aim to cooperate on rural water supply and the African Development Bank (ADB) is financing projects in urban areas (Kenya 2006b: 193).

According to Owuor et al. (2009),water sector interventions can take the form of local (intra-urban) initiatives, for instance to establish a water kiosk in a low-income neighbourhood with the (financial) assistance of an NGO. But interventions can also target a whole municipality or even a whole region, for instance the rehabilitation and/or improvement of the water (and sanitation) infrastructure. Perhaps the most far-fetching intervention project in urban Kenya is the Lake Victoria Region Water and Sanitation Initiative (LVWATSAN) being implemented by UN-HABITAT in association with the governments of Kenya, Tanzania and Uganda and with financial support from the government of the Netherlands.

The programme, which involves a mix of investments in the rehabilitation of existing infrastructure and capacity building at local level, is designed to assist the people in the Lake Victoria towns to meet water and sanitation related MDGs (UN-HABITAT 2007; 2008). The first phase, which focused on rehabilitation of water supply sources, extending water supplies to the poor areas and constructing sanitation facilities, was designed to have an immediate impact in improving water and sanitation services targeting seven towns of Homa-Bay and Kisii in Kenya, Masaka and Kyotera in Uganda, Bukoba and Muleba in Tanzania, and the border town of Mutukula (UN-HABITAT 2007).

With a clear pro-poor focus, the LVWATSAN programme is intended to generate desirable outcomes with a lasting impact on the lives of the poor. These outcomes include improved access to water, sanitation, solid waste management and drainage services in the project areas; functional and gender focused strategies for sustainable management and monitoring of rehabilitated systems; institutionalised capacity building; and a contribution to the reduction in pollutant loads entering Lake Victoria. It is also hoped that the programme towns will provide a model for national authorities and donors, including international financing institutions, to replicate in other towns within the region (UN-HABITAT 2008).

In a preliminary study tour of five towns in Kenya, namely Eldoret, Kisumu, Homa Bay, Kisii and Nakuru to assess the extent of interventions in the water and services provision, Owuor et al (2009) established that Eldoret municipality does not have any NGO, CBO or agency actively involved in water interventions at the local level. However, ELDOWAS may once-in-a-while depend on a Dutch NGO, SNV, for informed research. In 2007, for example, SNV conducted a survey of water vendors in the town and the results shared with ELDOWAS. Kisumu municipality has a number of NGOs working in various water sectors. The active NGOs in water and sanitation include Sustainable Aid in Africa International (SANA), Africa Now, World Vision and CARE Kenya. Wandiege Water Community Project is a water service provider registered by LVSWSB just like KIWASCO. Sustainable Aid in Africa (SANA) which started as a Dutch-Kenya bilateral programme (1982-2000) in rural water and sanitation in the then South Nyanza District of Nyanza Province deals with issues related to domestic water supply and targets the un-served urban and peri-urban informal settlements and the poor in general, besides dealing with environmental sanitation. The main source of water in Nakuru municipality is boreholes. The African Development Bank (ADB) has funded the drilling of 17 borehole: 5 in Baharini, 3 at Nairobi Road and 8 in Kabatini. The Lake Victoria Region Water and Sanitation Initiative in Homa Bay (LVWATSAN-Homa Bay) and a similar in Kisii has worked closely with the respective municipalities and Multi-Stakeholder Fora and initiated a number of short- and long-term water and sanitation interventions in the in LVWSB, especially under the jurisdictions of SNWSCO and GWASCO water service providers. It is clear that the stated interventions have in a way improved the services in terms of accessibility and quality. It is however, not clear why Eldoret which has no NGOs supporting water service provision, has less acute water problems than Kisumu which has several NGOs offering various services in the water provision sector. In all the five towns, the interventions have supported the establishment of water kiosks to be run in collaboration with various interest groups and/or local Community Based organizations.

The most interactive forum orchestrated through the interventions is best exemplified by Multi Stakeholder Forums (MSF) established by LVWATSAN in Homa Bay and Kisii towns. The MSFs ensure that the interventions under the LVWATSAN programme are developed and implemented in a manner that is informed by and responds to the needs of the local stakeholders. Through regular communication and feedback, the forums also ensure that stakeholders understand and support the achievement of goals and objectives of the programme. MSF worked together with the municipal councils of Homa-Bay and Kisii Municipalities and have been identified as a pro-poor governance mechanisms intended to include and involve poor people and all stakeholders in decision making on matters concerning them. It is a vehicle for a collective participatory approach to problem solving. These forums bring together all possible stakeholders, such as:

- representatives of local authorities,
- water and sanitation service providers,
- NGOs, CBOs and Faith Based Organisation (FBOs)
- private sector
- water vendor associations
- media and
- poor women and men, the elderly, youth, orphans and other vulnerable groups, among others.

The multi-stakeholder forums facilitate the active participation of a broad range of stakeholders at town level, in the design and implementation of the programme interventions (Owuor, et al. 2009). In a way, the MSFs have become a form of consumers' regulatory mechanism on the type and quality of services they deserve.

The interventions have in a way made various water service providers to establish some pro poor programmes in their areas of jurisdictions. KIWASCO is implementing a pioneer 'delegated management model' in Nyalenda – a densely populated slum area in Kisumu. This is a model where KIWASCO sells water in bulk and at a subsidized tariff to a private operator in the community, who in turn manages its distribution and other aspects. The selected operator acts as an agent of KIWASCO in terms of connecting customers, operating the sub-network, collecting revenue and fixing leaks. It is not only a performance- based contract but also a profit-making enterprise towards access to clean and affordable water. They have their own independent management, network, operations and tariffs. ELDOWAS has established ten (10) water kiosks provided but given to interest groups or individuals to operate. The UN-HABITAT's LVWATSAN programme is actively involved in both short-term and long-term interventions in water and sanitation in the municipality. This is being done in collaboration with the Municipal Councils of Homa Bay and Kisii, SNWASCO, GWASCO and the Multi-Stakeholder Forum (MSF-Homa Bay and Kisii).

Already, the LVWATSAN programme has constructed two water kiosks in Shauri Yako of Homa Bay estate to increase access to clean water in low-income areas. These two water kiosks have been left to MSF-Homa Bay to determine which of their group members to run them. NAWASSCO has constructed 7 water kiosks in Nakuru to serve the low-income estates of the municipality. Four of these kiosks are located in Rhonda and Kaptembwa but only 3 are operational. These water kiosks are managed by a CBO known as NAROKA (Owuor et al, 2009).

## 2.3 Privatization and public private partnerships in water sector

Privatization means the transfer of public sector assets, control and financing of enterprises to the private sector. Private sector participation and/or privatization of water supply often imply commercialization Bakker (2003b). Rakodi (2000) views commercialization as the creation of quasi market conditions in public service delivery through increased cost recovery and introduction of performance measurement systems. According to Alila et al (2007), the interaction between the government and business is generally an ongoing process and forms the basis for their interaction in nature and scope. He further stated that the key parameters shaping the interactions between government and business include market liberalization, privatization, good governance, public goods and services, development capital and policy implementation. The services can be provided of the government's own accord, and /or business owners and others in need can place demands for their provision.

Adapting the classification of Stottman (2000), Onjala (2002) and UN-Habitat (2003, O.A. K'Akumu (2006) identifies ten (10) types of PPP applicable to water enterprises. The range, is a continuum from Public Enterprises in which the asset ownership, management, tariffs regulation are all under statutory control, followed by Public limited company(PLC), Service contract, Management Contract, *Affermage* contract, Lease contract, Concession contract, Built-Operate-Transfer(BOT), Joint Venture, to Divestiture in the extreme end. In Divestiture, other than quality monitoring which is in the hands of the public, all other controls including asset ownership, capital, management, and tariff regulation are under private control.

Although some forms of PPP like contract and lease management, might resemble privatization, they are actually not the same thing. PPP falls in between public enterprises at one end of the continuum and divestiture to the very extreme end. It is divestiture, which for all practical purposes, involve privatization. Traditionally, water services have been provided by the public sector. A water institution is termed public if the ownership is in the public sector and the control is in the public sector. Control is in terms of responsibility for day-to-day management of the utility. Privatization will occur with any introduction of private sector participation in the ownership and/or control of a water service institution. Privatization of water therefore refers to the process and outcome of the introduction of the private sector in the ownership and/or control of water utilities.

## 2.4 Experience of ppp and privatization globally

In a study of privatization of water services in Kenyan Local Authorities, Asingo( 2005), identifies five governance issues that have affected the privatization of water sectors globally:

1.  Reasons for the privatization of water services;
2.  The identification of the service provider, and how the service provision is transferred from public to private providers,

3.    The impact of the water privatization on the poor people,
4.    The concern that the privatization of public utility service delivery tends to shift accountability of service providers to policy makers rather than the service users, particularly where privatization grants service monopoly to a private provider.
5.    The concern for cost recovery for privatized public good services like water.

The justifications for the privatizations wherever they take place have been pegged on the inability or failure of the central and local governments to provide services to the people. This is in most cases though not always attributed to financial factors. For example Nelspruit Local Authority (NLA) in South Africa, entered into public-private partnership to relieve it of the financial burden of upgrading water and sanitation services and ensuring efficient service provision (Asingo, 2005). However, some cities have addressed poor service delivery without necessarily altering ownership and management. For example the city of Bulawayo, Zimbabwe enhanced water service delivery by putting up mechanisms to minimize the use of unaccounted water through conservation (Asingo, 2005).

Different countries have used different methods to transfer service provision from public to private providers and registered different experiences. Nelspruit city, South Africa used open tendering method to identify a private company to manage water services on a concession basis for an initial period of 30 years. The Local Authority was to retain the role to regulate tariffs and set water and sanitation service quality standards according to the national government policy (Cardone and Fonseca, 2003).

In Guinea, water privatization proceeded through a franchise arrangement where the government transferred the ownership of urban water supplies to major cities including Conakry to a state owned national water authority, the *Societe Nationale des Eaux Giunea* (SONEG) in 1989.SONEG in turn invited private companies to bid for a franchise to operate and manage water services in the seventeen urban centres. The bid was won by *Societe de Exploitation des Eaux de Guinea (SEEG)*. SONEG continued to own water assets, undertake new investments, plan the sector, service debts, set tariffs and monitor the activities of SEEG.SEEG in turn was to operate and manage existing supply facilities, bill and collect payments in the 17 urban centres, undertake small scale investments and pay rental fee to SONEG (Bayliss, 2000). It is important to note that Guinea was not faced with water accessibility problems but poor quality water and low coverage of connected water. It had several alternative sources of water such as well water, connection through neighbours meter, and collected rain water. The main challenges included unaccounted for water, low collection rate from the public sector and high price of water (Menard, C and Clarke, G., 2000). In essence, Menard et al 2000, have concluded that despite the challenges and with the availability of water for 24 hours daily, the provision of water services improved under private than it could have been under public ownership.

In Mauritania, the government delegated Water Management in small towns to private providers called *Concessionaires* in 1993. Each *concessionaire* was expected to supply water to a community on a yearly basis for those with diesel powered systems and on a monthly basis for those with solar-powered systems under cost recovery principles where users pay for water consumed. In each case, the concessionaire only recovers maintenance and operation costs as the government meets the capital cost (Cardone and Fonseca, 2003).

Some Governments have also used devolution method to a lower level. For example, in Colombia, water service provision was devolved to local governments in 1994 and the federal government adopted an overall and regulatory role. A regional agency, *Acquontioqua* was set up to own and operate water services in the city council of Marinilla and other municipalities. In 1997, *Acquontioqua* awarded a management contract for the urban centres to a domestic private firm through a transparent process which involved citizen participation. Within 3 years, an additional 3,500 people were connected to the system, unaccounted for water decreased, service level and water quality have been increased, existing infrastructure was upgraded and 99% of the city population was provided with water 24 hours a day (Cardone and Fonseca, 2003). Dagdeviren, H. (2008) in a study of ten commercialization of water services in Zambia, observed that the World Bank sponsored management contract of water services in the copper belt mining towns of Zambia had to be reverted to public utility, Nkana Water Service Company because its performance was no better than those of public companies. The unaffordability rate of water tariff in Lusaka, as measured by the ratio of household monthly expenditure to household income, increased from 40% in 2002 to 60% in 2006 using 3% benchmark (Dagdeviren, H. 2008).

The accessibility to safe water rate also decreased from 73% in 1990 to 53% in 2005 (World Bank, 2006). Most households in peri-urban areas due to the informal nature of most of the settlements, depend on boreholes, communal or public taps built by commercial utilities, NGOs, and donors. The management of the schemes for the water service provision in the informal settlements have taken various forms. Some are managed solely by the community, several are managed by communities in co-operation with public utilities, while others are managed by vendors (Dagdeviren, 2008). Barraque' (2003) argument that the economic and political areas are a product of a country's social governance and that for any policy to be successful, social, economic and political dimensions need to be taken into account. Therefore, if the intended policy is not contextualized within the appropriate pattern of social governance, it is doomed to be rejected. This could explain the rejection of privatization of water supply policies in Cochabamba, Bolivia, and Nespruit, South Africa and successful scheme in City Council of Marinilla Colombia where the citizens directly participated in developing privatization terms and contract between the council and the private water companies. Cochabamba water concession projects in Bolivia were cancelled as a result of: Vested interests, combined with politics, lack of proper communication and street protests (Nickson and Vargas, 2002).

It is worth mentioning that regulatory mechanisms, whether through citizen participation or statutory, is crucial to the outcomes of privatization. It is widely recognized that regulation and regulatory governance are key elements of development-policy thinking in promoting pro-poor market-led development(Kirkpatrick and Parker, 2004).

## 2.5 Experience of ppp and privatization of provision of water services in Kenya

Privatization in Kenya has been carried out in two phases. The first round of privatization was executed on a sectoral basis. It took place mainly in the late 1980s and early 1990s and happened without a comprehensive national policy on privatization. It involved financial corporations and utility corporations like electricity, telecommunications and water. Privatization of the concerned enterprises were guided by privatization policy entrenched through the revision of statutes for the concerned sectors or corporations. Privatization in Kenya began with a divestiture exercise that saw the government sell proportions of its shares in the public enterprises.

The second round of privatization is yet to come following the publication of The Privatization Bill, 2004. The bill defines privatization as

a.   "The transfer of public entity's interests in a state corporation or other corporation" and
b.   "The transfer of the operational control of a state corporation or substantial part of its activities", to a non-public entity (Republic of Kenya, 2004: 55).

It envisages the following benefits to the Kenyan economy:

1.   Improvement of infrastructure and delivery of public services by the involvement of private capital and enterprise,
2.   Reduction of demand for government resources,#
3.   Generation of additional government revenue by receiving compensation for privatization initiatives,
4.   Improvement in regulations of the economy by reducing conflicts between the public sector's regulatory and commercial functions,
5.   Improvement in efficiency of the Kenyan economy by making it more responsive to market forces and
6.   Broadening of the base of ownership in Kenyan economy and the enhancement of capital market development.

The privatisation of water services in Kenya was ushered in by the sectoral reforms through Water Act of 2002 even before the Privatization Bill of 2004 was published. The Government of Kenya established Seven Water Service Boards (WSBs) under one Water Regulatory Board (WSRB).In each of the WSBs, the Water Service Providers registered as Public Limited Companies(PLCs). Each was expected to embrace the commercialization of services principle in the provision of water as a public good to customers at a profit.

The local authorities in Kenya introduced commercialization as a strategy for ensuring sustainable and efficient delivery of water and sanitation services (UNCHS, 1998b).Towards this end, most local authorities have formed or are in the process of forming Public Limited Companies (PLCs) to run on strict commercial lines under 'agency contracts' from the parent local authority. The emphasis by local authorities is towards ensuring that under the framework of commercialization, companies formed to provide water plough back the bulk of their earnings into improving service delivery while allowing local authorities to retain part of the earnings to cover costs such as personnel expense (O. A. K'Akumu and P. O. Appida, 2006).

## 3. Institutional framework for water management and monitoring mechanisms

The section presents the framework as provided by the water Act 2002, and describes it in relation to the envisaged roles as it examines the efficacy of the institutions in performing the perceived roles. It also examines the role of popular participation in the governance of water resources and distribution management.

### 3.1 Institutions for water management in Kenya

### 3.1.1 Institutional framework resulting from water act 2002

Water Resources Management issues are captured under Part III of the Water Act 2002. The governance institutions and instruments established for the water resources management under this part include the Ministry (Minister), Water Resources Management Authority (WRMA), Water Appeal Board (WAB) and the Catchment Area Advisory Committees (CAACs) while the instruments include the National Water Resources Management Strategies (NWRMS), National Monitoring and Information System on Water Resources (NMISWR), Catchment Management Strategy (CMS), permits and appeals. The institutional water management framework in Kenya is as presented in the figure below:

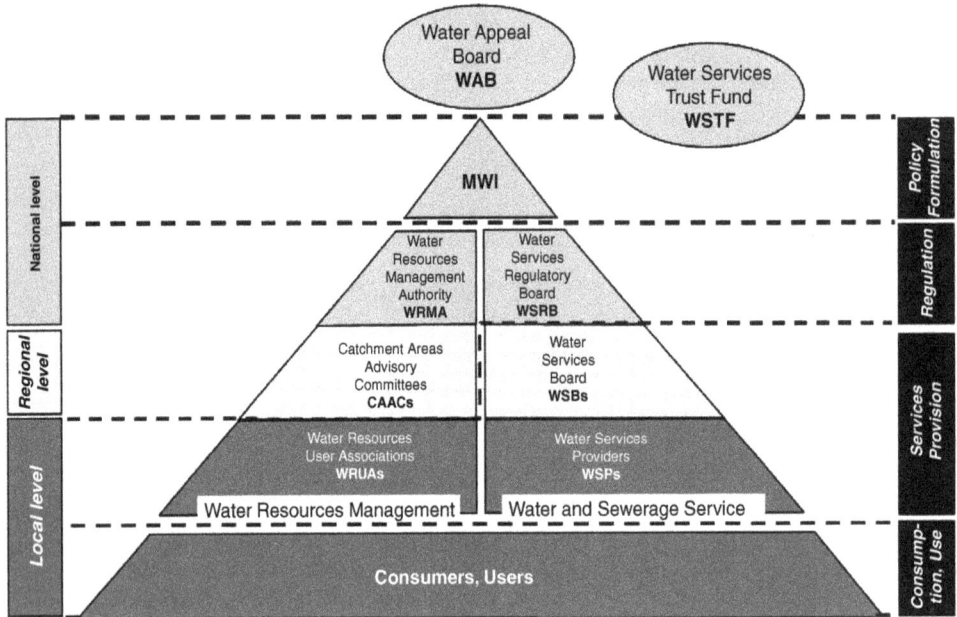

Source: Republic of Kenya (2007)

Fig. 1. Government's schematic representation of the institutional framework resulting from the Water Act of 2002

The body that formulates the strategy is the Water Resources Management Authority, (WRMA). WSRB registers or licences WSPs in each WSB. WRMA has its regional agencies called Catchment Area Advisory Committees (CAACs) which takes care of the designated catchment areas. The CAAC's key function is to advise their respective WRMA regional office on:

- water resource conservation, use and apportionment,
- the grant, adjustment, cancellation or variation of any permit and,
- any other matters pertinent to the proper management of water resources.

It is on this basis that WRMA shall issue or cancel permit for water use by a WSP. Water Appeals Board (WAB) hears appeals from mostly non state actors that have been aggrieved by action of some state actors in water governance and usage. The non-state actors are usually represented at the local level by Water Resources Users' Associations (WRUA's).

## 3.1.2 Description of the water management institutions

The minister causes the formulation of National Water Resources Management (NWRSMS), through public consultation. The strategy is based on the data obtained from the National

monitoring and Information Systems on water Resources (NMISWR). The body that formulates the strategy is the Water Resources Management Authority, (WRMA).The membership to this board are all appointees of the minster except the Chairman who is appointed by the president. This is the state agent that is charged with the governance of water resources. WRMA has its regional agencies called Catchment Area Advisory Committees (CAACs) which takes care of the designated catchment areas.

All the 15 members of each CAAC is appointed by WRMA in consultation with the minister. The membership is drawn from the following:

1.  representatives of ministries or public bodies,
2.  representatives of regional development authorities and local authorities,
3.  representatives of farmers,
4.  representatives of the business community,
5.  representatives of non-governmental organisations and
6.  other persons of demonstrable competence.

Each CAAC is supposed to develop its Water management strategy which is expected to:

1.  Take into account the classification of water resource and its quality objectives,
2.   Be consistent with the NWRMS,
3.  Prescribe the principles, objectives, procedures and institutional arrangement, use, development, conservation and control of water resources,
4.  Have water allocation plans that set out principles for allocating water and
5.  Provide mechanisms and facilities for enabling the public and communities to participate in managing the water resources.

The CAAC's key function is to advise their respective WRMA regional office on: water resource conservation, use and apportionment; the grant, adjustment, cancellation or variation of any permit; and any other matters pertinent to the proper management of water resources. It is on this basis that WRMA shall issue or cancel permit for water use.

The Water Appeals Board (WAB) at the national level comprises a membership of a chairman who is appointed by the President and two members appointed by the minister. The main function of WAB is to hear appeals from mostly non state actors that have been aggrieved by action of some state actors in water governance and usage. The non state actors are usually represented at the local level by Water Resources Users' Associations (WRUA's).

In as much water services provision appear separate from and parallel to water resources management at face value, in actual sense, they are so intertwined in the water governance set-up to the extent that they in terms of operations, have to go hand in hand. The Ministry at national level formulates policies for the institutions of Appeals board, Water services Trust Fund, and also Water Resources management Strategy. The policies so formulated are implemented through the Regulatory agencies like WRMA and WSRB, each having regional authorities in the names of CAACs and WSBs, respectively. CAACs and WSBs are also service providers at the regional levels, while WRUAs and WSP at service providers at the local level.

One of the objectives of the National Water Services Strategy is: "...to institute arrangements to ensure that at all times there is in every area of Kenya a person capable of providing water supply."(Republic of Kenya, 2002). This is meant to ensure that no area is left without a water supply Programme. The strategy would contain details such as existing water services; the number and location of persons who are not being provided with basic water supply; a plan for the extension of water services to under-served areas; a time frame for the plan and an investment programme. In the set up to achieving the objective, the government has set up contingency plans to enhance accessibility of water to all. These include; trust funds, social tarification, contractual clauses and alternative water providers.

Section 83 of the 2002 Water Act makes provision for the establishment of the Water Services Trust Fund. Funding is expected from the three principal sources: parliamentary appropriations, donations/grants/bequests and statutory payments. The objective of the fund is to help finance water provision in areas of Kenya without adequate water services. The trust fund will receive development money from the government as budgetary allocations, and it may also get money from taxing water users or providers. The money can then be used to finance investments to provide water for the poor and to neglected areas.

Social tarification refers to charging a "social" rather than a commercial tariff. It is a policy instrument that may be used to ensure that the poor get water in instances where charges based on full cost-recovery would be too expensive. It works for those poor citizens who are connected to the main water network but who may otherwise not be able to afford the market price. Contractual clauses or conditionality's as well as specifying tariffs in the contract. It is also possible to specify that part of the performance of the contract includes extending the water network to an area which is either not served or underserved – for instance, to informal settlements or peripheral communities. If such conditions are not adhered to, a breach would be implied and the licence could be withdrawn. The Water Services Regulatory Board will be in charge of supervision. The board is also mandated to take action against the licensee, which includes withdrawal of the licence.

In perspective, water services provision and water resources management are intertwined in the water governance set-up, in terms of operations, and have to go hand in hand. The Ministry at national level formulates policies for the institutions of Appeals Board, Water Services Trust Fund, and also Water Resources Management Strategy. The policies so formulated are implemented through the Regulatory agencies like WRMA and WSRB, each having regional authorities in the names of CAACs and WSBs, respectively. CAACs and WSBs are also service providers at the regional levels, while WRUAs and WSP are service providers at the local level.

### 3.2 Popular participation and governance of water services

Like other countries in the world, water governance policy is premised upon the Dublin statement on sustainable water and development principle number 2 stating that the management of water and development should be based on participatory approach of governance involving users, planners, and policy makers at all levels. The implication is that decision making involving water projects is made with full involvement and public consultation of all users in the implementation process. It is in this respect that the water reforms in Kenya, envisaged role will be performed by WRMA through its regional agency of CAAC and the grass root level representatives in the name of WRUA.

A WRUA is an association of water users, riparian land owners, or other stakeholders who have formally and voluntarily associated for the purposes of cooperatively sharing, managing and conserving a common water resource (Definition in WRM Rules 2007). A WRUA is an association of water users, riparian land owners, or other stakeholders who have formally and voluntarily associated for the purposes of cooperatively sharing, managing and conserving a common water resource (Definition in WRM Rules 2007).Has this been effectively done? What implication does this have on WSPs? If this is done what impact does it have on independent small scale water providers? Is the process tenable and at what cost? Can the small independent water service providers afford the process and at what cost? If not, is it not impacting negatively on the water service provision?

Sessional paper No. 1 of 1999 on Water Resources Management and development provides policy guidelines with four broad objectives addressing both water resources management and service delivery to:

a.  Preserve, conserve and protect water resources and allocate them in a rational, sustainable and economic manner,
b.  Supply quality water in sufficient quantities to meet various needs,
c.  Set up an effective institutional framework for water resource management and
d.  Develop water sector financing system.

The policy developed five principles:

1.  To advocate for an integrated water resource management strategy that can address the multiple water needs and uses.
2.  It clarifies the role of the government as a regulator and a manager of water resources, the public and private sectors as co-providers of water services; and community as contributors to water resource management.
3.  It proposes that water catchment area committees should be the water planning and management units that s should be formed to serve as principle advisors on water allocation decisions to enhance transparency and accountability.
4.  It calls for separation of water service delivery functions from the water resources regulatory and management functions.
5.  It recommends volumetric fees for water abstraction, and adopts a "Polluter Pays Principle" to control pollution.

WRMA created under section 7, while maintaining the above principles in pursuit of the stated objectives, was expected to further develop guiding principles and guidelines for water resource allocation, regulate water resource quality, manage water catchments and determine charges to be imposed on use of water from any source. Section 46 of the same Act creates WSRB to license water service providers, handle consumer complaints against licensees; develop guidelines for fixing water tariffs; and develop model agency agreements between local authorities and private water companies. Section 51 of the Act creates Water Service Boards (WSBs)whose main role is to ensure efficient and economic provision of water services. To do this, they are to enter into agency agreements with water providers, mainly private water companies and community Project Water Cycles.

The Water Act 2002 requires stakeholder participation around; 1. Public and stakeholder consultation in developing the NWRMS, CMS and Protected Areas; and 2.Public consultations in matters of water resource allocation. In all the WSBs in Kenya, there are various mechanisms for stakeholder consultation. These include formal institutional arrangements (CAAC) public notification, through newspapers and public announcements. However, the interaction with the WRUAs goes beyond just public consultation, but seeks to enhance participation of primary beneficiaries, the water users. Section 5.7 of the Water Act 2002 states that "WRMA shall endeavour to support WRUAs". In this respect provision shall be made to enable them to access the funds provided by the WSTF for management and development within their areas of jurisdiction".

Essentially the CAAC is one way in which stakeholders can participate and influence water resource management within catchment areas. However, The Water Act 2002 states that the role of CAACs should be *advisory*. This implies that WRMA is not bound by decisions of the CAACs, and that authority and responsibility of decisions remains squarely with WRMA. This therefore contradicts the presumed purpose of popular participation. CAAC is also not mandated to abide by the resolutions of WRUA. Furthermore WRUA is not a representative association of the people in a catchment but a club of interested stakeholders. It is worth noting that CAACs are not intended to be representative of individual water users of a particular area but rather of stakeholder groups. Essentially there is no direct relationship between a water user and a CAAC member. CAAC members are appointed by WRMA, so it is up to WRMA to make sure that CAAC members are genuinely *representative* of the respective stakeholder groups. It is worth noting that WRUAs are not specifically mentioned as one of the stakeholder groups to be included in the CAAC, although Section 16(3)(f) provides a clause that can be used to include competent WRM individuals who could arguably be drawn from the WRUAs. The official interaction from WRMA down to WRUA appears non binding.

## 4. Conclusion

This paper has demonstrated that Public Private Partnerships are crucial in enhancing water service delivery in Kenya. It is also noted that it is part of the privatization and/or liberalization of water services. However, it is more of a response to the strict sense of privatization. In the water sector in Kenya, Small Scale Water Service Providers have emerged as alternative means to water service provision to areas where residents either find water from the major Water Service Providers inaccessible or more expensive. This has manifested itself in the form of water kiosks, water vendors and reseller, water stand pipes, community water, truckers and even cart pushers. It is important to note that small scale water service providers have been appreciated not only in Africa and Latin America but also in Asia.

Despite the proliferation of small scale water providers in Kenya, the process has not been smooth. The institutional regulatory mechanisms has made it difficult for them to effectively participate in the market. The procedures of Water Reforms Act 2002 despite making attempts to enhance popular participation in the water governance, resource management and development, have been discouraging to small entrepreneurs. The process to be recognized as a WSP is lengthy and expensive for small scale water suppliers. Before a permit is issued, a water user submits application to WRMA, go through the technical assessment and public notification processes. It is after then that authorization for construction is issued and a certificate of completion thereafter. It is then that a permit for water provision is issued.

The effort to enhance public participation in the governance of water service provision and management of water resources is merely impressionistic and not easy to sequence practically. WRMA is not bound by decisions of CAAC which equally is not bound by recommendations of WRUA. None of the institutions are representative of the other's interest. Secondly, WRUA as is presently constituted, is a club and not a representative of water users in any particular Zone. Membership is through individual interests implying that there are so many stakeholders who could be left out because they have not indicated interest to join the WRUA.

Several small scale water service providers would therefore be easily blocked from accessing permits due to conflict of interest with WRUA members.

Another area of concern is the role overlap between WRMA and WSRB since both have powers under the Water Act 2002 to determine water charges. Section 73 of the same Act also allows licensees including WSBs to determine water service tariffs hence the argument by Asingo (2007) that it is WSP which knows the cost to be recovered and hence should be the one to fix water tariffs in consultation with WSBs and WSRB. Ministry of local governments' role in provision of Water services has been reduced to obscurity as private water companies are expected to work very closely with the Ministry of Water Development, yet in practice they control the provision of services by virtue of being the largest shareholders in the Public Limited Companies.

It can therefore be deduced that the results of any water supply service provision will depend on the interrelationships between the state, regulators, citizens as consumers of the services taking into account multi-dimensional interactions amongst the parties. In this respect the government's regulatory framework should be facilitative rather than controlling of small scale water service providers. For example, it is against market principles to place one player (Large Scale Water Service Providers) at an advantaged position so as to be not only a supervisor of a competitor (Small Scale Water Service Providers) at a cost, but also be the one to recommend its registration.

## 5. List of abbreviations and acronyms

| | |
|---|---|
| ADB | African Development Bank |
| AFD | French Agency for Development |
| CAAC | Catchment Area Advisory Committees |
| DANIDA | Danish International Development Agency |
| DFID | Department for International Development |
| ELDOWAS | Eldoret Water and Sewerage Company |

EU              European Union
FINNIDA         Finnish Development Agency
GWASCO          Gusii Water and Sewerage Company
JICA            Japan International Cooperation Agency
KEWASCO         Kericho Water and Sewerage Company
KFW/GTZ         German Development Agency
KIWASCO         Kisumu Water and Sewerage Company
LA              Local Authority
LVWATSAN        Lake Victoria Region Water and Sanitation Initiative
NMISWR          National Monitoring and Information System on Water Resources
NWC&PC          National Water Conservation & Pipeline Corporation
NWRMS           National Water Resources Management Strategies
PPP             Public Private Partnerships
SIDA            Swedish International Development Agency
SNWSN           South Nyanza Water and Sewerage Company
SPAs            Service Provision Agreements
UNICEF          United Nations Children's Fund
WAB             Water Appeal Board
WRMA            Water Resources Management Authority
WRUA            Water Resources Users Association
WSB             Water Services Board
WSP             Water Services Provider
WSRB            Water Services Regulatory Board
WSTF            Water Services Trust Fund

## 6. References

ADB Asian Studies. 2002.
         http://www.adb.org/documents/books/asian_water_supplies/chapter07.pdf.
         downloaded on 18/07/2011
Alila P. Et al "Business in Kenya: Institutions, Interactions, and Strategies" in Alila P. et al
         (eds), Business in Kenya: Institutions and Interactions University of Nairobi Press,
         Nairobi, 2007
Asingo, P.O. Privatization of Water Services in Kenyan Local Authorities: Governance and
         policy Issues. IPAR Discussion Paper no. 067/2005, Nairobi, 2005.
Clarke, G. et al. *Has Privatization of Water and Sewerage improved coverage? Empirical Evidence
         from Latin America*. World Bank policy Working Paper No. 3445,
Estache, A & Rossi, A. M."How Different is Efficiency of Public and Private Water
         Companies in Asia?" in The world Bank Economic Review, Vol 16, no.1. Oxford
         University Press:
         http://www.jstor.org/stable/3990169on 06/08/2010
Kjellen, Marianne"Complementary Water Systems in Dar es Salaam, Tanzania: the case of
         water vending." Water Resources Development, 16:143-154., 2000
Mogaka H. et al. "Impacts and Costs of Climate variability and Water Resources, 2003.

Njiru, C. "Utiliy-Small Scale water enterprise Partnerships:serving informal urban settlement s in Africa".Water Policy 6, 443-456., 2006

O. A. K'Akumu "Privatization of the urban water supply inKenya: policy options for the poor"in*Environment Urbanization* Vol 16 No 2 October 2004

"The political ecology of water commercialization in Kenya "in *Int. J. Environment and Sustainable Development,* Vol. 6, No. 3, 2007

"Privatization model for water enterprise in Kenya" in *Water Policy* 8, 2006.

O. A. K'Akumu and P. O. Appida "Privatization of urban water service provision: the Kenyan experiment" in *Water Policy* 8 (2006)

Onjala, J. "Essential services- Electricity and Water: The Challenges and options for Business in Kenya". In in Alila P. et al (eds), Business in Kenya: Institutions and Interactions University of Nairobi Press, Nairobi, 2007

Owuor S.l O. & Foeken D.W.J. *Water reforms and interventions in urban Kenya*: Institutional set-up, emerging impact and challenges, ASC Working Paper 83, 2009

Republic of Kenya *The Water Act Chapter 372, Laws of Kenya*: Revised Edition. Government Printer, Nairobi, 1972.

Republic of Kenya. *Report on next step in the commercialization of water and sanitation services in Eldoret, Kericho and Nyeri Municipal Councils.* Ministry of Local Government, Nairobi, 1996.

Republic of Kenya *The Kenya Gazette Supplement No. 107 (Acts No. 9): The Water Act 2002.* Government Printer,Nairobi, 2002.

Republic of Kenya. *The Kenya Gazette Supplement No. 14 (Bills No. 4): The Privatization Bill, 2004.* GovernmentPrinter, Nairobi, March 31.

Republic of Kenya *"The National Water Services Strategy (2007-2015)"*, Nairobi: Ministry of Water and Irrigation, 2007.

Republic ofKenya. Welfare Monitoring Survey II, 1994, Basic Report. Nairobi: Central Bureau of Statistics, Office of the Vice President and Ministry of Planning and National Development., 1996

Solo, Tova M. "Small scale entrepreneurs in urban water and sanitation market. "Environment and Urbanization. 11:7-31, 1999

UN-Habitat. *Water and Sanitation in the World's Cities: Local Action for Global Goals.* Earthscan, London, 2003.

Water Services Regulatory Board (2009): *A performance Report of Kenya's Water Services Sub-Sector Issue No 3.*,
http://www.wasreb.go.ke/index.php?option=com_content&task=view&id=70&Itemid=145downloaded on 15/1/2011

Water Sector Reform Secretariat *Proposed Delineation of Boundaries for Water Services Board.* Water Sector ReformSecretariat, Nairobi, 2003.

Water Sector Reform Secretariat: *A Handbook of the Water Sector Reforms.* Water Sector Reform Secretariat, Nairobi, 2005.

World Bank *World Bank Development Report* , New York, Oxford University Press, 1983.

World Bank (b). *Bureaucrats in Business: The Economics and Politics of Government ownership* , New York, Oxford University Press, 1995.

World Bank. *World Development Report*, World Bank. Washington DC, 1997

World Bank (a). "The Republic of Kenya: Towards a Water Secure Kenya". *Report No. 28398-KE.* World Bank: Washington DC, 2004.

World Bank (b). The World Development Report on Infrastructure and Development. World Bank: Washington DC, 2004.

# 11

# Bringing Water Regulation into the 21st Century: The Implementation of the Water Framework Directive in the Iberian Peninsula

Antonio A. R. Ioris
*University of Aberdeen*
*United Kingdom*

## 1. Introduction

Water is anything, but trivial. That observation is easily demonstrated by the intricate, often contested, nature of water use and conservation in Europe, which normally encapsulates operational challenges, intersector disputes and multi-level political expectations. If the traditional forms of water use were typically based on cooperation and mutual understanding (vis-à-vis subsistence irrigation and community water supply), the recent history of water development is more closely associated with large-scale interventions and growing rates of water demand. Mounting environmental pressures make the reconciliation of antagonistic interests even more difficult, especially in areas with relatively low stocks of water and an inadequate institutional organisation. Throughout the 20th century, both the Keynesian and the post-Keynesian phases of water management have tried to develop rational approaches to restore and maintain the integrity of freshwater systems.[1] If the Keynesian period was marked by large infrastructure projects and centralised planning, the post-Keynesian blueprint is now characterised by non-structural and more flexible responses. In that context, a succession of plans and regulatory efforts launched by the European Union in the last two decades have attempted to improve the institutional mechanisms for dealing with old and new water management problems. To a great extent, the end result of that salient water policy has been an 'organised anarchy' characterised by problematic preferences, unclear technology and fluid participation, whilst the overall trend of resource overuse and the uneven sharing of the environmental impacts remained mostly unchanged (Richardson, 1994). That is why the approval of the Water Framework Directive (WFD) in 2000 – currently in its first cycle of implementation – has been perceived as a promising opportunity to enhance the regulatory capacity of national governments and public agencies, as well as a central tool in the reform of the collective basis of social learning and bring water management in Europe to the 21st century (see Hedelin & Lindh, 2008).

---

[1] The post-Keynesian phase of water management began with the United Nations Mar del Plata conference in 1977 and, not by chance, coincided with the aftermath of the crash of the Bretton Woods monetary order, the oil crisis, and the declining role of the state. The connection between water management reforms and the larger politico-economic reorganisation has had major consequences for the assessment of problems and formulation of solutions, as discussed below.

The broad range of activities related to the implementation of the new Directive represents a very special episode in the history of environmental regulation in Europe. Likewise, the introduction of the WFD constitutes an important element of an affirmation of the political legitimacy and administrative authority of the evolving European Union statehood system. Because of its large-scale consequences, the complex reorganisation that follows the WFD epitomises a distinctive case in the sociology of water management, what van der Brugge & Rotmans (2007) describe as a transition from the previous focus on hydraulic infra-structure works to a new phase based on the adaptive, co-evolutionary coordination of improved responses. The new Directive is not only associated with technical and administrative expedients, but also relies on the affirmation of 'protonorms', such as watershed democracy, water marketisation, international river diplomacy and the notion of integrated management, that all compete to normalise the contemporary forms of water governance (Conca, 2006). The multiple components of the new European regulation related to the implementation of the new water directive certainly constitute one of the most comprehensive examples of a programme of environmental conservation around the world. Notwithstanding the ambitious nature of the WFD regime, the bulk of the official measures seem yet to be too centred on technical and bureaucratic procedures with limited consideration of the also important political and ideological dimensions of water management. That can represent a major implementation problem, because at the same time that the Directive encourages a more efficient allocation and use of scarce water resources, the success of the WFD also depends on dealing with some thorny social issues that influence the allocation and management of water, such as stakeholder inequality and environmental injustices (Surridge & Harris, 2007).

Our aim in this chapter is to investigate the introduction of new water management institutions and how it has influenced intersector and interspatial relations, particularly in terms of public water and sanitation services. More than a decade after the approval of WFD, it is the appropriate time now to discuss achievements and constraints of specific catchments and countries in order to assess the overall European progress. It needs to be examined whether the WFD agenda – essentially, the range of public and private activities related to implementation of institutional reforms around the allocation, use and conservation of water that have followed the approval of the new European Directive – provides a coherent set of guidelines to revert structural shortcomings and pave the road for more sustainable forms of water management. We will consider here some of the key dilemmas involved in the management of water in the Douro catchment (called Duero in Spain) with an emphasis on the Portuguese context (which is unusual, as most analysis of the catchment focus on the Spanish side). The study was initially inspired by the observation of Dominguez et al. (2004) that conflicts and problems in the Douro have been many times hidden from the public debate and, therefore, deserve to be properly examined. The empirical results show the socionatural complexity of the catchment and a situation of growing problems and evident regulatory shortcomings. In effect, because of its size and geographical complexity, the Douro represents one of the most challenging areas for the achievement of WFD objectives in southern Europe.

The Douro (Figure 1) is the largest Iberian river basin (97,290 km²) with 78,954 km² in Spain and 18,336 km² in Portugal (respectively 15.6% and 19.8% of each national territory), which corresponds to 17% of the peninsular area. According to Sabater et al. (2009), the main river channel is 572 km long. The first 72 km flows through steep valleys and the remaining 500 km of the river meanders through an open valley over soft tertiary sediments. The mean water temperature ranges from 11.2 °C in the headwaters to 14.0°C in the lower reaches. Mean precipitation in the Portuguese section is 1,016 mm/year and in the Spanish section is 625 mm/year (Maia, 2000, quoted in Dominguez et al., 2004). The catchment has a strong relationship between rainfall and river flow, with the maximal discharges occurring in the spring and minimum in the summer. High discharge periods are usually correlated with peaks in suspended solids. The mean flow at the river mouth is 903 m³/s.

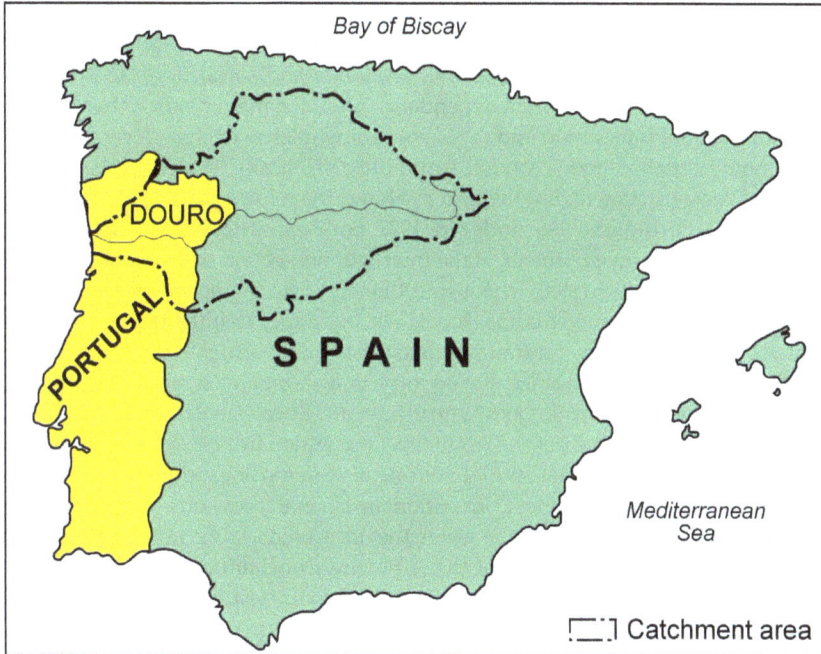

Fig. 1. The Douro Catchment in the Iberian Peninsula

Water use in the catchment is dominated by agriculture and, secondly, by hydroelectricity (one quarter of Spanish and more than half of Portuguese generation are located in the Douro), although industries, cities, navigation and mines are also important user sectors. Total water usage is between 26-31% of the natural mean flow and the storage capacity corresponds to 8.8% of the natural mean flow (7.7% in Spain and 1.1% in Portugal), according to Maia (2000, in Dominguez et al., 2004). In terms of the ratio between abstraction and availability, the level of water stress the Douro is not much different than the River Guadiana in the south of the Peninsula and with much lower rates of rainfall (European Commission, 2007). There exist more than 50 large dams built for hydropower and irrigation, with a particular concentration in the last 350 km of the river channel (Bordalo et al., 2006; Sabater et al., 2009), which has caused the extinction of ¾ of the local fish species (Azevedo, 1998). Because of untreated effluents coming from Spain, at the point of entry of the Douro in Portugal the level of pollution is considerably high (particularly in term of nitrate). Around 50% of the water bodies in the river basin in Portugal have chemical and biological standards at levels that are below the legal requirements (National Water Institute

[INAG], 2001), whilst the majority of the river stretches in Portugal and Spain present a less than good ecological condition due to irrigation abstraction, urban effluents, impoundments and riparian deforestation (Commission for the Coordination and Regional Development of the North [CCDR-N], 2000).

Following the holistic goals of new water regulation, both Portugal and Spain are now required to improve the scope of the responses and broaden the agenda of water management more in line with the expectations of those social groups not previously involved in the decision-making process. That should occur not only within national borders, but also between the two neighbouring countries. Nonetheless, if the two nations are profoundly connected by many cultural, economic and social ties – to a large extent, these are associated with the common dependence upon the main rivers –, there also exists a permanent dialectic of integration and repulsion, sometimes reaching a level of dispute that prevents genuine collaboration. Portugal is not only physically located in the downstream section of the Douro catchment, but the history shows the reluctance of Spain to consider the full extent of the Portuguese demands. In 1927 both countries signed an agreement to discipline hydroelectric developments in the international section of the Douro (later ratified by other treaties in 1964 and 1998), which split the river into segments instead of allowing a joint construction of hydropower dams. It was not by chance that the treaty coincided with the initial stage of the highly centralized dictatorship in Portugal (since the 1926 coup), which in the subsequent decades led the country to an isolationist, authoritarian model of economic development.[2] With the joint entry into the European Union in 1986, bi-lateral negotiations led to the signature of the Albufeira Convention in 1998, which determined that Spain had to guarantee a minimal annual volume of water at several points along the river. However, Spain has breached the Convention in several occasions, such as during the droughts of 2001-2002 and 2004-2005 when the thresholds were not respected. Further discussions produced an amendment of the Albufeira Convention in 2008, which has now quarterly and weekly flow thresholds, but still not put to the test.

Despite institutional developments at the national, Iberian and European levels (directly or indirectly related to the new Directive), the crux of the matter, not often grasped by the majority of existing assessments and discussion papers, is the myriad of political clashes and regulatory shortfalls that hinder the adoption of more effective and fairer management of water in the Douro. To overcome those limitations and fully understand the complexity of WFD, it is necessary to employ a multispatial and multisector analysis that articulates the higher (i.e. European) and lower (i.e. locality level) geographical scales, as well as situates the discussion beyond the technocratic parlance that still permeates most official documents and academic assessments. The following pages will offer a critical reflection about changes related to WFD by primarily focusing on the Portuguese section with some insights into the Spanish side of the catchment. That aims to provide a representative example of the controversies that characterise the current implementation of the new Directive. It will be

---

[2] Also in 1927, the river basin authority was created in Spain, which is called Duero Hydrographical Confederation (CHD) with responsibilities for water planning, water quality, flood prevention, and environmental protection. An advisory board (the River Basin Council) was established in Portugal in 1994, but it was only in 2007, with the creation of the Northern Hydrographical Region Administration (ARH-N), that an executive authority equivalent to CHD was formed in the downstream country.

necessary to consider the repercussion of official policies on different water users and the interchanges between the lower Douro (around the city of Porto) and what we generically define here as the upper Douro (the area around and upstream the demarcated area of port wine production).

The empirical part of the study consisted of two research fieldtrips to the Douro in the year 2008 (March-April and October-November) as visiting researcher at the University of Porto. The overall research strategy was the 'embedded case study', as described by Yin (1994), which starts with the consideration of embedded sub-units of social action and then scaled them up to identify common patterns at higher scales. The study explored interests and behavioural patterns of various geographical locations and stakeholder sectors, as well as about the institutional framework in which they operate. The research effort initially consisted of contacts with key informants, academics and policy-makers. Based on this preliminary information, we developed a database of public and non-governmental sectors that guided further interviews, the analysis of documentation and the collection of background information. By mapping the various organisations, their discourse and stated aims, it was possible to compare intra- and inter-group differences and the range of alliances or disputes. A total of 43 in-depth interviews were conducted with water users, regulators, and NGO and campaign activists. Interview respondents were identified from an array of organisations that represented multiple interests in the water management sector. Additional information was obtained in the libraries of the universities of Porto, Coimbra, Lisbon, Valladolid and Salamanca, in libraries of Vila Real, Miranda do Douro and Peso da Régua, and at the information centre of the National Water Institute (INAG) in Lisbon. Also public events sponsored by both governmental and non-governmental entities were also attended during the period of work in Portugal and Spain.

The chapter is organised as follows: the next section presents the institutional evolution of water management and regulation in Portugal and in the Douro, which will serve to inform the assessment of the implementation of WFD. The subsequence section deals with the achievements and constraints of the WFD regime, exploring evidences of innovation and continuity. The final parts summarise the analysis and offer some general conclusions.

## 2. Economic and institutional evolution: Portugal and the Douro

The attempts to reform the management of water in the Douro embody some of the most emblematic difficulties to translate the WFD regulation into practical improvement measures. The debate about the decentralization of water management – one of the tenets of the WFD regulatory regime – happens in tandem with a growing discussion about the transference of state duties to the regional spheres of public administration, as well as with broader claims for local autonomy, social inclusion and even economic development (vis-à-vis, for example, the series of conferences organised by the City Council of Porto in 2008). First of all, it is important to recognise that the use of water in the catchment had and still continues to play a strategic role in terms of regional development. The upper reaches have been the electric powerhouse of Portugal, due to the construction of large hydropower schemes since the 1950s, whilst the lower section of the catchment became associated with light-industrial production and the export of port wine. Until the early 20th century, wine was transported to the city of Porto in small boats (called 'rabelo'), but fluvial navigation started to decline with the inauguration of a railway line in 1887 and, more importantly, road transport in the early 20th century (Pereira & Barros, 2001).[3] At the same time, the transformations of the mechanisms of water use are closely related to the socioeconomic processes of change in the northern region of Portugal. Efforts to recover the regional economy have included actions related to increasing the use of freshwater resources, particularly in terms of new hydropower dams, fluvial tourism and the expansion of the water supply and sanitation network (CCDR-N, 2006).

The above points illustrate how the social and physical transformations around the use of water in the Douro reflect the broader 'choreography' of regional, national and international demands. Portugal started to intensify its economic and monetary integration with the rest of the continent in the 1960s, when joined the group of countries that founded the European Free Trade Association. That culminated in the full membership of the European Union (in 1986) and the adoption of the euro as the national currency (in 1999). The industrialisation and economic development of Portugal has been historically led by the national state, but such a condition has been increasingly criticised by national and international political forces. Crucially, the style of the WFD regulation is closely consistent with the neoliberal direction of European economic policies (see below), but neoliberalising reforms have neither guaranteed economic growth nor avoided the persistence of macroeconomic imbalances (Amador, 2003). It is important to emphasise that the evolution of environmental regulation in Portugal has followed the broader adjustments of public policies and the reconfiguration of the state according to a perspective of economic liberalisation and pro-market incentives. According to Queirós (2002), Portugal has made much progress in establishing a revised environmental legislative framework (largely but not solely in response to European Union directives), strengthening its environmental institutions (including the Ministry of Environment, Spatial Planning and Regional Development), developing national environmental planning (e.g. its first national environmental plan, in 1995) that covers the entire country (e.g. national coastal area protection plans, national nature protection plan, municipal land use plans). The introduction of the WFD in Portugal is an integral part of this institutional reorganisation and, in the words of a senior authority, the complexity is situated in the tension between the centenary tradition of the Portuguese law system and the formal requirements of the European legislation (see Ambiente Online, 2005). Considering the changes that took place in the last century, it is possible to schematically describe five successive phases of water use and development in the Douro, which echo national and international transformations (see Box 1). Note the transition from Keynesian forms of state intervention until around 1986 and the prevalence of post-Keynesian and neoliberal approaches ever since.

The impact of human activities on the water bodies in the Portuguese section of the Douro is evident one considering the trend of water quality classification. Different than other rivers in the south of the Iberian Peninsula, quantitative water impact does not represent the main management problem, but the pollution of the Douro and its lower tributaries. Water quality is seriously affected by household and industrial effluents (due to the lack of sewage collection and treatment), as well as diffuse pollution from agriculture that is mainly originated in Spain. The activity that consumes the largest volume of water in the Douro is

---

[3] According to the navigation authority (IPTM), the transportation of commodities in the Douro still remains a viable means of transportation and reached 140,000 tons in 2004 (94% of that total related to the export of granite to northern European countries).

| Precursory period (till early 20th century) | Navigation in the Douro increased significantly in the early 18th century with the transportation of port wine from the Peso da Régua region to the Porto docks (Pereira, 2008). The first hydropower generation site in the country was installed in a Douro tributary in 1894. Since the 1880s, water supply to the metropolitan area of Porto passed to rely on a treatment plant in the Sousa River, a tributary of the Douro under the operation of a French concessionary company (Amorim & Pinto, 2001). |
|---|---|
| Hydraulic period (1919-1986) | The Water Law of 1919 established a higher recognition of the importance of water for the socioeconomic development of the country (Cunha et al., 1980). The Law stipulated that water use required a prior authorisation from the state, which was later confirmed by the Decree No. 468 of 1971. It was during this phase that most of the large infrastructure works were built and key technical agencies were created and (the Hydraulic Services General Directory in 1949 and the Basic Sanitation General Directory in 1973). Some of the most strategic hydropower plants were built in the Douro, such as Picote (1958), Miranda do Douro (1960) and Bemposta (1964). The recently established dictatorship cancelled the contract wit the French concessionary in 1927 and municipalised the water services in the city of Porto. In 1940, a well field along the Douro (in Zebreiros) increased the supply of water to the metropolitan area. |
| Transitional period (1986-1993) | The regulatory context started to change after Portugal joined the European Union in 1986. During this period, a growing number of publications (e.g. Miranda, 1986) started to emphasise the need to adopt modern water management, in particular economic instruments based on the polluter-pays principle. A dedicated regulatory agency, the National Water Institute (INAG), was crated in 1990. Since 1985, the Crestuma-Lever reservoir, located at 21.6 km from the mouth of the Douro became almost the only suitable source for the production of potable water for approximately two million inhabitants of the Porto region (the same dam had impacted negatively the well field because it reduced the river flow and increased the rate of salinity in Zebreiros). The first tourism navigation ship started to operate in the Douro in 1986 and since then the industry has grown significantly (from 6,440 passengers in 1994 to 180,691 in 2004). |
| Water service liberalisation and river basin plans (1993-2005) | The approval in 1993 of the Decree No. 379 provided the legal basis for the gradual concentration of water services in the hands of regional companies. There has been a continuous trend towards regional water utilities, which is part of a movement from dispersed to concentrated sources of water supply, a tendency that has increased in recent years (Thiel, 2006). In 1994, a series of decrees reorganised the regulation of water use in Portugal and introduced the recognition of the economic value of water: No. 45 (on river basin plans), No. 46 (water user licence) and No. 47 (a charging scheme that included volumetric bulk water tariffs). Under that national legislation, the National Water Council and various river basin councils, including one for the Douro, were established as advisory boards and largely formed by civil servants. The Douro river basin plan was adopted in 2001, but it was only marginally implemented. |

Box 1. (continues on next page) Historical Evolution of Water Use and Water Development in Portugal and in the Douro

| WFD regulation (the current phase, since the approval of WFD in 2000) | The WFD was translated into national legislation in 2005 (Portugal, Law No. 58/2005) and attempts to forge improvements in several areas, including technical assessments, decision-making and regulatory enforcement. The WFD promotes the concept of water as an economic commodity and, therefore, the economic principles are main criteria in the determination of cost-effective mitigation measures and in assessing the case for derogation on grounds of disproportionate costs. The translation of the WFD into national legislation also launched the legal basis for the creation of water markets in Portugal (i.e. markets for the transaction of water use licences) that is claimed to allow the reduction of pollution through market transactions and at the minimal cost (D'Alte, 2008). The financial-economic regime, which introduced bulk water charges, was approved in 2008. In the end of that year, the Water Regulatory Agency (ARH) was preparing the production of river basin management plans, but it was expected that the deadline of end of 2009 would not be achieved. |
|---|---|

Box 1. (continued) Historical Evolution of Water Use and Water Development in Portugal and in the Douro

the irrigation (114,000 hectares cf. INAG, 2005). Industrial demand is another main user sector and its main environmental significance is the discharge of effluents into the river system, which aggravates the level of pollution. For the purpose of this analysis, we obtained data from the national surveillance system (available at http://snirh.pt), which has been used to inform the implementation of the new water directive. It can be seen in Figure 2 that there is an undefined trend of water quality in recent years (note a recovery of Class A, the best water quality condition, in 2007, together with a decline in Classes B, C and D, and a sudden increase in Class E situations). Environmental impact is, however, not restricted to pollution, by also include the negative influence of dams on native species, sediments and riparian habitats.

Probably the experience that best encapsulates the interface between social, economic and environmental demands in Portugal – before and under the WFD regime – has been the redesign of public water services.[4] For many years, the water industry had been systematically criticised for its fragmentation into small, localised companies, with high operational costs and limited investment capacity (e.g. Alves, 2005; Martins, 1998). The historical origin of the fragmentation of public water services was the delegation of responsibilities to municipal and sub-municipal administration, which still today are the main providers (among the 278 municipalities in Portugal, it is reported that exist 610 operators cf. Monteiro & Roseta-Palma, 2007).[5] Another characteristic of the Portuguese water industry is the operational separation between drinking water production (abstraction and treatment), called 'high services', and retail water distribution (supply of water to households and commercial customers), called 'low services'. To facilitate the understanding of the complex water industry currently in operation in Portugal, it is possible

---

[4] The country has one of the highest per capita footprints in the world (2,264 m3/year, cf. Malheiro, 2008), which is related not only to cost of water, but also with climatic conditions, technological stands, and patterns of production and consumption.

[5] Note that these numbers do not match the figures published by the regulator (IRAR, 2008).

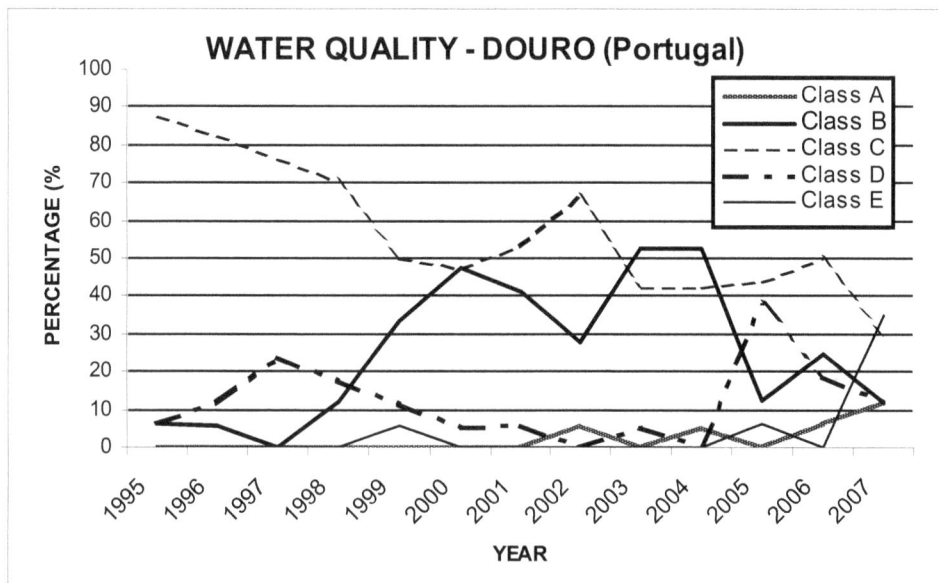

Fig. 2. Water Quality Trend in the Portuguese Section of the Douro Catchment

to classify the sector of water supply and sanitation as: *national state jurisdiction*: direct state management, delegated management (to public companies entirely owned by the national government) and concessions (to companies owned by the national government in partnership with municipal authorities, or between public and private companies); and *local authorities jurisdiction*: direct management (municipal, municipalised or intermunicipal services), delegated management (sub-municipal ['freguesia'], municipal or intermunicipal services) and concessions (to companies owned by the national government in partnership with municipal authorities, or between public and private companies). The distribution of water supply and sanitation operators is summarised in Table 1.

After the approval of a new legislation in 1993 (Law 379/1993), there has been a gradual movement towards the consolidation of high services in regional entities, which are supposed to provide gains of scale and rationalise water abstraction at the regional level. A national state-owned company was created in 1993 (Águas de Portugal), which has ever since formalised partnerships with local authorities in order to create regional companies (Águas de Portugal typically owns 51% of the regional company and the local authorities together own 49% of shares).[6] In the Douro, there are two such companies, the Águas do Douro & Paiva (in the Porto metropolitan area) and the Água de Trás-os-Montes & Alto Douro (in the upper river basin). Nonetheless, at the same time that the treatment of water is being transferred to regional utilities, some municipalities have contradictorily created their own companies to operate independently, such as city of Porto, which in 2006 established the Águas do Porto. Coherent with the current macroeconomc policies and the

---

[6] Águas de Portugal also became an international player involved in the privatisation of water services in other countries, such as in Brazil.

contemporary model of water governance, the reorganisation of the water industry has created important opportunities for private business, especially through the operation of municipal or multimunicipal concessionaries (in the form of public-private partnerships), whilst also stimulates private sector involvement in terms of outsourcing and operation and maintenance contracts (Water and Waste Regulatory Institute [IRAR], 2008).

| National government jurisdiction | | | | |
| --- | --- | --- | --- | --- |
| Type | Entity | Regional (high) or local (low) | Water supply (number of operators) | Basic sanitation (number of operators) |
| Direct state management | national state | High | 0 | 0 |
| | | Low | 0 | 0 |
| Delegated management | public company | High | 1 | 0 |
| | | Low | 1 | 0 |
| Concessions | multimunicipal concessionary | High | 13 | 16 |
| | | Low | 1 | 0 |

| Local authority jurisdiction | | | | |
| --- | --- | --- | --- | --- |
| Type | Entity | Regional (high) or local (low) | Water supply (number of operators) | Basic sanitation (number of operators) |
| Direct state management | municipal services | High | 62 | 67 |
| | | Low | 205 | 217 |
| | municipalised services | High | 5 | 5 |
| | | Low | 25 | 24 |
| | intermunicipal services | High | 2 | 0 |
| | | Low | 0 | 0 |
| Delegated management | municipal/ intermunicipal public company | High | 5 | 7 |
| | | Low | 16 | 18 |
| | sub-municipal public company | high & low | 155 | 0 |
| Concessions | Municipal concessionary | High | 8 | 11 |
| | | Low | 22 | 16 |

Table 1. Classification of Water Service Providers in Portugal (adapted from IRAR, 2008)

Whereas the new paradigm for water supply and sanitation in Portugal is consistent with policies that emphasise efficiency and rational management, the regulatory agency – the Institute of Water and Waste Regulation, IRAR, which was established in 1997 – still remains with a narrow remit and only deals with the concessionary companies, leaving the great majority of the municipal operators to self-regulate themselves.[7] In addition, the investment capacity and financial health of water utilities have deteriorated rapidly in the

---

[7] In 2008, there was a national debate about extending IRAR's duties to the other types of operators, but it was still difficult to see any firm movement in that direction.

last few years, demonstrated by a growing preoccupation with the level of debts, the ineffectiveness of many capital investments and the difficulty to raise money (IRAR, 2008). According to the national plan for the period 2007 to 2013, it will be necessary to invest € 3.8 bi (€1.6 bi in high services and €2.2 bi in low services) to secure 95% of public water supply coverage and 90% of public sanitation coverage (Ministerial Resolution 2339/2007). Different than in the recent past, European funds are expected to pay for only a fraction of that total amount, which means that the sector needs to find additional sources of investment and, probably, continue to increase the charges paid by the customers. Despite a constant effort to recover the costs and the controversy that it creates, between 1998 and 2005 tariffs increased below the rate of inflation and the charging scheme continued to be characterised by high levels of complexity and unfairness (Monteiro & Roseta-Palma, 2007). Furthermore, if it is undeniable that improvements in water services and environmental conservation require capital investments and incur high maintenance costs, the concentration of efforts around cost-recovery measures tends to diminish the attention to environmental and social dimensions of water services. The ongoing experience of the water industry has significant parallels and connections with the introduction of the WFD in the Douro, as discussed next.

## 3. The contested search for efficiency and the multiple tensions under WFD

As described above, the introduction of the WFD in Portugal has accelerated a process of institutional change initiated in the previous decades, particularly after the entry of the country into the European Union. Since the approval of the 2005 water legislation that translated Directive into national legislation, open events and regular media coverage have helped to broaden the debate about the new water regulatory regime. Nonetheless, underneath an apparent convergence of public opinion, there lays a stream of continuities and uncertainties not yet adequately considered. In several of our interviews it was mentioned that a major shortcoming is the insufficient opportunities available for the public to contribute during the regulatory transition. Historically, stakeholder engagement in water management and environmental issues has been very low in both Portugal and Spain, as much as between Portugal and Spain (Barreira, 2003). After the introduction of the WFD, the involvement of the public has remained restricted to consultations and formalist activities that offer little transparency and produce limited impact on decision-making (Veiga et al., 2008). In particular, the round of meetings organised in 2007-2008 by the government to discuss the new legislation ended up being something like a 'big imbroglio' because it has been limited to a small number of participants and merely ratified decisions made in advance by the government (interview with a NGO activist, 19 Nov 2008). Among the general members of the public, the criticism about the current water reforms has been related to a loose resistance against utility privatisation and in favour of vaguely defined 'water rights'. The superficial understandings of the conceptual underpinnings of the Directive permeate also the discourse of many environmental activists and academics that do not seem entirely aware of the politicised basis of the WFD regime.

Another significant evidence of continuity between past and present approaches is **the top-down assessment of environmental impacts and future scenarios**. A series of reports have been commissioned to estimate environmental pressures and impacts, as required to inform the implementation of the Directive, but by and large these assessments constitute little more than a compilation of generic data gathered from fragmented sources of information. The analyses tend to maintain the focus on pure hydrological modelling, paying scant attention to ecological conservation (Moura, 2007) or to traditional forms of water use practiced by local communities (Cristovão, 2006). The initial WFD report concluded that the Douro catchment has, among all the Portuguese rivers, the highest proportion (57.1%) of surface water bodies at risk of not achieving WFD targets (for the purpose of WFD, the river basin was classified according to 613 water bodies); there is an additional percentage of 23.4% of water bodies potentially at risk of not complying with the same objectives (INAG, 2005). The main sources of pressure seem to be pollution from agriculture and untreated sewage discharges, but it is not clear whether that proportion of impacted water bodies is reliable or the picture was exaggerated by the superficial nature of the assessment. The irony is that such assessments may affect negatively the resolution of water management problems: an overwhelmingly bad picture may have the perverse effect of diluting the focus away from the real problems and serve as justification for 'doing nothing' (i.e. under the assumption that the task is not feasible, the WFD regime allows an application for 'derogation' [exemption]).

At any rate, the narrow involvement of the public and the precarious scientific understanding of the socionatural complexity of the Douro catchment have not prevented the policy-makers from concentrating their attention on the aspects of WFD regulation that more directly correspond to the broader political and macroeconomic goals of the Portuguese government. Above all, a great deal of the ongoing regulatory effort has prioritised the achievement of higher levels of operational and economic efficiency, which represents the most emphasised aspect of the WFD regime in Portugal so far. The prevailing discourse claims that efficiency constitutes a 'win-win' game, insofar as the environmental pressure on aquatic systems can be reduced – in theory – by lowering the level of water demand and effluent discharge, which also represents economic savings to the water user (epitomised by Cunha et al., 2007).[8] That is illustrated by the ideas advocated by Professor Correia - the Secretary of State for the Environment – for whom the WFD regime is essentially a matter of cost reduction and higher efficiency. Although also mentioned in government documents, other dimensions of the new regulatory context are systematically overlooked. For instance, in Jun 2008, at the opening session of the National Association of Portuguese Municipalities, the minister argued that:

"Water demand in Portugal is estimated at 7,5 billion m³/year, of which agriculture is the main user sector, making use of 87% of the total, whilst urban supply demands 8% and the industrial sector, 5%. However, not all the water abstracted is effectively utilised, given that an important proportion is associated to inefficient use and losses. (...) There are various reasons to take the efficient use of water as a strategic goal. First of all, there is a growing consciousness in society that water resources are limited and, thus, it is necessary to protect and conserve (...). [Another reason] is the economic interest at the national level, inasmuch as potential savings related to water correspond to significant figures, estimated at around 0.64% of national GDP (...) The efficient use of water is still important in

---

[8] This argument obviously ignores that increases in efficiency can be easily minimised by additional water demand that, in the end, magnify the level of environmental impact.

regarding the rationalisation of investments, to the extent that it allows a better use of existing infra-structures, reducing or even avoiding the need to increase water abstraction systems (…). The efficient use of water corresponds to the economic interest of the citizens, to the extent that makes possible a reduction in the costs of water use."

The connection between efficiency, private gains and water management should come as no surprise, given that the minister has been himself one of the champions of the water reforms under the new paradigm of efficiency and economic rationality (cf. Correia, 2000). That is coherent with the tenets of environmental economics that underline the implementation of WFD, in particular the requirement to calculate the economic value of environmental impacts and the cost of mitigation measures. In practice, it has been translated into numerous applications of contingent valuation methodologies around Europe (e.g. Del Saz-Salazar et al., 2009) that unnecessarily reduce the complexity of socionatural water systems to the 'common ground' of money value. Although the chief water regulator in the Douro has expressed a more careful handling of the economic element of the new Directive (Brito et al., 2008), national policies constantly reinforce the idea that the main responsibility for improving water management lies in the hands of individual water users who should make their decisions in the light of a utilitarian economic thinking. The colonisation of the public debate by business expressions and the (material and symbolic) commodification of water is not an innocent occurrence, but reinforce the association of the WFD regime with government efforts in other policy areas (e.g. reduction of state enterprise and establishment of public-private partnerships). The emphasis on treating water as a commodity is illustrated in Figure 3 (poster of an event held at the time of our fieldwork).

Another step in the direction of exacerbating the economic dimension of WFD is the persistent claim that water is increasingly scarce and, as a result, should attract a monetary charge equivalent to its level of shortage. The corollary is that the scarcity of water can only be universally discerned by the stakeholders if the resource is quantified in monetary terms (i.e. the economic value). In other words, the access to water should be priced and charged, regardless of the existence of cultural and social expressions of value. The introduction of bulk water charges (Article 9 of the Directive) is the regulatory instrument that more concretely translates this ideological equivalence between water value and money value.[9] Water charges have represented the main controversy related to the WFD in Portugal, particularly in the period between 2005 and 2008. After three years of debate, it was eventually decided that the charges should be calculated taking into account also the volume of effluent discharge, extraction of inert material, land use area, public water projects and the level of regional water scarcity. It is unfortunate that the regular clashes between stakeholders and public authorities ended up giving the impression to the general public that the regulatory regime under WFD is ultimately about monetary costs and tariffs, rather than about environmental conservation (cf. our interviews with local stakeholders).

---

[9] In addition, the imposition of bulk water charges helps to enforce the new regulation: the income of the charges will serve to pay for at least 2/3 of the regional water administrations (ARH) and will feed into a national fund, which will serve to pay for environmental restoration measures. Note that several stakeholders complained during our interviews that the environmental benefits that may arise from the revenues from the charges are doubtful and uncertain.

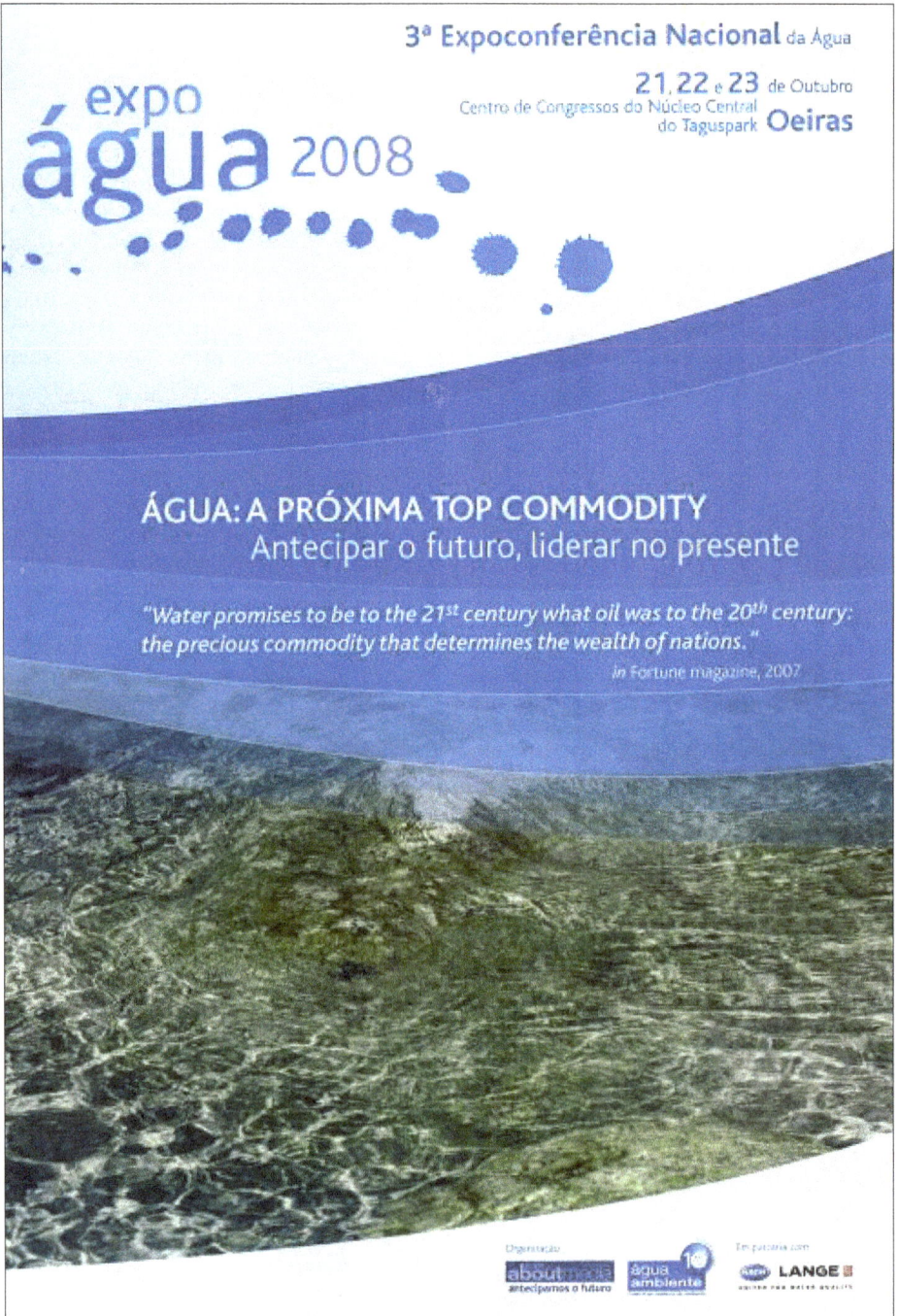

Fig. 3. A Congress Flyer Where Water was Directly Depicted as a Commodity

Agriculture is probably the water user sector that best encapsulates the anxieties in relation to the new water charges and the WFD in general. Farmers are now expected to pay the second higher charges (€ 0.003/m³ plus the other charging factors), but their resistance to the new user charges are not simply proportional to the financial burden. On the contrary, other political and cultural factors also interfere in the disputes, although not normally recognised.

According to the last river basin plan (INAG, 2001), there are 200,000 hectares of irrigation in the Douro catchment, the great majority being small, intensive farming units located between Porto and Vila Real. These farmers have been described in official documents as responsible for the highest rate of water demand and the lowest rates of user efficiency, which imply that investments are needed for the development of backstage technical capabilities and adequate planning procedures (INAG, 2001). That is reaffirmed in the first WFD report (INAG, 2005), which estimates that the tariffs paid by agriculture prior to the new Directive (i.e. which was adopted in some public agriculture projects in the Douro) only used to recover 9% of the total costs of water supply (note that equivalent urban tariffs used to cover 82% of the same costs). It mans that the difference was paid in the form of government subsidies to the farming sector and that is now increasingly seen as unpopular and unacceptable by water regulators. In addition to lowering the subsidies, the water Directive introduced the universal payment for bulk water charges as a mechanism to 'steer the behaviour' of the water users (as declared by government representatives in a seminar organised by the Portuguese Farmers Confederation on 08 Jul 2008). It should come as no surprise, then, that the majority of farmers believe that the new environmental regulation is an extra-burden to a sector that is already under serious pressure due to declining governmental support (under the Common Agriculture Policy [CAP]) and the transfer of European funds to the Eastern side of the continent.[10] In our interviews, both enterprise and small farmers were unanimous in criticising the charges and blaming the northern European countries, where irrigation is less critical, for imposing the new water regulation. Four months after the introduction of charges (on 01 Jul 2008), members of the agribusiness argued that water has a huge 'value' for the farmers, but it should not have a monetary 'price'. In an interview on 21 Nov 2008, it was declared that:

"I consider a distortion of competition the application of a new fee on water used by agriculture in the Mediterranean countries. Why? Well, if you live in Scotland, or in Brussels, you have much higher and more often precipitation, whilst in Portugal it rains less and for shorter periods of time. A farmer in Portugal has to invest in water storage and pipelines, pay for the irrigation equipment, energy and in ten years has to replace the equipment. The costs are very high and already restrain water use. In this context, comes the European Union and says 'we all need to pay for water in order to improve efficiency and environmental quality. (...) The farmers don't need to pay for water to use it more efficiently... You know, the farmer already has a deep relation with the water cycle. Now, the main risk is that this charge becomes [merely] a new tax that will not contribute to improve the

---

[10] Farmers also criticise the delays and mismanagements in other areas of government intervention, such as the protest expressed by the Fruit Association of Armamar about the fact that the Temilobos dam (in the middle section of the Douro), which was planned to provide water for 1,200 hectares of irrigated apple groves, was still not operational in the end of 2008, two years after its completion.

environment. (…) I strongly believe that in situations of water scarcity the user should pay less, not more for water".

It is evident that such argument subverts the logic of environmental economics, which postulates that scarce resources should attract higher user charges. Farmers in the area also mentioned that there is limited room for improving efficiency (at least at low costs), since they are the first to want to save water and reduced operational costs with electricity and irrigation equipment (which they claim to have done already). That indicates how the economic value of water, instead of a straightforward figure, is in effect a highly contested and contestable concept. By their turn, representatives from the small farmers community complained that the charges were adopted in Portugal before the definition of environmental management targets, which ultimately serves to demonstrate that the new water policies are centred on the 'commercialisation' of water and not on the protection of nature. The following passage summarises the feeling among the small, family agriculture:

> "[M]any times the farmers and the agriculture sector are sees as reckless users of water. These discussions fail to consider the reality of the Portuguese agriculture, as well as ignore the deep, even passionate, relationship of the farmer with water (…) [T]his law liberates the state from the responsibility to look after the conservation of water, given that it leaves it open to the market. About the social relevance of water, little or nothing is said. (…) [the consequences of the new charges] are inevitably the increase in production costs and, as a result, the elimination of those that don't have financial means to pay for it" ([National Agriculture Confederation [CNA], 2006).

In addition, sector representatives protested that the bulk water charges in Portugal are three times higher than equivalent figures in France and that it was adopted by the Portuguese government two years earlier than in Spain (i.e. 2008 in Portugal and 2010 in Spain). Nonetheless, in the Spanish side of the Douro the controversy about volumetric charges to agriculture has also dominated the public discussion about the impacts of the new water regulation. The sector is responsible for 93% of the water demand in Spain because of 563,105 hectares under irrigation, specially concentrated along the main river channel and in some of the larger tributaries (Gómez-Limón et al., 2008). The use of water in the river basin is claimed to be one of the least efficient in Spain, which has again become a strong justification for modernization and search for efficiency (Domingues et al., 2004; Gómez-Limón & Gómez-Ramos, 2007). As in Portugal, economic modelling based on multi-criteria objectives suggests that water pricing could exert significant influence on the behaviour of farmers in terms of water use due to shifts to better equipment, less water demanding and rainfed crops (Gómez-Limón & Martínez, 2006), but because of the declining profitability of agriculture only low or very low volumetric charges can be arguably borne by farmers. It seems also that the impact of bulk charges would be mainly on incoming irrigators, because those already established will have major difficulties to adjust their practices and would probably abandon or reduce their activity, with consequent loss of jobs in the region (Gómez-Limón et al., 2008). On the top of that, because of the climatic conditions of Castilla y León, productivity is relatively low and, according to agronomic research in the University of Valladolid changes in irrigation equipment are unlikely to significantly improve economic and technical efficiency (personal communication from university researchers). Nonetheless, the official position of the CDH, the water regulator, remains firmly in favour of replacing surface irrigation with spray irrigation equipment in order to save water. Contradictorily, there are also plans to build new water storage dams in the headwaters of several Douro tributaries to increase the irrigation area. Both the new dams and the volumetric charges have received some level of opposition from the 400 associations of irrigators in the Spanish Douro. In the case of the community of Bajo Carrión (visited during our research in the Douro), the directors had recently resigned and new elections were called exactly because of disagreement about the modernization targets required by the regulator (i.e. the majority of the members voted against the acceptance of efficiency-centred regulatory demands).

The underlying problem with policies that try to induce higher efficiency through charges and the sudden incorporation of external costs is that it ignores existing social and spatial inequalities, which can be aggravated if not properly considered, as mentioned by Tsakalotos (2004: 29), "...while the expansion of the market, and market-type arrangements, are often defended on the grounds of efficiency, they are also often implemented in a manner that goes well beyond the discourse of efficiency. (...) Such a strategy makes alternative conceptions much more difficult to conceptualize, let alone carry out". If the introduction of bulk water charges has represented a major controversy among small and large farmers, an analogous situation happened among companies responsible for public water supply and sanitation. Despite the fact that a full privatisation (i.e. divestiture) seems out of the political agenda – in large measure, because of fierce public opposition – the association between water and money remains present in the collective imaginary of the population (illustrated in a Portuguese newspaper cartoon in Figure 4). It has been widely stated in official documents, reports and guidance that public water services in Portugal were and continue to be thwarted by inefficiency and that the introduction of WFD should be associated with cost-recovery measures and higher water user charges.[11] In particular, local water providers ('low' companies) are blamed for their backward thinking as a "hindrance to the development of water supply sector" (that is exactly the expression used in the cover page of the main magazine of water services in Portugal, Água & Ambiente, June 2005). Rather than being politically neutral, those claims for cost recovery have provoked tensions and uneasiness between the various water utilities that operate in the same geographical area (i.e. the 'high' and 'low' companies; see details above). For example, in 2008 the municipal company formed to serve the city of Porto (Águas do Porto) was able to reduce the purchase of water from the Águas do Douro & Paiva in 80,000 m³/day (out of a total of 280,000 m³ distributed daily), according to its chief-manger (interview on 14 Nov 2008). That corresponds to a net saving of € 216,000/month in terms of payment made to the regional company or around 12% of her income (in 2008). As a result, Águas do Douro & Paiva tried unsuccessfully to raise their tariffs by 8% in 2008, but the government allowed an increase of 5.5% (note that the rate of inflation in the year 2008 was 2.7% in Portugal).

If a large company such as Águas do Porto was able to confront the regional water authority, other municipal entities are left in a much weaker position to negotiate costs and conditions with the regional water utilities. In our interviews with managers, engineers and politicians responsible for the water services in the cities and towns in the upper Douro, we

---

[11] Nonetheless, as in the case of the low elasticity price-demand of agriculture mentioned above, the increase of user charges in the last few years has had limited influence on the level of water demand (Monteiro & Roseta-Palma, 2007).

Fig. 4. The Transformation of Water into Money in Portugal and under the Influence of the European Union (by Luís Afonso, "O Público", 22 Feb 2004)

detected a considerable level of resentment about the pressures exerted by the central government in favour of the regionalisation of the service. Some municipalities that passed to buy water from the regional companies are even contemplating a return to local water abstraction and treatment. It was constantly mentioned that the purchase of water from the regional company normally costs more than twice the local costs with abstraction and treatment. Part of this difference can be explained by the investments made by the larger company to comply with drinking water legislation, something that many local authorities fail to observe. Moreover, there is also a clear resentment with the fact that heavy public investments were made by the national government in the Porto metropolitan area in the past, but today's investments are expected to be borne by the local water companies via customer charges (i.e. the cost-recovery policy). More than the regional companies, local water operator face major political barriers to transfer higher charges to the population and that has led to growing protest and some cases of physical violence (as in the invasion of the Peso da Régua Council in 2002). It is therefore not unexpected that a similar criticism took place after the announcement of the WFD bulk water charges in 2008 (vis-à-vis newspaper articles published in the period). As in the agriculture sector, public reaction lacks proportionality with the additional financial burden (i.e. the impact of the WFD charges on each household is relatively low, estimate at around € 0.20 per month, which corresponds to 2.5-3.0% of the average tariff). Interestingly, the cost of the tariff is likely to be relatively low for the majority of urban water users, as much as it is for the farmers, which suggests that the opposition expressed is not really about the financial levy *per se* but rather a deep antipathy toward the interference in long-established water use practices. It suggests that public opposition is not just about the charge, but it reacts against a vague sense of lost ownership and the disruption of established forms of relation between society and nature.

While the general population reacts – in spontaneous or organised ways – against additional charges in agriculture and urban water supply, other more coordinated protests intensify against the construction of large dams in the Douro (something that the WFD regime has been so far unable to prevent, because of political pressures). The new dams are part of the attempt to secure 60% of electricity from renewable sources by 2020, which has been strongly confirmed by the Prime Minister, as in a public event when he stated that "Portugal is the European country with more hydropower reserves to be exploited" (RTP News, 20 Nov 2008). As mentioned above, the Douro is the main powerhouse of Portugal and is again where six (out of ten) new large hydropower schemes will be built, according to the National Programme of Dams with High Hydroelectric Potential (INAG, 2007). If in the past the dams were erected across the main channel, the focus of the construction of hydropower dams is in the tributaries, such as in the Rivers Tua and Tâmega. The Citizenship Movement for the Development of the Tâmega has challenged the activities of the energy companies responsible for the new dams (the Portuguese EDP and the Spanish Iberdrola). It is still vividly present in the memory of the local residents the controversy about a dam planned in the River Côa in the 1990s and firmly resisted because of the impact on archaeological sites with rupestrian paintings. In the Tâmega, the campaign against the Fridão dam and to protect the town of Amarante started in 1995. Probably the largest mobilisation today is against a dam in the River Sabor, a large structure (123 metres high) that will flood 2,820 hectares and also impacts on archaeological sites (see Figure 5 regarding a protest event in Apr 2008). Despite the likely impact on important conservation reserves, the government gave the go-ahead for the project, which was then appealed to the European Commission. The anti-dam activists lost the appeal in 2007, but were planning to resort to the European Court of Justice on grounds of what they see as 'serious mistakes' of the environmental impact assessment (interview with NGO activist, 19 Nov 2008).

Apart from environmental impacts, another source of criticism about dams in the Douro is the general feeling that the hydropower schemes build in the last decades have contributed little to improve the live of the communities of the Upper Douro. After the construction, the operation of the dams only generates a small number of jobs in the region and brings only marginal contribution to local communities (cf. our interviews with residents and city councillors). The fact that electricity is generated in the same area of the dams and then transferred to other parts of the country, reinforces a sense of dual citizenship between the coast and the inland. For long time now, the rural areas of the Upper Douro have been suffering from depopulation, loss of small-scale agriculture, abandonment of cultivable land, and lack of viable economic perspectives (see CCDR-N, 2007). The economic decline of the rural areas has been taking place for decades and recent development initiatives focused on diversification and market integration (most with European Union support) have not reversed that trend. On the contrary, it has resulted in a higher level of dependency, uncertainty and lower self-sufficiency (Moreno, 2003). The economic and cultural transformations taking place in the Douro have largely operated under the influence of foreign investments (Roca & Oliveira-Roca, 2007), but such policies have had little

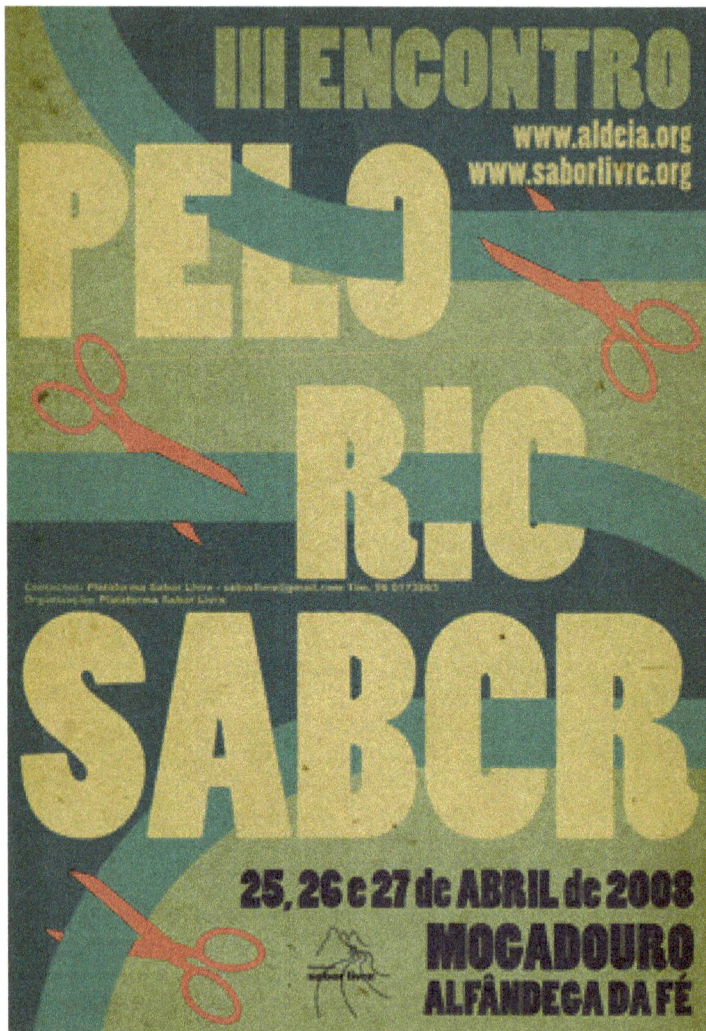

Fig. 5. Leaflet of the Mobilisation Against the Sabor Dam [i.e. the leaflet says 'In favour of the Sabor River' and invites the population for a protest in April 2008]

effectiveness in promoting the changes require by small and medium-size enterprises (Bateira & Ferreira, 2002). That context of perceived remoteness and misfortune is reflected in socio-economic and interpersonal relations, which includes **a disregard for traditional forms of collective water management**. We had the opportunity to visit a number of sites that where until the recent past (around 30 years ago) used to practice a community form of irrigation. These are areas of family agriculture where, in the past, each day of the week a different farmer used to divert water to his/her piece of land, with full transparency and accountability among the community regarding the amount of water used. That is the case in the rural communities of Vila Real, where the irrigation infrastructure had to be carved in a hilly landscape, which also involved family and community work. Because of changes in agriculture production and concentration of landed property, such forms of water use are disappearing fast. That is an example of how changes in water management practices intensely encapsulate local and international dynamics, but unfortunately there has been almost no space to consider those issues that fall out of the mainstream ethos of the WFD regime.

## 4. Discussion: Spatial rigidity and monotonic categorisation

The implementation of the Water Framework Directive represents a decisive moment in the institutional history of water management in Europe, Portugal and the Douro. The WFD regime, including methodological improvements and more stringent targets, constitutes what can be called a 'metarregulation' with wide range impacts and lasting consequences. The higher level of concern for environmental impacts and the wasteful patterns of water use can be identified as positive steps in the direction of resolving lifelong problems. At face value, the detailed timetable of the new Directive seems to offer a robust mechanism for the assessment of ecological trends and the formulation of cost-effective solutions. However, the implementation of the Directive has served to consolidate an interpretation of problems that favours specific political and macroeconomic interests. The prevailing approaches systematically conceal that water reforms are an integral part of broader social transformations in the mechanisms of production and consumption of tradable goods and in the interpersonal relations. Likewise, mainstream procedures tend to ignore that the WFD regulation brings water management further into the sphere of money circulation and power political forces, which happens in important and contingent forms. Under a hegemonic approach informed by such technical and economic translation of problems, an array other important aspects of water management have received almost no attention, such as inter-catchment integration, the delegation of decision power and the balance of power behind the technological fix. WFD creates new opportunities to raise management issues (such as the increasing degradation of surface and ground water bodies) but there remains a tension between continuity and innovation that essentially reflect political clashes. The new Directive is implemented by invoking an apparent consensus about water issues, but under surface remains a series of intricate complexity of intersector and geographical inconsistencies. Making use of a universalising symbolism of 'common' challenges and 'shared' responsibilities, the implementation of the WFD never avoided being itself a locus of disputes and power affirmation.

It can be accepted that the WFD conveys improvements in many areas, such as a holistic approach to catchment issues, the consideration of cumulative impacts and the cyclical (adaptive) response to environmental degradation pressures. Even so, serious controversies persist in relation to the priorities of state action, which operates in favour of certain interests at the expense of broader, and more legitimate, social expectations. It should be remembered that the state includes a range of government bodies, regulatory agencies, parliaments and courts, a large entity that extends from the local to the global with fluid boundaries and exposed to the disputes between groups, classes and geographical areas (Jessop, 2008). The complexity of the state apparatus is even greater in the contemporary world, where a multiplicity of goals and liabilities frequently create significant confusion among members of the general public. It is not clear to everybody that statehood is being qualitatively reformulated according to a wild interplay between homogenisation and particularisation, which unfolds towards higher levels of business competition, market liberalisation and economic growth (Brenner, 2004). The hegemonic reorganisation of the state according to neoliberal demands constitutes a multifaceted, non-linear and multiscalar process that tends to engulf all areas of social action and, crucially, to reshape socionatural relations according to the political and economic priorities of global markets (see Finlayson et al., 2005). The difficult challenges involved in that progression towards an Europe of interconnected localisms and pervasive market rationality is yet more acute in its semi-peripheral countries, such as Portugal and Spain, which are expected to breach the development gap with northern regions whilst also cope with democracy deficits and growing environmental threats.

Our current assessment of the WFD experience builds upon a previous analysis that identified the overly ambitious goals of the Directive and the (often neglected) need to carefully consider the historico-geographical features of the Douro. The internal contradictions of the new regulatory landscape was then defined as a 'techno-bureaucratic shortcut', which means a tendency to produce superficial adjustments in practices and procedures whilst the overall trend of (bureaucratised and exclusionary) management remains largely unchanged (Ioris, 2008). Based on the points discussed above, it is now possible to further argue here that the 'techno-bureaucratic shortcut' has effectively two main ontological foundations, namely a spatial rigidity and the monotonic categorisation of water management issues. The first source of constraint – spatial rigidity – is related to the static understanding of how ecological and socionatural processes interact and evolve. The Directive has been territorialized (to the catchment scale) by ignoring the constant and perpetual remaking of the catchment's spatial configuration (i.e. the social and socionatural relations that produce space). The new regulation has progressed inflexibly across rigid geographical axes – above all, the nested spheres of governance of the European Union – with limited opportunity for deviating from a priori established management directions. Under the assumption that all Europe requires the same form of water management and regulation, the national state is powerfully inserted in a dialectics of inertia and modernisation that is predetermined by the transnational centres of political power. In that context, the regulatory principles of water management emanate concentrically from the top (the EU apparatus controlled by the stronger groups of interest) to the member states and from that to catchments and locations. The result of this rigid management of water is a pressure for the homogenisation of water management and regulation, which happens, first and foremost, through a narrow set of scientific methodologies typically developed in the northern European countries and reproduced with almost no modifications in Portugal (e.g. Bordalo et al., 2006).

Second, the interpretation of management problems and the formulation of possible solutions have followed the monotonic categories of the new European regulation, in particular the myriad of environmental economics tools that colonise the nucleus of WFD regulation, such as water charges, water markets, and the payment for ecosystem services. Under this quest for technical and operational efficiency local knowledge and the indigenous understanding of the hydrological system are being rapidly lost. The introduction of new semiotic basis for water management leads to the translation of local water issues into a technical vocabulary that is only shared by a small number of stakeholders (i.e. regulators, professional activities, engineers, and consultants). Because of this monotonic understanding of water problems, the direction of water management is decided upfront, with limited scope for innovation and creativity at the local level. It is true that the erosion of autochthonous wisdom did not start in the period of WFD implementation, on the contrary, it has been the outcome of larger processes of social and economic change, in particular the abandonment of traditional agriculture practices and depopulation. Nonetheless, the new Directive accelerates that process, given that the national states enjoy limited flexibility to decide about technical thresholds and regulatory instruments. Due to the spatial rigidity and monotonic assessments, there is a tendency to bypass the more time consuming steps of the new regulation (in particular, public participation and information sharing) and, unsurprisingly, opt for the aforementioned 'bureaucratic shortcut'. Overall, the shortcut tendency is itself an outcome of the very structure of the new regulation, which allows limited room for the detailed understanding of local circumstances and the genuine engagement of stakeholders.

## 5. Conclusions

This brief examination of the local experience of water institutional reforms in the Douro demonstrates the persistent mismatch between regulatory objectives and the actual procedures and relations taking place in different parts of the river basin. The process of water regulatory reforms started in the 1990s, following macroeconomic and politico-institutional changes, and was translated into new legislation and increasing calls for an integrated management of catchments. However, it was really the opportunity created by WFD that provided the opportunity to introduce an new, more holistic regulatory rationality. Yet, underneath the new institutions, which include the introduction of water charges, public consultations, preparation of plans and scientific assessments, there is a constant reaffirmation of a centralised and selective basis of dealing with water management questions. Those problems have seriously limited the prospects of the new water institutional framework. Behind the hectic agenda of activities related to the introduction of WFD, it is possible to discover the persistence of old established practices that had marked the history of water management of the European Union in previous decades. Attempts to improve water management in the catchment under the WFD regime have often revived long-established cleavages and the inconsistencies of public policies related to the allocation, use and conservation of shared resources, which have typically privileged certain groups of stakeholders and geographical areas.

It was shown how a rationalistic approach to water problems has prevailed and pervaded most of the recent reforms. The narrow focus on engineering constructions has been replaced by more subtle attempts to manage water through economic incentives and impact mitigation, but without ever addressing the underpinning contradictions of water use and economic development. Although there is a shift from single processes to water regulation, there remains a clear line of continuity between the past and the present of water use and conservation in the Douro. If WFD helps to draw attention to water problems and mobilises private and public resources, at the same time it unravels silent conflicts, creates competition and not necessarily facilitates the participation of the weaker social groups. This paradox is not resolved within the water regulatory framework only, but requires broader political basis for dealing with shared problems. If in the past, public investments were in water infrastructure, the current top-down approaches to water management basically reproduce this engineering-based model of development and management, without questioning the causes of environmental degradation and the main beneficiaries. Under WFD, water is emerging (or re-emerging) as a locus of political disputes involving a myriad of stakeholder groups and spatial relations. The Douro is an emblematic example of how water management should be understood as not only a technical and economic matter, but also directly related to political questions of social exclusion. What is still lacking is a genuinely innovative way of dealing with water problems, one that resolves the uneven balance of power between spatial areas and social groups, as well as incorporates traditional wisdom and the contribution of local people in the development of innovative solutions to old and new water management challenges. Unless social differences and the reproduction of social inequalities are addressed, water management problems will remain unchanged.

The ultimate result is that, notwithstanding legal and discursive improvements, the long-term causes of water problems – namely, political pressures for maximising the economic outcomes and minimising the investments in social equity and environmental conservation – have been left out of the process of regulatory change. The limited availability of long-term monitoring data and detailed technical studies have contributed to reinforce the two fundamental hindrances of the regulatory regime under WFD (namely, spatial rigidity and monotonic categorisation of problems), leading to an evasion of references about the political origins and the socioeconomic consequences of environmental impacts. In the end, WFD remains a contested experience of environmental regulation that oscillates between attempts to commodify nature (e.g. bulk water charges, valoration of ecosystem services, calculation of disproportionate costs) and the affirmation of techno-bureaucratic mechanisms of law enforcement (i.e. that neglect the demands and needs of large proportion of water stakeholders). The asymmetry of political power also operates in the interstices of the regulation, given that the water reforms promoted through WFD have served to implement a particular worldview and serve specific interests under a universalising discourse and a naturalisation of hegemonic agendas. On the other hand, the imposition of techno-bureaucratic approaches to water management has prompted the emergence of various forms of opposition, either at the local level or in coordination with other national and international forms of contestation (as the criticism of water privatisation and the campaigns against the new dams in the Upper Douro). The success of the next stages of the implementation of WFD will depend on the ability to perceive the broader socionatural complexity of water management, the pursuit of effective forms of social inclusion and a more equal balance of negotiation power.

## 6. Acknowledgements

The author warmly thanks the interviewees in Portugal and Spain, the University of Porto for the logistical assistance and the financial support received from The Carnegie Trust for the Universities of Scotland.

# 7. References

Alves, J.F. (2005). *Águas do Douro e Paiva S.A.: Dez Anos 1995-2005*, Águas do Douro e Paiva, ISBN 972-99717-0-6, Porto, Portugal.

Amador, J. (2003). The Path Towards Economic and Monetary Integration: The Portuguese Experience. *Finance a Uver - Czech Journal of Economics and Finance*, Vol. 53, No.9-10, pp. 413-429, ISSN 0015-1920.

Ambiente Online. (2005). Águas de Portugal Caminha para Abertura de Capital a Privados, 25.05.2005, Available from
http://www.ambienteonline.pt/noticias/detalhes.php?id=2597.

Amorim, A.A. & Pinto, J.N. (2001). *Porto d'Agoa: Serviços Municipalizados de Águas e Saneamento do Porto*, ISEP, ISBN 972-953-43-2-2, Porto, Portugal.

Azevedo, J. (Ed.). (1998). *Entre Duas Margens: Douro Internacional*, Tipografia Guerra, Viseu, Portugal.

Barreira, A. (2003). La participación pública en la Directiva Marco del Agua: Implicaciones para la Península Ibérica, *Proceedings of II Congreso Ibérico sobre Gestión y Planificación del Agua*, Porto, November 2000.

Bateira, J. & Ferreira, L.V. (2002). Questioning EU Cohesion Policy in Portugal: A Complex Systems Approach. *European Urban and Regional Studies*, Vol.9, No.4, (October 2002), pp. 297-314, ISSN 0969-7764.

Bordalo, A.A.: Teixeira, R. & Wiebe, W. (2006). A Water Quality Index Applied to an International Shared River Basin: The Case of the Douro River. *Environmental Management*, Vol.38, No.6, (December 2006), pp. 910-920, ISSN 0364-152X.

Brenner, N. (2004). *New State Spaces: Urban Governance and the Rescaling of Statehood*, Oxford University Press, ISBN 978-0-199-27005-7, Oxford, UK.

Brito, A.G.; Costa, S.; Almeida, J.; Nogueira, R. & Ramos, L. (2008). A Reforma Institucional para a Gestão da Água em Portugal: As Administrações de Região Hidrográfica, *Proceedings of VI Congreso Ibérico sobre Gestión y Planificación del Agua*, Vitoria-Gasteiz, Spain, December 2008.

CCDR-N. (2000). *Atlas Ecológico do Rio Douro: Divisão em Troços Ecológicos do Rio*, CCDR-N & Junta de Castilla y León, ISBN 972-734-236-1, Porto, Portugal & Valladolid, Spain.

CCDR-N. (2006). *Norte 2015: Competitividade e Desenvolvimento: Uma Visão Estratégica*, Comissão de Coordenação e Desenvolvimento Regional do Norte, Porto, Portugal.

CCDR-N. (2007). *Programa Operacional Regional do Norte 2007-2013*, Comissão de Coordenação e Desenvolvimento Regional do Norte, Porto, Portugal.

CNA. (2006). A Água e a Agricultura: Novas Realidades. *Voz da Terra*, Vol.49, (October November), pp. 33-44, ISSN 0870-5356.

Conca, K. (2006). *Governing Water: Contentions Transnational Politics and Global Institution Building*, MIT Press, ISBN 0-262-53273-5, Cambridge, USA & London, UK.

Correia, F.N. (2000). O Planeamento dos Recursos Hídricos como Instrumento da Política de Gestão de Água. *Recursos Hídricos* Vol.21, No.1, pp. 5-12, ISSN 0870-1741.

Cristovão, A. (2006). Douro-Duero: Em Busca de Caminhos para o Desenvolvimento Transfronteiriço, *Proceedings of the Congresso em Homenagem ao Douro/Duero e Seus Rios: Memória, Cultura e Porvir*, Zamorra, Spain, April 2006.

Cunha, L.V.; Gonçalves, A.S.; Figueiredo, V.A. & Lino, M. (1980). *A Gestão da Água: Princípios Fundamentais e Sua Aplicação em Portugal*, Fundação Caloustre Gulbenkian, Lisbon, Portugal.

Cunha, L.V.; Serra, A.; Costa, J.V.; Ribeiro, L. & Oliveira, R.P. (Eds.). (2007). *Reflexos da Água*, Associação Portuguesa de Recursos Hídricos, ISBN 978-972-99991-4-7, Lisbon, Portugal.

D'Alte, T.S. (2008). O Mercado de Águas em Portugal: O Comércio de Títulos na Lei da Água. *Fórum de Direito Urbano e Ambiental*, Vol.38, pp. 83-96, ISSN 1676-6962.

Del Saz-Salazar, S.; Hernández-Sancho, F. & Sala-Garrido, R. (2009). The Social Benefits of Restoring Water Quality in the Context of the Water Framework Directive: A Comparison of Willingness to Pay and Willingness to Accept. *Science of the Total Environment*, Vol.407, No.16, (September 2009), pp. 4574-4583, ISSN 0048-9697.

Dominguez, D.; Manser, R. & Ort, C. (2004). *No Problems on Río Duero (Spain) – Rio Douro (Portugal)? The Science and Politics of International Freshwater Management*, ETH, Zurich, Switzerland.

European Commission. (2007). *Addressing the Challenge of Water Scarcity and Droughts in the European Union: Impact Assessment, Com(2007)414 final*, Commission of the European Communities, Brussels, Belgium.

Finlayson, A.C.; Lyson, T.A.; Pleasant, A.; Schafft, K.A. & Torres, R.J. (2005). The "Invisible Hand": Neoclassical Economics and the Ordering of Society. *Critical Sociology*, Vol.31, No.4, (July 2005), pp. 515-536, ISSN 0896-9205.

Gómez-Limón, J.A. & Gómez-Ramos, A. (2007). Opinión Pública sobre la Multifuncionalidad del Regadío: El Caso de Castilla y León. *Economía Agraria y Recursos Naturales*, Vol.7, No.13, pp. 3-25, ISSN 1578-0732.

Gómez-Limón, J.A. & Martínez, Y. (2006). Multi-criteria Modelling of Irrigation Water Market at Basin Level: A Spanish Case Study. *European Journal of Operational Research*, Vol.173, No.1, pp. 313–336, ISSN 0377-2217.

Gómez-Limón, J.A.; Morales, A.; Alonso, M. & Diéguez, S. (2008). *Caracterización del Uso del Agua en la Agricultura. Plan Hidrológico de la Parte Española de la Demarcación Hidrográfica del Duero*, Confederación Hidrográfica del Duero, Valladolid, Spain.

Hedelin, B. & Lindh, M. (2008). Implementing the EU Water Framework Directive: Prospects for Sustainable Water Planning in Sweden. *European Environment*, Vol.18, No.6, (November December 2008), pp. 327-344, ISSN 0961-0405.

INAG. (2001). *Plano de Bacia Hidrográfica do Rio Douro – Relatório Final*, Ministério do Ambiente e do Ordenamento do Território, Lisbon, Portugal.

INAG. (2005). *Relatório Síntese sobre a Caracterização das Regiões Hidrográfi cas Prevista na Directiva-Quadro da Água*, Ministério do Ambiente e do Ordenamento do Território, Lisbon, Portugal.

INAG. (2007). *Programa Nacional de Barragens com Elevado Potencial Hidroeléctrico (PNBEPH)*, INAG/DGEG/REN, Lisbon, Portugal.

Ioris, A. A. R. (2008). Regional development, nature production and the techno-bureaucratic shortcut: the Douro River Catchment in Portugal. *European Environment*, Vol.18, No.5, (September October 2008), pp. 345-358, ISSN 0961-0405.

IRAR. (2008). *Relatório Annual do Sector de Águas e Resíduos em Portugal – 2007*, Instituto Regulador de Águas e Resíduos, Lisbon, Portugal.

Jessop, B. (2008). *State Power: A Strategic Relational Approach*, Policy Press, ISBN 978-07456-3320-6, Cambridge, UK.

Malheiro, P. (2008). Portugal Apresenta Quinta Maior Pegada de Água per Capita do Mundo. *Água & Ambiente*, Vol.119, (October 2008), pp. 8-10.

Martins, J. P. (1998). *Serviços Públicos de Abastecimento de Água e de Saneamento: Opções de Financiamento e Gestão dos Municípios Portugueses*, AEPSA, Lisbon, Portugal.

Miranda, J. C. (1986). Para uma Política da Água em Portugal. *Recursos Hídricos*, Vol.7, No.3, 5-7, ISSN 0870-1741.

Monteiro, H. & Roseta-Palma, C. (2007). *Caracterização dos Tarifários de Abastecimento de Água e Saneamento em Portugal*, ISCTE, Lisbon, Portugal.

Moreno, L. (Ed.). (2003). *Guia das Organizações e Iniciativas de Desenvolvimento Local*, ANIMAR, Lisbon, Portugal.

Moura, R. M. (2007). A Importância do Vale do Rio Douro na Conservação da Paisagem e os Problemas de Gestão Decorrentes. *População e Sociedade*, Vol.13, pp. 107-123, ISSN 0873-1861.

Queirós, M. (2002). O Ambiente nas Políticas Públicas em Portugal. *Finisterra*, Vol.37, No.73, pp. 33-59, ISSN 0430-5027.

Pereira, G. M. (Ed.). (2008). *As Águas do Douro*, Águas do Douro e Paiva, ISBN 978-972-36-0992-9, Porto, Portugal.

Pereira, G.M. & Barros, A.M. (2001). *Memória do Rio: Para uma História da Navegação no Douro*, Afrontamento, ISBN 978-972-36-0547-1, Porto, Portugal.

Richardson, J. (1994). EU Water Policy: Uncertain Agendas, Shifting Networks and Complex Coalitions. *Environmental Politics*, Vol.3, No.4, pp. 139-167, ISSN 0964-4016.

Roca, Z. & Oliveira-Roca, M. N. (2007). Affirmation of Territorial Identity: A Development Policy Issue. *Land Use Policy*, Vol.24, No.2, (April 2007), pp. 434–442, ISSN 0264-8377.

Sabater, S.; Feio, M.J.; Graça, M. A. S.; Muñoz, I. & Romaní, A. M. (2009). The Iberian Rivers, In: *Rivers of Europe*, K. Tockner; U. Uehlinger & C. T. Robinson, (Eds.), 113-149, Academic Press, ISBN 978-0-12-369449-2, London, UK.

Surridge, B. & Harris, B. (2007). Science-driven Integrated River Basin Management: A Mirage? *Interdisciplinary Science Reviews*, Vol.32, No.3, (September 2007), pp. 298-312, ISSN 0308-0188.

Thiel, A. 2006. Institutional Changes in Water Management in the Algarve: Actors and Crises Due to Contradictions of Capitalism, *Proceedings of the V Congreso Ibérico sobre Gestión y Planificación del Agua*, Faro, Portugal, December 2006.

Tsakalotos, E. (2004). Social Norms and Endogenous Preferences: The Political Economy of Market Expansion, In: *The Rise of the Market: Critical Essays on the Political Economy of Neo-Liberalism*, P. Arestis & M. Sawyer, (Eds.), 5-37, Edward Elgar, ISBN 978-1-84-376725-1, Cheltenham, UK and Northampton, USA.

van der Brugge, R. & Rotmans, J. (2007). Towards Transition Management of European Water Resources. *Water Resources Management*, Vol.21, No.1, (January 2007), pp. 249-267, ISSN 0920-4741.

Veiga, B.G.A.; Chainho, P. & Vasconcelos, L. T. (2008). A Directiva-Quadro da Água Enquanto Elemento Potenciador dos Processos de Participação Pública: Casos de Portugal e França, *Proceedings of the V Luso-Moçambicano Congress of Engineering*, Maputo, Mozambique, September 2008.

Yin, R. K. (1994). *Case Study research: Design and Methods*, SAGE, ISBN 978-080-39-5663-6, Newbury Park.

# Part 5

## Water Demand / Water Pricing

# 12

# Water Soft Path Analysis – Jordan Case

Rania A. Abdel Khaleq
*Ministry of Water and Irrigation*
*Jordan*

## 1. Introduction

In water management today, there are two primary ways of meeting water-related needs, or two "paths". One path can be called the "hard" path which relies almost exclusively on centralized infrastructure and decision making: building dams and reservoirs, pipelines and treatment plants, and establishing water departments and agencies. It delivers water, mostly of potable quality, and takes away wastewater. The second path or the "soft" path may rely on centralized infrastructure, but complements it with extensive investment in decentralized facilities, efficient technologies, and human capital. It delivers diverse water services matched to the users' needs and works with water users at local and community scales. –

The purpose of this chapter is to present the "Soft Path Analysis" as an approach that improves the overall productivity of water use, rather than the business as usual approach with the endless search of new sources of water supplies.

The water soft path is modelled on the highly successful approach to energy known as the soft energy path. The "soft path" planning approach for fresh water differs fundamentally from supply focused planning. It starts by changing the conception of water demand. Instead of viewing water as an end product, the soft path views water as the means to accomplish certain tasks. The role of water management changes from building and maintaining water supply infrastructure to providing water related services, such as new forms of sanitation, drought-resistant landscapes, urban redesign for conservation and rain-fed ways to grow crops (Brandes et. al, 2005)

What calls for new approaches are also the inadequacies by which water planners and policymakers are addressing the new challenges that are further complicating the traditional approaches to solving the water problems. Issues such as regional and international water conflicts, the dependence of many regions on unsustainable groundwater use, the growing threat of anthropogenic climate change, and our declining capacity to monitor critical aspects of the global water balance are all currently inadequately addressed by water planners and policymakers. If these challenges are to be met within ecological, financial, and social constraints, new approaches are needed (Gliek, 2003).

### 1.1 Concept of soft paths

Soft paths can be defined briefly as approaches to natural resources management that rely on a multitude of distributed, relatively small-scale sources of supply coupled with ultra efficient ways of meeting end-use demands (Brooks et al, 2004). Gleick (in Brooks ,2004) provides a more comprehensive definition:

What is required is a "soft path," one that continues to rely on carefully planned and managed centralized infrastructure but complements it with small-scale decentralized facilities. The soft path for water strives to improve the productivity of water use rather than seek endless sources of new supply. It delivers water services and qualities matched to users' needs, rather than just delivering quantities of water. It applies economic tools such as markets and pricing, but with the goal of encouraging efficient use, equitable distribution of the resource, and sustainable system operation over time. In addition includes local communities in decisions about water management, allocation, and use.

## 1.2 Twentieth century policy and planning

The predominant focus of water planners and managers has been identifying and meeting growing human demands for water. Their principal tools have been long-range demand projections and the construction of tens of thousands of large facilities for storing, moving, and treating water. The long construction times and high capital costs of water infrastructure require that planners try to make long-term forecasts and projections of demand. Yet these are fraught with uncertainty. Three basic futures are possible: (i) exponential growth in water demand as populations and economies grow, (ii) a slowing of demand growth until it reaches a steady state, and (iii) slowing and ultimately a reversal of demand (Figure 1).

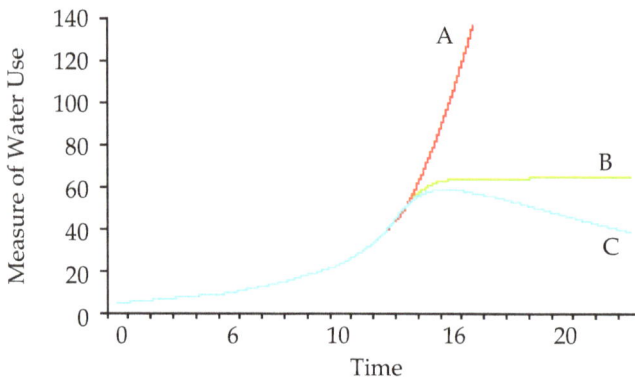

Fig. 1. Scenarios of future water use. (Gliek, 2003)

The three curves represent continued exponential growth in demand (A), a leveling off of demand to steady state (B), and declining demand (C).

Reviewing the last several decades of projections shows that in the developing world planners consistently assumed continued, and even accelerated, exponential growth in total water demand. Some projections were that water withdrawals would have to triple and even quadruple in coming years, requiring additional dams and diversions on previously untapped water resources in remote or pristine areas once declared off-limits to development (Gliek, 2003).

However, instead, total water withdrawals began to stabilize in the 1970s and 1980s in , and construction activities began to slow. More recently, the economic costs of the traditional hard path have also risen to levels that society now seems unwilling or unable to bear.

Similarly, as large-infrastructure solutions have become less attractive, new ideas are being developed and tried and some old ideas are being revived, such as rainwater harvesting and integrated land and water management. These alternative approaches must be woven together to offer a comprehensive toolbox of possible solutions (Gliek, 2003).

## 1.3 Experiences with actual soft path analyses

Soft path analysis as a detailed and rigorous method was initially developed around energy alternatives. Soft path analysis as a methodology was initially developed in the 1970s in a search for alternatives to conventional energy policy by Lovins. Its normative base was clear from the start. The early work was done within Friends of the Earth USA, partly as a way to counter the then-growing drive to build nuclear power plants (Brooks, 2005).

By the end of the 1970s, articles on soft energy paths were appearing in professional journals, and several books had been published. By the middle of the 1980s, the methodology could be considered proven; some 35 soft energy path studies had been published for various nations or regions of the world. Canadians were among the leaders in seeing the potential for soft energy paths, and Friends of the Earth Canada provided the base for some of the developmental work and later for four iterations of soft energy path analysis on Canada, including a 12 volume report by Brooks, Robinson and Torrie in 1983, and a more popular book by Bott, Brooks and Robinson in 1983. Methodological guides were also made available (Brooks, 2004).

Though no nation or state whole heartedly accepted soft path conclusions as guiding principles, their impact was quite evident in policies that began to lean toward soft technologies and in results that showed more "new" energy coming from gains in energy efficiency than from all new sources of supply together (Brooks, 2005).

In comparing and contrasting energy and water we can notice water and energy exhibit many analogies, not just as physical substances, but also in the ways in which human beings have developed them as resources. The gradual shift from simple to combined to highly complex technologies, and from individual to local to highly centralized systems, has typified these two key resources for human development. However, the shift has proceeded much further for energy than it has for water. In many respects, water systems and water policies are not so far from the soft alternative today as energy policies and systems already were 25 years ago. Water supply is, for example, typically municipal or at most national in scope, and much of it is publicly owned; energy supply is commonly global in scope, and much of it is privately owned (Brooks, 2003). But looking at water or energy as a bundle of services, rather than as a commodity, many more options can be conceived to satisfy demands (Brooks, 2005).

## 1.4 Water soft path analyses

From the first, analysts agreed that the soft path methodology could be applied to other natural resources, but analytical models have only appeared for energy, and, more recently and more partially, for water. Still today, it is fair to say that, to date, there has been no full water soft path study – at least not if by "full" we mean a semi-quantitative model of water scenarios based on soft path methods and relying on soft technologies.

The early proposals for and experiments with water soft path studies published prior to 2000 are described by Brooks (2003). Since that report, there have been further publications. By far the most important is another report from Peter Gleick and his colleagues at the Pacific Institute (Gleick et al., 2003). This report provides a review of urban water use in California, and of cost-effective methods to reduce consumption. This report is both more detailed and more rigorous than anything else to date. Happily, its conclusions are equally impressive: Without any change in water end-uses, economic structure or expected growth, at least one-third of all water use could be saved by the application of technologies that are cheaper than the costs of new supply. Should these technologies be adopted (at reasonable rates of implementation), projected economic and population growth in California could be accommodated without a single additional water supply project.

In Canada, The POLIS Project on Ecological Governance at the University of Victoria has created an Urban Water Demand Management group. Since 2003, this group has published a series of reports (Brandes, 2003; Maas, 2003; Brandes and Ferguson, 2004). The first report used information in the Statistics Canada Municipal (Water) Use Database (nicknamed MUD) to identify wide variation in both total and domestic per capita water use in Canadian municipalities. With some exceptions, it also identified an association of lower rates of use with the presence of water metres and with higher water prices. This report notes the opportunity to reduce water use in Canadian cities just by bringing the higher water consuming cities down to best practices elsewhere in Canada, and the latest report suggests the policies that would be effective at achieving this goal (Brooks et al, 2004).

## 1.5 Methodology of soft path studies

The concept for water soft paths is clearly attractive. Wolff and Gleick (2002) listed a number of characteristics of soft paths, but the key principles can be reduced to three:

- The first principle is to resolve supply-demand gaps in natural resources as much as possible from the demand side. Beyond the 50 litres per person-day commonly cited as the minimum for an adequate quality of life, there are many ways to satisfy human demands for water. The approach depends upon applying least-cost choices to every stage from water withdrawal to wastewater disposal and (ideally) reclamation plus emphasis on the need for the actual "services" desired, as opposed simply to providing quantities of water.
- The second principle is to match the quality of the resource supplied to the quality required by the end-use. It is almost as important to conserve the quality of a resource as to conserve its quantity. High-quality resources can be used for many purposes; low-quality resources for only a few. In contrast, we only need small quantities of high-quality resources but vast amounts of low-quality resources. Of course, those uses requiring high-quality resources are critically important, as with drinking (for water) and certain industrial processes (as for manufacturing semiconductors).
- The third principle is to turn typical planning practices around. Instead of starting from today and projecting forward, start from some future water-efficient time and work

backwards to find a feasible and desirable way ("a soft path") between that future and the present. The main objective of planning is not to see where current directions will take us, but to see how we can achieve desired goals. This step is called "backcasting" (to make an obvious contrast with forecasting). It is at this stage that appropriate transition technologies must be identified to bridge the time between full implementation of soft technologies. Finally, at the end of the process politically and socially acceptable policies and programs must be defined to bring about the desired changes.

## 1.6 Differences between soft and hard paths

The soft path can be defined in terms of its differences from the hard path. The two paths differ in at least six ways according to Wolff, Gary and Peter H. Gleick (2002):

1.  The soft path redirects government agencies, private companies, and individuals to work to meet the water-related needs of people and businesses, rather than merely to supply water. For example, people want to be clean or to clean their clothes or produce certain goods and services using convenient, cost-effective, and socially acceptable means. They do not fundamentally care how much water is used, and may not care whether water is used at all. Water utilities on the soft path work to identify and satisfy their customers' demands for water-based services. Since they are not concerned with selling water per se, promoting water-use efficiency becomes an essential task rather than a way of responding to pressure from environmentalists. The hard path, in contrast, fosters organizations and solutions that make a profit or fulfill their public objectives by delivering water—and the more the better.
2.  The soft path leads to water systems that supply water of various qualities, with higher quality water reserved for those uses that require higher quality. For example, storm runoff, gray water, and reclaimed wastewater are explicitly recognized as water supplies suited for landscape irrigation and other non potable uses. This is almost never the case in traditional water planning: all future water demand in urban areas is implicitly assumed to require potable water. This practice exaggerates the amount of water actually needed and inflates the overall cost of providing it. The soft path recognizes that single-pipe distribution networks and once-through consumptive-use appliances are no longer the only cost-effective and practical technologies. The hard path, in contrast, discounts new technology, and over-emphasizes the importance of economies of scale and the behavioral simplicity of one-pipe, one-quality-of-water, once-through patterns of use.
3.  The soft path recognizes that investments in decentralized solutions can be just as cost-effective as investments in large, centralized options. For example, there is nothing inherently more reliable or cost-effective about providing irrigation water from centralized rather than decentralized rainwater capture and storage facilities, despite claims by hard-path advocates to the contrary. Decentralized investments are highly reliable when they include adequate investment in human capital, that is, in the people who use the facilities. And they can be cost-effective when the easiest opportunities for centralized rainwater capture and storage have been exhausted. In contrast, the hard path assumes that water users— even with extensive training and ongoing public education—are unable or unwilling to participate effectively in water-system management, operations, and maintenance.

4.  The soft path requires water agency or company personnel to interact closely with water users and to effectively engage community groups in water management. Users need help determining how much water of various qualities they need, and to capture low-cost opportunities. In contrast, the hard path is governed by an engineering mentality that is accustomed to meeting generic needs.

5.  The soft path recognizes that ecological health and the activities that depend on it (e.g., fishing, swimming, tourism, delivery of clean raw water to downstream users) are water-based services demanded, at least in part, by their customers, not just third parties. Water that is not abstracted, treated, and distributed is being used productively to meet these demands. Water is part of a natural infrastructure that stores and uses water in productive ways. The hard path, by ignoring this natural infrastructure, often reduces the amount and quality of water available for use. The hard path defines infrastructure as built structures, rather than separating it into built and natural components.

6.  The soft path recognizes the complexity of water economics, including the power of economies of scope. The hard path looks at projects, revenues, and economies of scale. An economy of scope exists when a combined decision-making process would allow specific services to be delivered at lower cost than would result from separate decision-making processes. For example, water suppliers, flood control departments, and landuse authorities can often reduce the total cost of services to their customers by accounting for the interactions that none of the authorities can account for alone. This requires thinking about landuse patterns, flood control, and water demands in an integrated, not isolated, way.

## 1.7 Comparing different management approaches

When viewed on a spectrum, all three water management approaches – supply management, demand management, and the soft path – represent incremental steps toward sustainability. However, far from being a simple progression some key characteristics distinguish them. The most significant difference is the view of the limits of water available for human use and of the nature of the choices that should determine how we manage water. Figure 2 is an idealized sketch of the different paths that will result from following each of the three approaches.

Water demand management seeks primarily water efficiency, and is often focused on the implementation of cost-effective ways to achieve the same service with less water. Demand management options have been known for years, but with water prices kept artificially low, little incentive existed for widespread adoption (Brandes et al, 2005).

Though demand management has always been part of how water system operate, it is typically treated as a secondary or temporary measure needed until additional supplies are secured. Changing our water management paradigm requires that demand management become the primary focus. With rampant growth and the uncertainty of climate change, reducing the demand for water is our best "source" of "new" water (Brandes et al, 2005).

The soft path approach changes the conception of "water." Instead of being viewed simply as an end product, water becomes the means to accomplish specific tasks, such as sanitation or agricultural production. Conventional demand management asks the question "How" –

How can we get more from each drop of water? Water soft paths also ask the question "Why" – Why should we use water to do this at all? (Brandes et al, 2005).

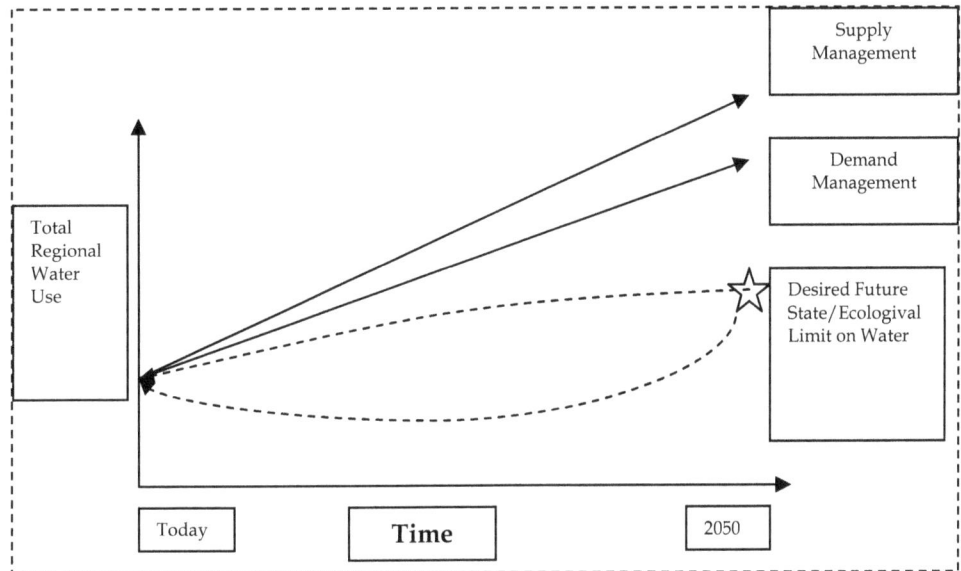

Fig. 2. Planning for the future with a soft path approach. (Brandes et al, 2005)

## 1.8 One continuing gap in soft path analyses

Probably the most legitimate criticism of energy soft path studies was that they neglected issues of equity. This led to many comments about the need to introduce environmental justice as an explicit element of policy, regardless of the nature of the policy. If that criticism was true of energy, it is even more so of water. The inequities in water and land distribution around the world are sizable and, as a result of misguided policies that promote centralization and privatization, they seem to be growing. As it is, poor people in urban areas commonly pay 10 times as much per litre for water of questionable quality as do richer people for water of good quality; and poor subsistence farmers sometimes (especially if they are women) get no water at all when commercial farms are adequately supplied (Webb et al , and Koppen et al in Brooks et al 2004). Though it is almost unquestionably true that water soft paths would improve the situation for poor people around the world, water soft paths by themselves are not sufficient. As emphasized by staff at the International Water Management Institute (IWMI) in Sri Lanka, water policies must be explicitly "pro-poor" and "pro-women". Urban water systems in developing countries are notoriously leaky if one compares the difference between water put into the system and water that reaches registered consumers. Some of those leaks are true losses, but some (highly indeterminate) portion is "stolen" or redirected to illegal taps, which may serve hundreds of poor residents. Fixing the "leaks," another common recommendation, should be undertaken only if coupled with additional, and possibly free, taps in low-income neighborhoods. In short, greater efficiency for water needs to be tempered with concern for equity, and this concern must be introduced explicitly in soft path analyses.

## 2. Where are we: Jordan water situation today

### 2.1 Introduction

Jordan is an arid to semi-arid country with land area of 92,000 sq km, located to the east of the Jordan River. Jordan's topographic features are variable. A mountainous range runs from the north to the south of the country. To the east of the mountain ranges, ground slopes gently to form the eastern deserts, to the west ground slopes steeply towards the Jordan Rift valley. The Jordan Rift valley extends from lake Tiberias in the north, at ground elevation of –220 m, to the Red Sea at Aqaba. At 120 km south of lake Tiberias lies the Dead Sea with water level at approximately –405 m. The southern Ghors and Wadi Araba, south of the Dead Sea, form the southern part of the Rift Valley. To the south of Wadi Araba region lies a 25 km coastline which stretches along the northern shores of the Red Sea. Due to the variable topographic features of Jordan, the distribution of rainfall varies considerably with location.

### 2.2 Climate

The climate in Jordan is characterized by a long, dry, hot summer, and a rainy winter. The temperature increases towards the south, with the exception of some southern highlands. Rainfall varies considerably with location, due mainly to the country's topography. Annual rainfall ranges between 50 mm in the eastern and southern desert regions to 650 mm in the northern highlands. Over 90% of the country receives less than 200 mm of rainfall per year, and 70% receives less than 100 millimeters per year. Figure 3 represents spatial variation of rainfall in Jordan.

Long term average annual rainfall for the country as a whole gives a total volume of 8352 million cubic meter (MCM). The minimum value of annual rainfall registered was 4802 MCM at the water year 1946/1947and the maximum annual value registered was 17797 MCM at the water year 1966/1967. Approximately 92.48% of the rainfall evaporates back to the atmosphere, the rest flows in rivers and wadis as flood flows and recharges groundwater. Groundwater recharge amounts to approximately 5.16 % of the total rainfall volume, and surface water amounts to approximately 2.36% of total rainfall volume. (Ministry of Water and Irrigation records)

### 2.3 Water situation

Jordan is considered to be a highly water-stressed country, with only 153 cubic meter per capita per year available in 2006 compared to an average of 1,200 m3 per capita for the whole of the Middle East (FAO, 2007).

The availability of water is classified as very low on the Water Stress Index, which indicates the degree of water shortage or scarcity. Water Stress Index is the value of annual rainfall that charges surface and groundwater divided by the total population ($m^3$/capita/year). Countries with less than 1,700 $m^3$/capita/year are regarded as countries with "existing stress", while countries with less than 1,000 $m^3$/capita/year are regarded as having "scarcity" and countries with less than 500 $m^3$/capita/year are regarded as having "absolute scarcity". With 153 $m^3$/capita/year Jordan falls into the category of "absolute scarcity"– a category comprising only 12 countries (UNEP 2002 in Fisher, 2005).

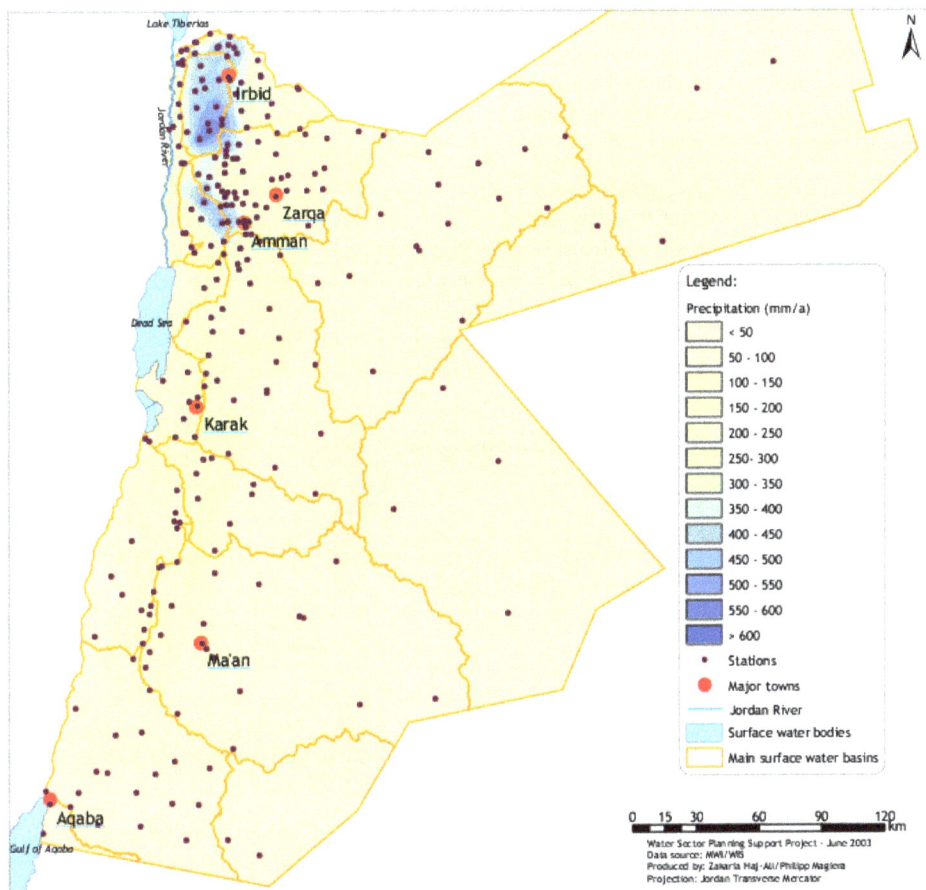

Fig. 3. Spatial distribution of rainfall in Jordan (National Water Master Plan, 2004)

The water challenge in Jordan stands as a major threat confronting human development and poverty alleviation. For this reason, the enhancement of water resource management is featured as a high priority in the National Agenda.

A description of how serious the water situation is in Jordan is presented in a paper written by Beautmont (2002) as follows:

Of all the countries in the Middle East it is Jordan which faces the greatest water problems (Salameh & Bannayan,; Beaumont in Beautmont, 2002). To meet its predicted urban water demand of 832 million cubic meters by 2025 would require 113% of its current irrigation use (1990s). In other words even if it reallocates all the irrigation water which was being used in the 1990s there would not be sufficient water to meet the expected demand. When figures on renewable water resources are examined the position becomes even more serious. It can be seen that Jordan has an internal renewable water resource base of 680 million cubic meters and a total natural water resource base of 880 million cubic meters. Yet in the 1990s withdrawals were 984 million cubic meters, which is well in excess of the total natural water resource base. Although a limited amount of reuse of water was occurring in Jordan, the explanation of this fact is that large quantities of water were being withdrawn from groundwater reserves at a rate faster than that of natural recharge. Jordan is, therefore, a country which will soon experience serious water shortages. Indeed, it is the **only** country in the Middle East which faces such a serious situation.

Later, Beautmont (2002) suggests that the only long-term solution would be for Jordan to embark on a policy of desalinated water supply for at least some of its major urban centers. However, it could be carried out from Aqaba. The great problem, though, with Aqaba is that the desalinated water would have to be transported over distances of at least 250 km, and pumped 1000 metres in height to reach the urban centres of Amman and Zerqa. In summary, there are no easy solutions to the water problem for Jordan. In the short term the reallocation of at least some of the irrigation water will buy time, but in itself it will not solve the water scarcity issue.

The following sections describes the situation in more details.

## 2.4 Water resources

Water resources consist primarily of surface and ground water sources. In recent years wastewater has increasingly been used for irrigation.

### 2.4.1 Surface water resources

Surface water resources in Jordan vary considerably from year to year. The long-term average surface water flow is estimated at 706.91 MCM/year, comprising of 451.40MCM/year base flow, and 255.51 MCM/year flood flow. Of these only an estimated 473MCM/year is usable or can be economically developed.

Surface water resources are unevenly distributed among 15 basins. The largest source of external surface water is the Yarmouk river, at the border with Syria. The Yarmouk river accounts for 40% of the surface water resources of Jordan, including water contributed from the Syrian part of the Yarmouk basin. It is the main source of water for the King Abdullah canal and is thus considered to be the backbone of development in the Jordan valley. Other major basins include Zarqa, Jordan river side wadis, Mujib, the Dead Sea, Hasa and Wadi Araba. Internally generated surface water resources are estimated at 400 million m /year (FAO, 1997). Figure 4 presents the main surface water basins in Jordan.

### 2.4.2 Groundwater resources

Groundwater is a major water resource in Jordan and the only water resource in many regions of the Kingdom. Twelve groundwater basins have been identified in Jordan, these include two fossil aquifers: Al-Disi and Al-Jafar. Some of these basins have more than one aquifer. The annual safe yield of the renewable groundwater supply is estimated to be 277MCM. An additional 143 MCM per year are considered available from non-renewable fossil aquifers that are sustainable for between 40 and 100 years. In 2005, the over-draft was about 144 MCM, consequently, the water level in several basins are declining and some aquifers are showing some deterioration of their water quality due to increased salinity. Figure 5 presents groundwater basins and sustainable abstraction per groundwater basin.

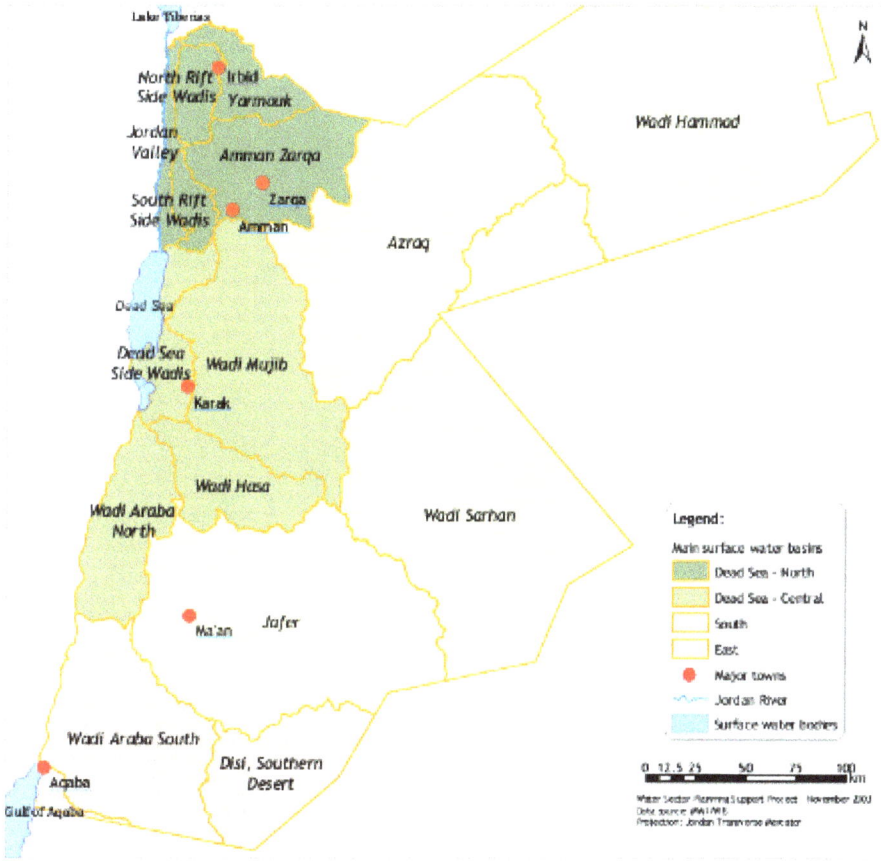

Fig. 4. Main surface water basins in Jordan. (National water Master Plan)

## 2.4.3 Wastewater

In a water-short country such as Jordan, wastewater is an important component of the Kingdom's water resources. Generally, fully treated wastewater is suitable for unrestricted use in agriculture and for aquifer recharge. "Jordan's National Water Strategy" (1997), argues that population pressure in Jordan has caused a chronic deficit in available freshwater, which has resulted in over abstraction of groundwater. Furthermore, there are limited opportunities to develop new freshwater sources and these are expensive, with high operating costs. Given this, the strategy states that treated wastewater is to be considered as a resource that, with due care for health and the environment, should be reused for agriculture, industry and other non-domestic purposes, including groundwater recharge.

The reuse of treated wastewater in Jordan reaches one of the highest levels in the world. The treated wastewater flow of the major wastewater treatment plant in the country is discharged to Zarqa River and the King Tall dam, where it is mixed with the surface flow and used in the pressurized irrigation distribution system in the Jordan Valley. Reused wastewater is becoming increasingly an essential element of Jordan's water budget.

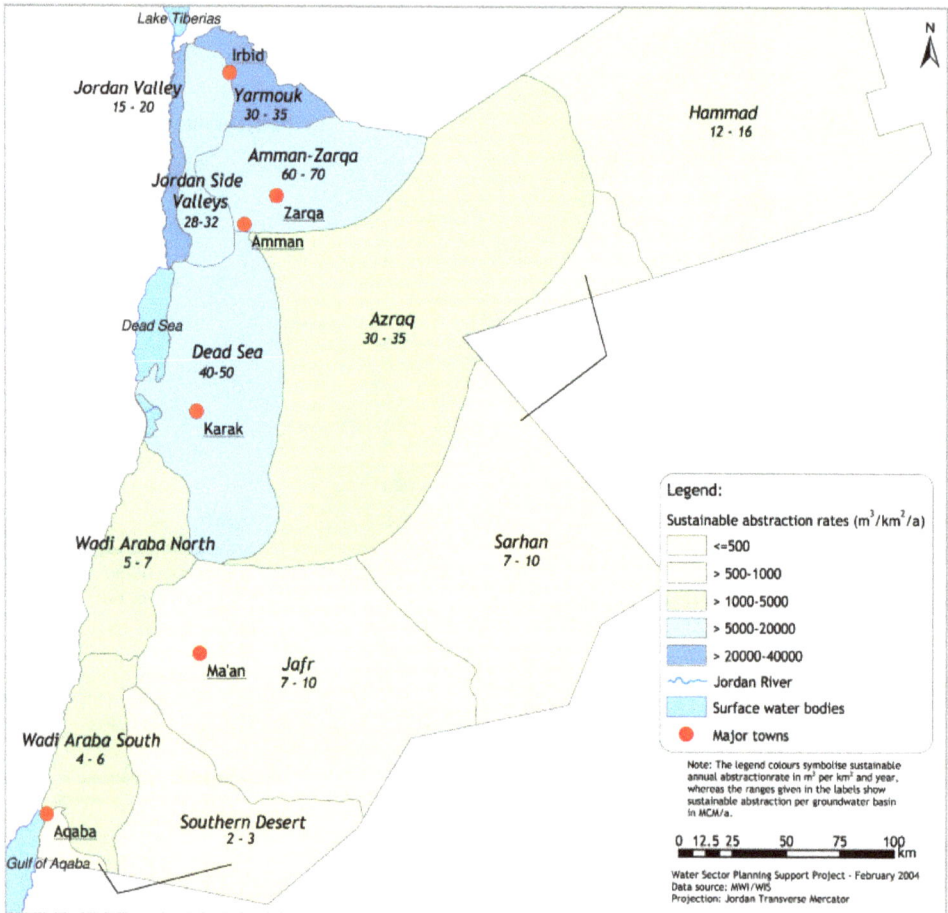

Fig. 5. Ground water basins and sustainable abstraction rate per ground water basin

In Jordan, about 84 MCM of wastewater were treated in 2005 and discharged into various water courses or used directly for irrigation, mostly in the Jordan Valley. Currently, approximately 60% of the urban population is provided with sewerage services.

Standards 893/2002 "Water-Reclaimed Domestic Wastewater" controls wastewater reuse in agriculture. The National Wastewater Management Policy (1998) allows for the Jordanian Standards on water reuse to be periodically examined to account for ambient conditions, end uses, socio-economics, environment and local conditions.

## 2.5 General water budget

In 2005 the total use of water in Jordan was 941 million cubic meters (MCM) or 164 $m^3$/capita/year t the total 2005 country's population of 5.47 million people. This usage included 77 MCM of nonrenewable groundwater (groundwater mining) and 83.5 MCM of treated wastewater. The total renewable freshwater resources in Jordan are estimated at 850

MCM, however the presence of groundwater mining and wastewater reuse in 2005 indicates that the demand already exceeds the availability of renewable water during that year. Table 1 shows the most recent statistical data on water use in Jordan by user sector and water source.

| Source | Uses in MCM | | | | Total Uses |
|---|---|---|---|---|---|
| | Municipal | Industrial | Irrigation | Livestock | |
| Surface Water | 74.7 | 4.5 | 265.21 | 7.0 | 351.41 |
| Ground Water | 216.66 | 33.903 | 254.629 | 0.826 | 506.01 |
| Treated Wastewater | 0.0 | 0.0 | 83.545 | 0.00 | 83.545 |
| Total | 291.36 | 38.403 | 603.384 | 7.826 | 941 |

(In Million Cubic Meters per Year, MCM/year)

Table 1. Sources of Sectoral Water Use in Jordan in 2005

From the Table 1 it is clear that Jordan is facing a chronic imbalance in the population-water resources equation. The per capita use of water will continue to decline at a rate equal to that of population increase. The renewable water resources falls short of meeting actual demand, which translates into the increase of food imports where the deficit in food balance reached $110 per capita in 1996 (Strategy, 1997).

## 3. Amman governorate

Amman Governorate enjoys a special position in Jordan because of its size and population, as well as its importance as the having the capital city Amman, the center of governmental institutions, communication, commerce, banking, industry, and cultural life.

Jordan is administratively divided into 12 Governorate. Figure 6 represents a map of Jordan showing the twelve Governorates. Amman Governerate is one a middle governerate and has an area of 7,579 km2. This represents 8.5% of Jordan's area. However, the population of Amman was 2125400 in 2005 representing 38% of the population in Jordan (DOS, 2005). The population densitiy is 280.4 persons per km2 (DOS, 2005).

During the last 10 years the amount of new building within the city has increased dramatically with new districts of the city being founded at a very rapid pace (particularly so in West Amman), straining the very scarce water supplies of Jordan as a whole, and exposing Amman to the hazards of rapid expansion in the absence of careful municipal planning.

Amman enjoys four seasons of excellent weather as compared to other places in the region. Summer temperatures range from 28 - 35 degrees, but with very low humidity and frequent breezes. Spring and fall temperatures are extremely pleasant and mild. The winter sees nighttime temperatures frequently near zero, and snow is not unknown in Amman, as a matter of fact it usually snows a couple of times per year. It typically will not rain from April to September, with blue skies prevailing. But lately it started to rain in April and the beginning of May. In fact about half the quantity of rain Amman and Jordan received in 2006 fell in April.

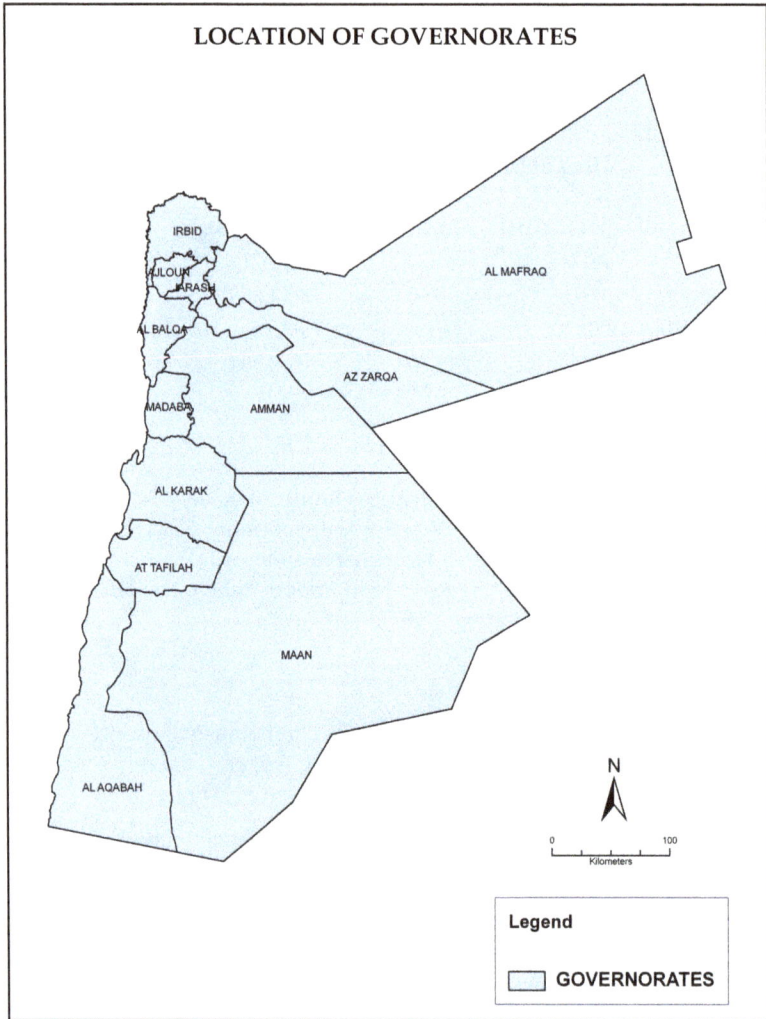

Fig. 6. Map of Jordan showing location of Amman Governorate

In Greater Amman, Lyonnaise des Eaux - Montgomery Watson - Arabtech Jardaneh (LEMA) has been in charge of public water supply since the signing of the major Water and Wastewater Supply Management Contract of Greater Amman in August 1999, till the end of 2006. Starting 2007, a governmental company "Meyahuna" was established to be in charge of the water supply in Amman.

## 3.1 Water in Amman Governorate

Water uses in Jordan are mainly defined as agricultural water use, municipal and industrial water use. Municipal consumption refers to the water consumed in a given year by the domestic, commercial and pastoral sectors in addition to the light industries.

In 2005, Amman Governorate total water uses amounted to 143.52 million cubic meters (MCM). The water use for the municipal sector was 119.87 MCM, the industrial water use was 1.26 MCM while water use for the agricultural sector was 22.4 MCM. The percentages of water use in each sector is as follows: 83.52% for the municipal sector, .88 % for the industrial sector 15.6 % for the irrigation sector.

Amman's total municipal and touristic uses witnessed significant increase during the past decades in both absolute and relative terms. This was mostly due to the growth in municipal consumption. Increased income and changes in way of life have also contributed to such an increase of water consumption, especially in the urban areas of Greater Amman. (Master Plan, 2004).

One third of water requirements to satisfy the municipal demands for Amman Governorate, are currently met from internal resources of Groundwater and surface water of the Governorate, while the two thirds need to be transferred from resources external to the governorate. The ability to increase water supply potential is further limited due to the over pumping of ground water resources, falling ground water levels and deteriorating groundwater quality. Thus, in order to satisfy the projected water demand in Amman Governorate, major water supply and transfer projects such as Mujib-Zara-Zarqa-Ma'in Saline Water Desalination Project and Disi Water project are needed

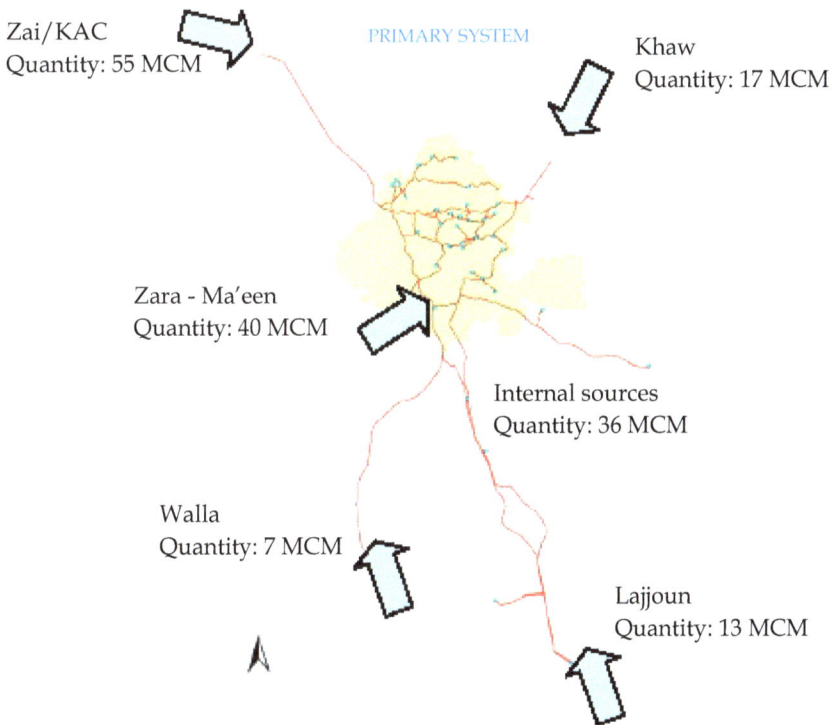

Fig. 7. Schematic location of bulk water sources in 2004. (Ministry of Water and Irrigation and USAID, 2006)

The most important water sourcefor Amman Governorate is the Yarmouk River and water collected from 10 other sources located in the northern part of the Jordan Valley that feed the King Abdullah and ultimately the Zai water treatment plant. The intake at Deir Alla is located at 230 meters below sea level and the water is pumped through a system of 4 pumping stations to 880 meters above sea level. The Zai water treatment plant provides conventional treatment (flocculation, sedimentation, rapid filtration and chlorination. (MWI et al, 2006).

An important new water source (Zara Ma'een project) is being completed and was operational in August 2006. The project comprises a 55 MCM per year reverse osmosis treatment plant for upgrading class III raw water of salinity between 1400-2000 mg/l to no more than 250 mg/l. The plant produces 47 MCM per year of drinkable water as defined in Jordanian Standards. (MWI et al, 2006). Figure 7 is a schematic location of bulk water sources in 2004

### 3.2 Water services

In 2005, about 97 % of the total population in Amman Governorate were served through some 362,500 service accounts. Due to lack of water resources, water supply is rationed in most of the service area; in 2005 for instance, the average hours of supply were 66 per week. However, about 60% of customers receive water more than 36 hours per week and 55 water districts within restructured CIP area are receiving water continuously as of May 2006. (MWI et al, 2006).

The water supply through public network to Amman Governerate was 119.86 MCM in 2005 (WAJ records). However, water billed according to LEMA company was 66.3. MCM only. This translates into non-revenue water of 53.56 MCM. The nonrevenue water can be split to apparent losses and real losses

Apparent losses are however considered to be part of the consumption, since they are due to illegal abstractions, inaccurate or erroneous meter readings , non-operational meters and/or un-metered connections

In order to estimate the actual water use in Amman an assumption was made that water losses of 53.56 MCM can be split equally to real losses and apparent losses. So real losses were estimated as 26.78 MCM and apparent losses also as 26.78 MCM. Table 2 establishes a standard water balance for Amman

| Own Sources 37.74 MCM | | Water Exported 8.65 MCM | Authorized Consumption | Billed Authorized Consumption 66.3. MCM | Revenue Water |
|---|---|---|---|---|---|
| | System Input 128.52 | | | Unbilled Authorized Consumption | |
| Water Imported 90.77 MCM | | Water Supplied 119.87 MCM | Water Losses 53.56 MCM | Apparent Losses 26.78 MCM | Non Revenue Water |
| | | | | Real Losses 26.78 MCM | |

Table 2. Establishing a standard water balance for Amman

### 3.3 Water uses

In order to use the soft path method, and in view of conflicting numbers reported in different publications analysis of billing recoded to estimate water use services.

Municipal Water was estimated based on the above water balance as (66.3+ 26.78 = 93.08 MCM)

### 3.3.1 Residential water use

To estimate the residential water use in Amman, The following steps where followed:

1.  Since the apparent losses adds to both residential and non residential uses. The percent of billed water residential use to the total billed water was calculated.

    Percent of billed residential water use =residential billed water use / total billed water
    $$= 58.5 / 66.3 * 100 = 88.2 \%$$

    Total apparent losses was calculated from an earlier section as 26.78 MCM

2.  To obtain the apparent losses for the residential sector, an assumption was made that this amount is proportional to the percent of residential use. Thus, the percent of the billed residential water use was multiplied by the total apparent losses.

    Apparent losses for the residential sector= 0.882* 26.78 MCM = 23.62 MCM

3.  The amount of apparent losses obtained was added to the billed residential water use. The total residential water use is estimated at

    Residential water use =Residential billed water use + apparent losses
    $$= 58.5 + 23.62 = 82.12 \text{ MCM}$$

Further, the residential water was used to calculate the per capita water use. The population of Amman according to DOS (www.dos.gov.jo) was 2125400 in 2005.

$$\text{Per capita water use } = \text{ residential water use} / \text{ population}$$
$$82.12 * 10^9 * / ( 2125400 * 365) = 106 \text{ liters} / \text{ capita} / \text{ day}$$

### 3.3.2 Non-residential water use

The non residential water use or as defined in the literature as Industrial, Commercial and Institutional water use (ICI), according to LEMA water billing data, was 7.8 MCM. An additional 3.16 MCM can added as result of the apparent losses using the same logic applied to residential water use. This result ICI water use of 10.96 MCM representing 7.6 % of the total water use in Amman.

The Water Efficiency and Public information for Action Project (WEPIA) project gathered some data or the subscription base and broken them down by sector. Their breakdown is as in Table 3.

This breakout is helpful. It shows that the recorded water deliveries are primarily to residential household and that a conservation program should address this sector. However, 10 MCM of water deliveries for the non-residential sector most likely does not represent the true consumption in these sectors, as they are likely to be receiving supplementary deliveries by tanker truck. The overall usage for the non-residential sector is likely to be considerably higher. Hospitals, for example, rarely rely on municipal deliveries, and therefore their recorded data are short by magnitudes of scale

| Sector | Annual Consumption (m3) | % Annual Consumption |
|--------|-------------------------|----------------------|
| Residential | 95,459,600 | 82.9% |
| Schools | 1,388,023 | 1.2% |
| Hospitals | 1,170,361 | 1.0% |
| Commercial | 14,626,426 | 12.7% |
| Industry | 351,399 | 0.3% |
| Governmental | 2,188,538 | 1.9% |
| Total | 115,184,347 | 100.00% |

Table 3. Water use broken down by sector

### 3.3.3 Industrial water use

Industrial water use here refers to the amount of water consumed by big industries, which utilise water produced locally, and mainly from groundwater wells. The industrial facilities in Amman Governorate include few food industries, Iron industries, Tiles industries and Pharmaceutical industries.

### 3.3.4 Agricultural water use

The water use for irrigation in Amman Governerate from groundwater wells and springs was 22.42 MCM according to the Ministry's Water Information System (WIS).

The Department of Statistics (DOS) implemented several agricultural surveys in 2004. The surveys included the following: cultivated area of fruit trees, cultivated areas of vegetables, and cultivated areas by field crops by type of crop in Amman governorate. For cultivated area with field crops of 321,186 dunums, it was estimated that 320,712 was non-irrigated, while the remaining 475 dunums, were irrigated using surface methods (DOS, 2004).

## 4. Soft path analysis for Amman

In Amman as in all Jordan, the water situation is critical. Population Growth, low rainfall, and increased economic activity are causing additional stress on the natural resources and in particular on water. In this section a soft path analysis will be developed for Amman. Three scenarios will be considered.

### 4.1 Scenarios for the soft path

Three Scenarios are considered for the purpose of developing a soft path analysis for Amman:

1.  Maintaining water supply at the same level of 2005, but providing water services at the same level of a water use of 106 liter/capita/day. Year 2005 is considered as the base year.
2.  Maintaining the per capita water use at the level of 2005 of 106 litre/capita/day, while accounting for the projected population growth, this will result in a higher water supply.
3.  Maintaining water supply at the same level of 2005, but providing water services at the same level of a water use of 135 liter/capita/day.

## 4.1.1 Official demand projection

Table 4 represents the official water demand projection for Amman Governorate. Details of this projection will be explained in a later section.

| Year | Municipal | Industrial | Touristic | Irrigation | Total Demand |
|------|-----------|------------|-----------|------------|--------------|
| 2005 | 147.1 | 1.21 | 2.79 | 74.5 | 225.6 |
| 2010 | 158.2 | 1.5 | 3.18 | 73.8 | 236.7 |
| 2015 | 176.1 | 1.87 | 3.6 | 73.3 | 254.87 |
| 2020 | 195.2 | 2.33 | 4.02 | 72.1 | 273.65 |

Table 4. Official water demand projection for Amman Governorate

## 4.1.2 Scenario one

In this scenario water supply is maintained at the same level as 2005. Residential water use will be maintained at 105,379,575 m$^3$ through the planning horizon, this will make the per capita water use drop. To supplement the drop of per capita water use per day from the level of 106 liters in 2005 to 67 liters in 2030 several options will be included in this soft path analysis. These options shall provide a total difference of 61,871,482 m$^3$ to keep the same level of water services.

The non-residential and industrial sectors will also maintain the same level of water use. Several options need to be developed to account for the growth in these sectors while maintaining the same level of water use. The agricultural sector will not expand any further in this scenario.

## 4.1.3 Scenario two

In this Scenario an assumption was made that the per capita water use will remain the same level of 2005, that is 106 liter/capita/day and population will grow according to official population growth figures. The residential water use will thus be 167,251,057 m$^3$ in 2030. In this Scenario, the assumption will be that the water requirement in 2030 will be 135 liter/capita/ day that is a total of 208,155,650 m$^3$ and to supplement this difference several option need to be considered. These options will need to supplement the difference of 40,904,594 m$^3$.

The non-residential and industrial sector will also grow at the same level of population growth. Options will be suggested to increase the efficiency in these sectors. The agricultural sector will not expand any further in this scenario.

## 4.1.4 Scenario three

In this scenario water supply is maintained at the same level as 2005. Residential water use will be maintained at 105,379,575 m$^3$ through the planning horizon, but the water services will be kept at the same level of having 135 litre/cap/day. To make this possible, a soft path will be developed to provide the additional 102,776,076 m$^3$ to reach the level 208,155,650 m$^3$ necessary for a water use of 135 litre/cap/day and a population of Amman of 3665483 in 2030.

Figure 8 presents the official demand projection for Amman, along with the propped three soft paths.

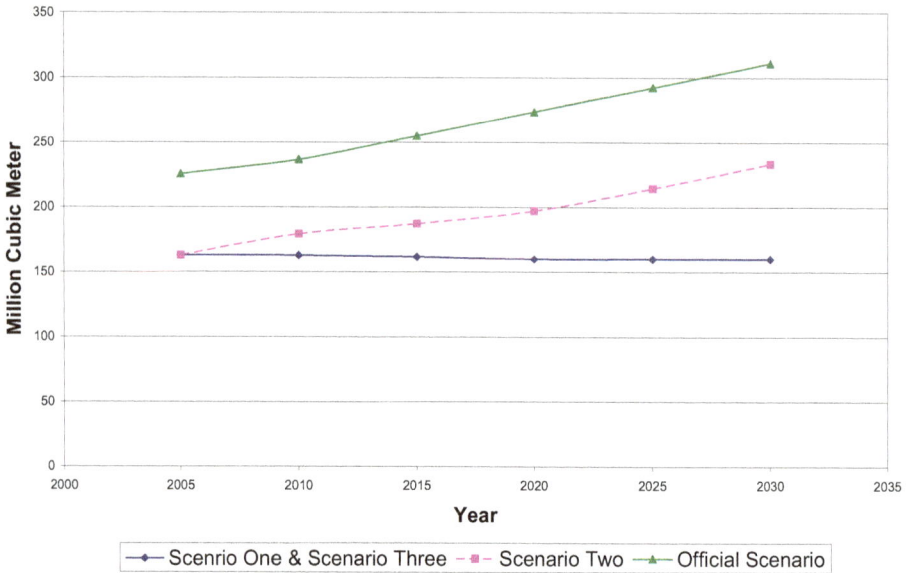

Fig. 8. Official projection and soft path scenarios for Amman Governorate

## 4.2 Methodology

The following basic steps for a soft path plan will be followed to develop the three Scenarios, these steps are after (Brandes et al, 2005)

1. Identify Water Services: List all services provided by water (e.g. residential indoor and outdoor, municipal parks, cooling). Some questions to answer include: Who is going to need water? For what purpose or goal water is needed? What kind of water is needed to meet a specific goal? How much water of a particular quality is needed to meet given goal?

2. Adopt a Projection for the Governorate- Look 25 years in the future of pre-existing official documents, demographic projections, and expectations of economic growth. Next, apply existing water use patterns to this projection (on a use-by-use basis), thus enabling a "business as usual" baseline.

3. Establish a Desired Future Condition: Create a desired future pattern for water supply and use. For example, a governorate might assume all future growth will be offset by conservation or efficiency and no new water sources will need to be developed.

4. Analyze Water Quantity and Quality: Establish the quantity of water required to provide the service identified for this projected future (Step 3) by applying as many of the water conserving options as can be adopted within the given time frame. Determine which uses require high water quality water notably (drinking, cooking and bathing) and which uses can proceed with lower quality water (toilet flushing, gardening, most forms of Agriculture, industrial applications, etc.)

5.  Review Water Supply options: Identify all current sources of water (surface and ground) and determine whether any is being over-used or degraded. Reduce withdrawals of fresh water or releases of wastewater that threaten long-term renewable use, and reject any new sources that cross major shed boundaries or create serious threat to ecological, cultural or social values. Indicate the relevant range of future supply adjustment that may result from climate change.
6.  Backcast (in contrast with forecast): Create various soft paths by designing incremental policies and programs to get from here to the future. Check each option to see whether it seems economically feasible, socially acceptable and politically achievable. This is an iterative process of backdating.

## 4.3 Step 1: Identify water services

The first step in preparing a soft path analysis is to establish a base line of current water use services. Looking at the water services allows us to evaluate the effect of improvements in end use technology and water demand management while maintaining the purpose for which the water is required.

### 4.3.1 Disaggregating consumption of sectors

### 4.3.1.1 Residential water use

According to earlier calculations in this section water use per capita in Amman was estimated at 106 liters/day.

| Household Fixture | % Total Consumption | Total Consumption (m3) |
|---|---|---|
| Drinking | 2.0% | 2,003,728 |
| Toilets | 17.8% | 17,833,179 |
| Showers | 27% | 27,050,328 |
| Clothes Washers | 11% | 11,020504 |
| Kitchen and Bath Faucets | 27% | 27,050,328 |
| Outdoor | 6.7% | 6,712,489 |
| Others | 7% | 7,013,048 |
| Car | 1.5% | 1,502,796 |
| Total | 100% | 100,186,400 |

Table 5. Indoor water consumption by end use

Understanding how the customer uses water is important to understanding how where to get conservation savings. End use information – that is, information about water flow at each specific point that the customer uses water – is important not only to analyzing conservation potential, but also to forecasting increases in water demand in the future.

No data exists of exact measures of water use services in Amman. To develop basis of estimate for water use services, prior studies in Ammam, and the results of a model developed by Rosenberg (personal contacts) were used. This model of household water use was based on a survey; questionnaire; and data of billed water use. The resultant analysis found the following percentages of various end use consumption for an average household.

The disaggregating here is a useful indicator of water use. Toilets, faucets and showers are according to Table 5 have the highest end uses in the household – 71.8% overall – and therefore represent clear conservation program opportunities. Residential water are estimated below are for an averages household. Water used here is only potable.

### 4.3.1.2 Urban agriculture

A statistical survey of urban agriculture was conducted by the Department of Statistics in (1999). Thirteen thousand households were surveyed in various regions around the greater Amman area. Extrapolated to the whole population, the survey's findings show the following:

- 50,097 households are practicing urban agriculture.
- The total cultivated area is an astounding 6,483,952 square meters.
- 86% of these households use the public network as a source of irrigation.
- Only 30% of the households suffer from water scarcity, thus leaving 70% who must be irrigating adequately or even excessively.

These findings suggest that an urban landscape program, particularly in the Amman region, may provide some water conservation potential. Although not likely to be a large quantity of water, it nonetheless would represent an area of savings worth exploring further.

### 4.4 Step 2: Adopt a projection for a region

The results displayed as the official demand projection part of the National Water Master Plan for Jordan of 2004, which has been produced by the Ministry of Water and Irrigation, in cooperation with the German Technical Cooperation (GTZ). Table 6 is a summary obtained reviewing different sections of the National Water Master Plan.

| Year | Municipal | Industrial | Touristic | Irrigation | Total Demand* |
|------|-----------|------------|-----------|------------|---------------|
| 2005 | 147.1 | 1.21 | 2.79 | 74.5 | 225.6 |
| 2010 | 158.2 | 1.5 | 3.18 | 73.8 | 236.7 |
| 2015 | 176.1 | 1.87 | 3.6 | 73.3 | 254.87 |
| 2020 | 195.2 | 2.33 | 4.02 | 72.1 | 273.65 |

Official Ministry of water and irrigation scenario

Table 6. Amman Governorate total water requirement (MCM)

Demand is defined as the amount of water required by a user. On the other hand, consumption represents the amount actually used "at the end of pipe". Both demand and consumption do not include any kind of physical losses. However, estimating Gross demand to be supplied need to include physical losses. In Jordan and as a result of water scarcity and rationing, water demand as defined earlier, is higher than actual water consumption. This is why the official demand projection starts at a higher value than other scenarios in figure 6.

Thus, gross municipal demand figures were based on a physical loss reduction program through systematic network rehabilitation program as presented in Table 7.

| Governorate | 2005 | 2010 | 2015 | 2020 |
|-------------|------|------|------|------|
| Amman | 28 | 22 | 18 | 15 |

Table 7. Physical losses per Governorate Assumption (%)

## 4.5 Step 3: Establish a desired future condition

### 4.5.1 Scenario one

The desired future for scenario one will be based on the assumption that Amman water requirements will be met with no need for extra water supplies than what is available in the year 2005. In order to maintain the water services at the same level, programs need to be designed to account for the difference of an average a per capita water use of 106 liters per capita per day (lpcd) at 2005 and 67 lpcd at 2030. This amounts to a total of 52,296,750 m3 for a population of Amman of 3665483 at 2030. Details of this scenario were presented in section 4.1.2.

### 4.5.2 Scenario two

The desired future for scenario two will be based on the assumption that Amman per capita water use will be the same as it was in the year 2005 that is 106 lpcd. In this Scenario, the assumption will be that the water requirement in 2030 will be 135 lpcd that is a total of 196,475,459 m³ and to supplement this difference a soft path several need to be developed. Details of this scenario were presented in section 4.1.3.

### 4.5.3 Scenario three

The desired future for this scenario will be based on maintaining a water supply at the same level as 2005. Residential water use will be maintained at 105,379,575 m³ through the planning horizon, but the water services will be kept at the same level of having 135 litre/cap/day. To make this possible, a soft path will be developed to provide the additional 102,776,076 m³ to reach the level 208,155,650 m³ necessary for a water use of 135 lpcd and a population of Amman of 3665483 in 2030.

## 4.6 Step 4: Analyze water quantity and quality

### 4.6.1 Scenario one

To be able to provide the per capita use of 67 l per capita per day at 2030 and be able to provide the same water services at the current level of 106 l/capita/day of 2005, the following programs need to be implemented at the residential sector. These programs will save a total of 61,871,482 m3 which is the difference between the projected gross residential use for scenario one of 105,379,575 m3 and the projected gross residential water use of 167,251,057 m3 if the water use will be maintained at 106 litre/capita/day. Savings are mainly based on savings from end uses summed across the total population according to the following assumptions as in Table 8. The programs include: retrofitting 50 % of the toilets, retrofitting 50% of the showerheads, retrofitting 50 % of kitchen and bath faucets, finding and fixing leaks, installing efficient washing machines with a percent of 20% , turning 20% of the irrigation systems in gardens to drip irrigation systems, growing low water consuming landscapes or crops, installing greywater systems for outdoor use with 20% coverage, installing rainwater harvesting systems for outdoor use with 20% coverage, installing greywater systems for indoor use with 3.7% coverage, installing rainwater harvesting systems for indoor use with 7% coverage.

1- Retrofit Toilets

| Population | Per capita | Percent used | Savings | Coverage | Water Savings |
|---|---|---|---|---|---|
| 3665483 | 106 | 0.178 | 0.2 | 0.5 | 6916032.675 |

2- Showerhead Retrofit

| Population | Per capita | Percent used in showers | Savings | Coverage | Water Savings |
|---|---|---|---|---|---|
| 3665483 | 106 | 0.27 | 0.2 | 0.5 | 10490611.36 |

3- Aerator Retrofits for Kitchen or bath faucets

| Population | Per capita | Percent used | Savings | Coverage | Water Savings |
|---|---|---|---|---|---|
| 3665483 | 106 | 0.27 | 0.2 | 0.5 | 10490611.36 |

4- Find fix and leaks

| Population | Per capita | | Savings | Coverage | Water Savings |
|---|---|---|---|---|---|
| 3665483 | 106 | 0.07 | 0.1 | 0.2 | 7770823.231 |

5- Install efficient water washing machines

| Population | Per capita | Percent used | Savings | Coverage | Water Savings |
|---|---|---|---|---|---|
| 3665483 | 106 | 0.11 | 0.1 | 0.2 | 854790.5554 |

6- Install Drip Irrigation System

| Population | Per capita | Percent used in | Savings | Coverage | Water Savings |
|---|---|---|---|---|---|
| 3665483 | 106 | 0.067 | 0.1 | 0.2 | 520645.1564 |

7- Install low water consuming landscape or crops

| Population | Per capita | Percent used in | Savings | Coverage | Water Savings |
|---|---|---|---|---|---|
| 3665483 | 106 | 0.067 | 0.1 | 0.2 | 520645.1564 |

8-install Graywater Collection System for outdoor use

| Population | Per capita | Percent used | Savings | Coverage | Water Savings |
|---|---|---|---|---|---|
| 3665483 | 106 | 0.067 | 1 | 0.2 | 5206451.564 |

9-Install Rainwater Collection System for out door use

| Population | Per capita | Percent used | Savings | Coverage | Water Savings |
|---|---|---|---|---|---|
| 3665483 | 106 | 0.067 | 1 | 0.2 | 5206451.564 |

10-Install Graywater Collection System for indoor use

| Population | Per capita | Percent used | Savings | Coverage | Water Savings |
|---|---|---|---|---|---|
| 3665483 | 106 | 0.178 | 1 | 0.037 | 2558932.09 |

11-Install Rainwater Harvesting System for indoor use

| Population | Per capita | Percent used | Savings | Coverage | Water Savings |
|---|---|---|---|---|---|
| 3665483 | 106 | 0.91 | 0.5 | 0.07 | 12375035.99 |
| | | | Total Savings | | 62,911,031 |
| | | | Required Savings | | 62,903,753 |

Table 8. Programs for soft path one

## 4.6.2 Scenario two

In this Scenario an assumption was made that the per capita water use will remain the same level of 2005, that is 106 l/capita /day and population will grow according to official population growth figures, that is the residential water use for Scenario 2 will be 167,251,057 m³ in 2030. In this Scenario, an assumption will made be that the water requirement in 2030 will be 135 liter/capita/ day that is a total of 208,155,650 m³. The difference of 40,904,594 m³ will be gained from implementing the programs described in Table 9. The programs include: retrofitting 50 % of the toilets, retrofitting 50% of the showerheads, retrofitting 50 % of kitchen and bath faucets, finding and fixing leaks, installing efficient washing machines

1- Retrofit Toilets

| Population | Per capita | Percent used | Savings | Coverage | Water Savings |
|---|---|---|---|---|---|
| 3665483 | 106 | 0.178 | 0.2 | 0.5 | 6916032.675 |

2- Showerhead Retrofit

| Population | Per capita | Percent used in showers | Savings | Coverage | Water Savings |
|---|---|---|---|---|---|
| 3665483 | 106 | 0.27 | 0.2 | 0.5 | 10490611.36 |

3- Aerator Retrofits for Kitchen or bath faucets

| Population | Per capita | Percent used | Savings | Coverage | Water Savings |
|---|---|---|---|---|---|
| 3665483 | 106 | 0.27 | 0.2 | 0.5 | 10490611.36 |

4- Find fix and leaks

| Population | Per capita |  | Savings | Coverage | Water Savings |
|---|---|---|---|---|---|
| 3665483 | 106 | 0.07 | 0.1 | 0.2 | 7770823.231 |

5- Install efficient water washing machines

| Population | Per capita | Percent used | Savings | Coverage | Water Savings |
|---|---|---|---|---|---|
| 3665483 | 106 | 0.11 | 0.1 | 0.2 | 854790.5554 |

6- Install Drip Irrigation System

| Population | Per capita | Percent used in | Savings | Coverage | Water Savings |
|---|---|---|---|---|---|
| 3665483 | 106 | 0.067 | 0.1 | 0.2 | 520645.1564 |

7- Install low water consuming landscape or crops

| Population | Per capita | Percent used in | Savings | Coverage | Water Savings |
|---|---|---|---|---|---|
| 3665483 | 106 | 0.067 | 0.1 | 0.2 | 520645.1564 |

8-install Graywater Collection System for outdoor use

| Population | Per capita | Percent used | Savings | Coverage | Water Savings |
|---|---|---|---|---|---|
| 3665483 | 106 | 0.067 | 1 | 0.03 | 780967.7347 |

9-Install Rainwater Collection System for out door use

| Population | Per capita | Percent used | Savings | Coverage | Water Savings |
|---|---|---|---|---|---|
| 3665483 | 106 | 0.067 | 1 | 0.1 | 2603225.782 |

10-Install Graywater Collection System for indoor use

| Population | Per capita | Percent used | Savings | Coverage | Water Savings |
|---|---|---|---|---|---|
| 3665483 | 106 | 0.178 | 1 | 0 | 0 |

11-Install Raionwater Harvesting System for indoor use

| Population | Per capita | Percent used | Savings | Coverage | Water Savings |
|---|---|---|---|---|---|
| 3665483 | 106 | 0.91 | 0.5 | 0 | 0 |
|  |  |  | **Total Savings** |  | 40,948,353 |
|  |  |  | **Required Savings** |  | 40,904,594 |

Table 9. Programs for soft path two

with a percent of 20%, turning 20% of the irrigation systems in gardens to drip irrigation systems, growing low water consuming landscapes or crops, installing greywater systems for outdoor use with 3 % coverage, and installing rainwater harvesting systems for outdoor use with 10 % coverage.

### 4.6.3 Scenario three

To be able to provide the per capita use of 67 l per capita per day at 2030 and be able to provide the same water services at the level of 135 l/capita/day of 2005, the following programs need to be implemented at the residential sector. These programs will save a total of 102,776,076 m3 which is the difference between the projected gross residential use of year 2005 of 105,379,575 m3 and the projected gross residential water use of 208,155,650 m3 if the

1- Retrofit Toilets

| Population | Per capita | Percent used | Savings | Coverage | Water Savings |
|---|---|---|---|---|---|
| 3665483 | 106 | 0.178 | 0.2 | 0.5 | 6916032.675 |

2- Showerhead Retrofit

| Population | Per capita | Percent used in showers | Savings | Coverage | Water Savings |
|---|---|---|---|---|---|
| 3665483 | 106 | 0.27 | 0.2 | 0.5 | 10490611.36 |

3- Aerator Retrofits for Kitchen or bath faucets

| Population | Per capita | Percent used | Savings | Coverage | Water Savings |
|---|---|---|---|---|---|
| 3665483 | 106 | 0.27 | 0.2 | 0.5 | 10490611.36 |

4- Find fix and leaks

| Population | Per capita | | Savings | Coverage | Water Savings |
|---|---|---|---|---|---|
| 3665483 | 106 | 0.07 | 0.1 | 0.2 | 7770823.231 |

5- Install efficient water washing machines

| Population | Per capita | Percent used | Savings | Coverage | Water Savings |
|---|---|---|---|---|---|
| 3665483 | 106 | 0.11 | 0.1 | 0.2 | 854790.5554 |

6- Install Drip Irrigation System

| Population | Per capita | Percent used in | Savings | Coverage | Water Savings |
|---|---|---|---|---|---|
| 3665483 | 106 | 0.067 | 0.1 | 0.2 | 520645.1564 |

7- Install low water consuming landscape or crops

| Population | Per capita | Percent used in | Savings | Coverage | Water Savings |
|---|---|---|---|---|---|
| 3665483 | 106 | 0.067 | 0.1 | 0.2 | 520645.1564 |

8-install Graywater Collection System for outdoor use

| Population | Per capita | Percent used | Savings | Coverage | Water Savings |
|---|---|---|---|---|---|
| 3665483 | 106 | 0.067 | 1 | 0.2 | 5206451.564 |

9-Install Rainwater Collection System for out door use

| Population | Per capita | Percent used | Savings | Coverage | Water Savings |
|---|---|---|---|---|---|
| 3665483 | 106 | 0.067 | 1 | 0.2 | 5206451.564 |

10-Install Graywater Collection System for indoor use

| Population | Per capita | Percent used | Savings | Coverage | Water Savings |
|---|---|---|---|---|---|
| 3665483 | 106 | 0.178 | 1 | 0.17 | 11757255.55 |

11-Install Raionwater Harvesting System for indoor use

| Population | Per capita | Percent used | Savings | Coverage | Water Savings |
|---|---|---|---|---|---|
| 3665483 | 106 | 0.91 | 0.5 | 0.25 | 44196557.12 |
| | | | **Total Savings** | | 103,930,875 |
| | | | **Required Savings** | | 103,808,347 |

Table 10. Programs for soft path three

water use was increase to 135 litre/capita/day. Savings are mainly based on savings from end uses summed across the total population according to the following assumptions as shown in Table 10. The programs include: retrofitting 50 % of the toilets, retrofitting 50% of the showerheads, retrofitting 50 % of kitchen and bath faucets, finding and fixing leaks, installing efficient washing machines with a percent of 20%, turning 20% of the irrigation systems in gardens to drip irrigation systems, growing low water consuming landscapes or crops, installing greywater systems for outdoor use with 20% coverage, installing rainwater harvesting systems for outdoor use with 20% coverage, installing greywater systems for indoor use with 17 % coverage, installing rainwater harvesting systems for indoor use with 25 % coverage.

## 4.7 Step 5: Review water supply options

### 4.7.1 Current water supplies

**Groundwater resources**

The estimated safe yield of renewable water resources in Amman governorate is in the order of 34 MCM/a. About 63 MCM is being abstracted from potential aquifers in the governorate (2005). Ground water quality in the area is generally good to fair quality (Total Dissolved solids (TDS) is in the range 500-1000 gm/l).

Deterioration of groundwater quality is vulnerable along the Seil region where some of the industrial waste is being disposed and as a result of the overdraft conditions that have been experienced in the governorate.

**Groundwater resources from other governorate**

Groundwater Resources From other Governorate are transported through pipelines to Amman Governorate from well fields in wadi Wala- Heidan (Madab Governorate), Katraneh and Lajoun (Karak Governorate) and Azraq and Corridor (Zarqa Governorate)

**Surface water resources**

Surface water resources in the Governorate are limited to rainfall/runoff in wadi Swaqa and al Botum. In addition, to Ras el Ain and Wadi Sir spring flows. The total potential of surface resources is estimated at about 7.4 MCM. Currently most of this is used for municipal purposes and the rest is used for irrigation.

**External surface water resources**

External surface water supplies is being conveyed to the Governorate from Yarmouk river via King Abdulla Canal/Dier Alla intake. About 60.3MCM of water supplies have been conveyed to Amman. The Deir Alla Zai Coveyor has the capacity of 90/a MCM.

**Non conventional water resources**

**Wastewater**

There are two existing treatments plants in Amman, Au Nsier and Wadi Sir.

- Abu Nsseir: is an activated sludge plant with a capacity of 4,000 m3/day or 1.5 MCM/year; and
- Wadi Essier: is an aerated lagoon plant with a capacity of 4,000 m3/day or 1.5 MCM/year

In addition to a small wastewater treatment plant at Queen Alia Airport.

The effluent of Abu Nsseir Wastewater Treatment Plan was about 2240.3 m3/day in 2005 and is currently used for landscaping of the medians adjacent to the treatment plant, while the effluent of Wadi al Sir was 2762 m3/day in 2005.

WAJ is developing three new wastewater treatment plants:

- As Samra secondary treatment plant, being built under a BOT scheme. This plant has four treatment trains with a total capacity of 267,000 m3/day (97 mm3/year). This plant will start operation in 2007. Additional capacity of 267,000 m3/day is planned for a later stage.
- South Amman secondary wastewater treatment plant. This plant has a capacity of 31,000 m3/day (11.3 MCM/year) and is expected to be operational in early 2008. South Amman wastewater project intended to serve more than 290,000 people living in this area.
- Giza-Talbiea secondary treatment plant. This plant has a capacity of 2,300 m3/day (0.8 MCM/year) and is expected to become operational by mid 2007.

**Zara main project**

Zara Main desalinated brackish water project, can make about 40 MCM of water available to Amman by the end of 2006.

**Disi project**

This project includes raising of water from an aquifer in the Disi-Mudawarra area in the south of Jordan and the conveyance of the water to the greater Amman area, a distance of approximately 325 kilometers. The conveyance system will have a capacity of transporting 100 MCM water per year. This project is currently under tendering.

### 4.7.2 Water supply options for scenario one

| Year | Total Demand MCM | Local Sources (MCM) | | | Possible Available Sources (MCM) | | | |
|------|------------------|---------------------|-------------|--------------|------|-----|--------------|----------------------|
|      |                  | Surface Water | Ground-Water | WW Effluent | Disi | Zai | Zara Main | From Other Governorates |
| 2030 | 160.18 | 10 | 34 | 16.18 | --- | 60 | 40 | ----- |

### 4.7.3 Water supply options for scenario two

| Year | Total Demand MCM | Local Sources (MCM) | | | Possible Available Sources (MCM) | | | |
|------|------------------|---------------------|-------------|--------------|------|-----|--------------|----------------------|
|      |                  | Surface Water | Ground-Water | WW Effluent | Disi | Zai | Zara Main | From Other Governorates |
| 2030 | 224.1 | 10 | 34 | 16.18 | 64 | 60 | 40 | ----- |

### 4.7.4 Water supply options for scenario three

| Year | Total Demand MCM | Local Sources (MCM) | | | Possible Available Sources (MCM) | | | | |
|------|------|------|------|------|------|------|------|------|------|
| | | Surface Water | Ground-Water | WW Effluent | Disi | Zai | Zara Main | From Governorates | Other |
| 2030 | 160.18 | 10 | 34 | 16.18 | --- | 60 | 40 | ----- | |

### 4.7.5 Summary

The water requirements could be met by reducing the groundwater pumping to the safe yield of 34 MCM/year. Irrigation from ground water shall be reduced and supplemented by irrigation from reclaimed water. For Scenario One and Three water supplied and Zai and Zara Main are of critical importance and these sources need to be used to the fullest extent possible. However, for Scenario 2, meeting the water requirement will need additional water that can be only provided by Disi Project.

### 4.8 Step 6: Backcast

In Step 3 the desired future has been identified as scenario one or two scenario three; in Step 4 ways to make that future work were identified; and in Step 5 supply constraints where defined. In this step we need to explain how to get to that future. Each option need to be checked to see whether it seems economically feasible, socially acceptable and politically achievable

### 4.9 Step 7: Write, talk and promote

The last step in soft path analysis, but can be considered the most important, is to get those conclusions to the public and especially to people who influence and make key decisions about fresh water. Considerable efforts should be put into promoting water soft path results.

## 5. Conclusion

The traditional approach to water supply led to enormous benefits. The history of human civilization is intertwined with the history of the ways humans have learned to manipulate and use water resources. The earliest agricultural communities arose where crops could be grown with dependable rainfall and perennial rivers. Irrigation canals permitted greater crop production and longer growing seasons in dry areas, and sewer systems fostered larger population centers (Gliek, 2002)

During the industrial revolution and population explosion of the nineteenth and twentieth centuries, the demand for water rose dramatically. Unprecedented construction of tens of thousands of monumental engineering projects designed to control floods, protect clean water supplies, and provide water for irrigation or hydropower brought great benefits to hundreds of millions of people. On the other hand, half the world's population still suffers with water services inferior to those available to the ancient Greeks and Romans. According to the World Health Organization's most recent study, more than 1 billion people lack access to clean drinking water, and nearly 2.5 billion people do not have improved sanitation services. Preventable water-related diseases kill an estimated 10,000 to 20,000 children each day, and the latest evidence suggests that we are falling behind in efforts to solve these problems (Gliek, 2002).

Further more, Groundwater aquifers are being pumped down faster than they are naturally replenished and more than 20 percent of all freshwater fish species are now threatened or endangered because dams and water withdrawals have destroyed the free-flowing river ecosystems where they thrive.

In the twenty-first century we can no longer ignore these costs and concerns. The old water development path—successful as it was in some ways—is increasingly recognized as inadequate for the water challenges that face humanity. We must now find a new path with new discussions, ideas, and participants. The Soft path offers this alternative. The adjective *soft* refers to the nonstructural components of a comprehensive approach to sustainable water management and use, including equitable access to water, proper application and use of economics, incentives for efficient use, social objectives for water quality and delivery reliability, public participation in decision making, and more (Gliek, 2002).

This chapter aimed at investigating the possibility of implementing this approach to Jordan and in particular in Amman Governorate. A soft path analysis was developed considering three different scenarios. Applying this analysis framework to Jordan, has demonstrated the urgent need of implementing strategies today that can reduce our dependence on more expensive supply side developments in the future. We have to start soon on establishing comprehensive water demand management program, particularly in urban areas and for the residential, commercial and institutional sectors.

Toilet retrofits rogram, showerhead retrofits program, aerator retrofits program, clothes washer retrofits program, audit leak detection, installing drip irrigation system, indoor and out door greywater reuse, rainwater harvestig for indoor and outdoor uses, public information programs, modifying water user behavior, reclaimed water use and recycling, a comprehensive leak detection and reduction program, and a more efficient agricultural sector.

The analysis proved that the need to improve the management of fresh water is great, and soft paths offer a way to design alternative management strategies. It also demonstrated that

- Jordan must shift emphasis from only expanding water supply to moderating water demand.
- We must learn how to get along with less water in total, and much less water per capita.
- As a water poor country we must learn how to become even more efficient than they already are.
- Making more efficient use of existing water resources through demand management is an economical and environmentally responsible way to meet growing demand for water.
- If Jordan is committed to aggressive pursuit of demand management, it would help preserve Jordan's existing valuable and limited natural water resources, and provide a readily available and low cost water resource for the coming years.

- Water savings allow new customers and new demands in Jordan to be supplied with water without taking more water from nature. Conservation of non consumptively used water create benefits that exceed the costs of conservation.

Realistically, both supply and demand approaches will be necessary, as has been demonstrated; however, the better approach will be from the demand side.

Finally, implementing this soft path requires a social choice to invest in the people, businesses, and cooperative arrangements that are needed for the maximum cost-effective water savings to become reality (Gliek, 2002). Government agencies or water suppliers must implement comprehensive, integrated economic, educational, and regulatory policies that remove the barriers and achieve the socially desirable level of water savings.

Unless demand management is fully integrated with water-supply planning, it will remain an underused and misunderstood part of our water future (Gliek, 2002).

## 6. References

Beaumont, P., (2002). "Water Policies for the Middle East in the 21st Century: The New Economic Realities" In: Water Resources Development. Vol.18, No.2, 315–334.

Brandes, O. and Brooks, D., (2005). "The Soft Path for Water in A Nutshell", A joint Publication of Friends of Earh Canada, Ottawa, ON, and the POLES Project on Ecological Governance, University of Victoria, Victoria, BC, Canada.

Brooks, David B., (2003). "Another Path Not Taken: A Methodological Exploration of Water Soft Paths for Canada and Elsewhere". Report to Environment Canada. Friends of the Earth Canada, Ottawa, ON.

Brooks, D., de Loë, R., Patrick, R. and Rose, G., (2004). "Water Soft Paths for Ontario: Feasibility Study". Report to the Walter and Gordon Duncan Foundation. Friends of the Earth Canada, Ottawa, ON.

FAO's Information System on Water and Agriculture. (1997).

Fisher, F. and Hber-Lee, A., (2005). "Liquid Assets: An economic Approach for Water Management and Conflict Resolution in the Middle East and Beyond". Washington, DC: Resources for the Future.

Gleick, P. , Loh, H., Gomez, S., Morrison, J., (1995). *California Water 2020: A Sustainable Vision.* Paci.c Institute for Studies in Development, Environment and Security, Oakland, CA.

Gleick, P., D. Haasz, C., Henges-Jeck, V. Srinivasan, G., Wolff, K., Cushing, K. and A. Mann. 2003. *Waste Not, Want Not: The Potential for Urban Water Conservation in California.* Pacific Institute for Studies in Development, Environment, and Security. Oakland, CA.

Gleick, P., (2003). "Global Freshwater Resources: Soft-Path Solutions for the 21st Century". In: *Science.* Vol 302, www.sciencemag.org.

National Water Master Plan, (2004), Ministry of Water and Irrigation and German Technical Cooperation – GTZ , Amman, Jordan.

Ministry of Water and Irrigation and USAID (2006), "Amman Water Management/ Commercialization Assessment Phase Two Report: Feasibility Analysis Of New Company Volume 2* - Annexes, Amman, Jordan.

# 13

# Cities and Water – Dilemmas of Collaboration in Los Angeles and New York City

David L. Feldman
*University of California, Irvine, California*
*U.S.A.*

## 1. Introduction

This chapter examines the different ways megacities manage water by comparing how Los Angeles and New York - two U.S. metropolises that divert water from distant sources - have worked with their surrounding regions to acquire, allocate, and manage public supplies. Early in their histories these cities, in their quest to acquire water, adopted a hegemonic relationship with their neighbors. In effect, they sought to control regional sources that could satisfy current as well as projected water needs (Hundley, 2001; New York City, 2011; Koeppel, 2000; 2001). Over time, and under external pressure, both cities embraced collaboration with adjacent communities to address water supply and quality issues whose scope and impact required regional accommodation and sharing of authority. What they have done to achieve accommodation in light of water stress, and how they have done it, may afford lessons for megacities across the globe that face comparable challenges.

New York and Los Angeles diverged in their motives for and methods of collaboration, in part because their water challenges differ. New York's central challenge currently revolves around managing water quality and the safety of its drinking water. Meeting this challenge is virtually impossible without cooperation with non-governmental actors in other political jurisdictions from whence its water supply comes - and who would be severely burdened financially if the city had to build a large regional water filtration plant. For Los Angeles, by contrast, water (and air) quality issues in the Owens Valley - the source, since 1913, of one-third of the city's water - have driven efforts to partner with valley stakeholders to negotiate gradual reductions in flow and restoration of the watershed. While both cities were initially concerned with water supply, however, over time they both became increasingly worried over water quality and the need for integrated approaches to managing supply and quality.

## 2. Method and approach

Our approach is four-fold. We: 1) analyze the hydrological and political factors influencing water decisions; 2) compare these cities' water policy histories; 3) examine their current collaborative challenges; and, 4) draw out their most important similarities and their lessons for other cities. For Los Angeles, we focus chiefly upon the Owens River with briefer discussion of newer (i.e., mid 20th Century) issues, including the *State Water Project* which diverts water from the Sacramento-San Joaquin - Bay Delta, and the Colorado River

Aqueduct, completed in 1940. While the Owens Valley case revolves around a powerful, growing city initially diverting water from a modest agrarian region in order to support future growth, and then restoring a portion of that region's water under federal order, the latter cases revolve around endangered species protection and climate variability, respectively, as factors that compel change in urban water policy.

For New York City, the chief focus of our discussion is the Croton and Catskill watersheds - the former is the city's original regional water source, dating to the 1840s, while the second was developed in the late 19th Century. In more recent years, both watersheds have been part of the so-called *New York City Watershed Protection Plan* designed to protect the city's water supply from sewage and runoff-induced contamination through adopting cooperative land use controls and other measures. These watersheds are the source of fully one-half of the city's water supply. Additional case material from the Delaware River, an interstate stream which New York relies upon for the other half of its water supply, is also discussed.

Section 3 sets the stage for comparison by first considering two vital questions: 1) how do megacities affect water supply and quality in their nested regions; and, 2) why are Los Angeles and New York good cases for studying these issues? Despite being located in a highly-developed society, and perceived as having safe, well-managed water systems, this was not always the case. Beyond this, as we shall see, Los Angeles and New York share important challenges with regards to infrastructure, the need to conserve water, and climate change which may translate into lessons for other megacities facing similar problems.

## 3. Policy context – Megacities and water

A number of accounts suggest that global freshwater supplies are increasingly facing severe *stress*: a growing imbalance between available supplies within various regions on one hand, and demands on those supplies by multiple users on the other. Water stress is generally attributed to population growth, climate variability (including extreme drought), and inadequately maintained and/or deteriorating water supply and treatment infrastructure. Experts view stress as caused by demographic, climatological, and socio-economic factors intersecting in various ways (World Meteorological Organization, 1997; World Resources Institute, 1998; Alcamo, et. al., 2003; World Water Council, 2005).

While these three factors are compelling sources of stress, a more nuanced cause is rapid *urbanization*, exemplified by the phenomenal growth of so-called "megacities" composed of tens-of-millions of people. Megacities are a sprouting phenomenon in developing nations, especially, where cities and towns already comprise some 80% of the planet's urban populace. More than two-thirds of the world's urban residents live in cities in Africa, Asia, and Latin America. Moreover, since 1950, the urban population of these regions has grown five-fold, while in Africa and Asia alone, urban population is expected to double by 2030 (Satterthwaite, 2000; UNPF, 2007).

Large cities generally, and megacities in particular, contribute to water stress in two ways: 1) they are often located some distance from the water sources needed to maintain their growth; thus they must divert water from outlying rural areas which, in turn, often produce the food and fiber to support them; and, 2) soaring birthrates and in-migration (the latter often from these same outlying areas) place extra burdens upon water infrastructure, and generate severe health and hygiene problems. Both of these contributors underscore the complex ways demography, economics and climate factors interact.

Urban-related water stressors can be more precisely de-constructed as three-fold problems. First, large cities generate huge volumes of wastewater which are costly to treat and, if left untreated, can contaminate local wells and streams. Second, the *spatial* "footprint" caused by sprawling horizontal urban development and annexation imposes numerous water-related problems, including paving of city streets and commercial districts (contributing to pollutant runoff and diminished groundwater recharge), and consumption of water for parks and outdoor residential use (increasing evapo-transpiration and taxing local supplies).

Third, while greater concentration of people in cities may lower unit costs for many forms of water infrastructure (Satterthwaite, 2000) the need to expand water supply and treatment networks over vast distances increases the likelihood of distribution system leaks and other failures. All these problems have been observed in a number of Third World megacities, and underscore how urbanization exacerbates climate change impacts on scarce water supplies; imposes extraordinary pressures on surrounding regions; and, outraces infrastructural capacity (UN, 2009: 32; Adekalu, et. al., 2002; Downs, Mazari-Hiriart, Dominguez-Mora, & Suffet, 2000; Gandy, 2008; Tortajada and Casteian, 2003; Yusuf, 2007; and Zérah, 2008).

### 3.1 Hydrology and Geography as prologue – A Los Angeles and New York overview

So, what can the experiences of Los Angeles and New York teach us about water stress and large cities? Conventional wisdom might suggest that being located in highly-developed societies both are far better in managing water supply and quality than their counterparts in less-developed nations. In reality, however, their longer-standing experience as large urban conurbations makes them instructive cases for other megacities. This is so for three reasons.

First, early in their histories, both cities faced many of the same challenges to public health and wastewater management that their Third World counterparts face today. These challenges included confronting the role foul and unhealthful water plays in the spread of infectious disease (a particular problem for New York City which, in 1832, suffered a severe cholera epidemic attributed to contaminated drinking water, Koeppel, 2000, 2001; American Museum of Natural History, 2011). Another includes the need to take decisive, yet adaptable, action to upgrade public works in order to provide residents with abundant water, and determining whether satisfying the need for safe, secure, and dependable supplies was best left to the "efficiencies" provided by private sector investment, or better suited to management by governmental entities. Los Angeles and New York confronted this latter challenge early in their histories, as we will see (Glaeser, 2011: 99; Mulholland, 2002). To a large extent both challenges drove these cities to divert water from outlying regions.

Second, in diverting water from outlying hinterlands, Los Angeles and New York generated well-documented, but vastly different, environmental and social impacts upon these adjacent regions. In the case of Los Angeles, diversion of water imposed reductions of both in-stream flow and groundwater in Owens Valley. These reductions, in turn, degraded local fisheries and wildlife habitat (McQuilkin, 2011), while acquisition of adjacent lands overlying aquifers deprived Owens Valley communities of the ability to pursue real estate development for commercial and residential use (VanderBrug, 2009).

For its part, by acquiring much of the open space surrounding its reservoirs in the Catskill and Croton watersheds, a positive economic outcome generated by New York City was retention of low-density residential development that preserved the region's rural character

(Westchester County Department of Planning, 2009). Later, sewage plant outfalls and non-point pollution around these same reservoirs released contaminants into the city's water supply which generated further, less popular land acquisition measures to avert pollution through eminent domain and condemnation suits - a strategy that continued through the early 1990s (New York State Department of Environmental Conservation, 2010a, 2010b).

Finally, although these cities have very different hydrological features, New York and Los Angeles share *two* remarkably similar water problems. First, both cities have experienced an outracing of available supply as a result of locally-generated demands. Second, while the former is located in a wet and humid region, while the other is dry and semi-arid, both cities have needed to look outside their political boundaries for additional supply. They have also employed similar strategies to acquire water and land rights to ensure control over the watersheds from whence their water comes. By their wide range of conditions, we might suggest that these cities bound the impacts faced by most of the world's large urban centers.

That both cities share these problems in common underscores an important point about water stress: the traditional distinction between arid and semiarid regions on the one hand and more humid areas on the other, as a means of maintaining that arid regions' water problems mostly revolved around inadequate water quantity while humid areas' problems are water quality related, is not a valid claim. Water scarcity can occur in any *urbanized* region if demands cannot be attenuated (Feldman, 2009).

Los Angeles is located in a flat, triangular-shaped semi-arid basin bounded on its north and east by mountains and on the west by the Pacific. Its Mediterranean climate experiences some 39.54 cm (15.58") of average annual precipitation, all in the form of rain, which is collected by two major streams that rise in the San Gabriel portion of the Transverse Range dividing Southern from central California - the Los Angeles and San Gabriel Rivers. The Los Angeles River was the city's major source of water for nearly a century, providing drinking water and serving as the irrigation source for local vineyards and orange groves - both through an elaborate system of ditches and channels called *zanjas* (Gumprecht, 2001: 3; Los Angeles Department of Water and Power, 2010b).

New York City, by contrast is located on the Atlantic Coastal Plain in a slender portion of land bounded by the outfall of the Hudson and East Rivers, and referred to as the Atlantic slope drainage. The humid continental climate, fed by the Gulf of Mexico and Atlantic weather systems, produces some 127 cm (50") of precipitation per year, some 63.5 - 76.2 cm (25 - 30") of which falls as snow. In its initial period of settlement, local water supplies were provided through ponds, streams, and springs located on the island of Manhattan (American Museum of Natural History, 2011; New York State Department of Environmental Conservation, 2010b). These hydrological differences help explain how both cities initially managed water supply, while their phenomenal demographic growth helps us understand the remarkably similar path both took in seeking hegemony over regional supplies. Table 1 depicts major features of the water supply systems of New York and Los Angeles.

## 4. Comparing policy history – The evolution of regional dominion

While Los Angeles and New York developed along different trajectories, especially early on (i.e., New York grew at a faster rate much earlier), in regards to water supply they followed two strikingly comparable patterns of development. First, both sought to fully exploit locally-available resources through collective effort. Second, when these sources proved insufficient to support further growth, they acquired more distant sources. Acquisition of these sources was predicated on concerns with water security, safety, and plentifulness.

| Water supply characteristic | Los Angeles | New York |
|---|---|---|
| Major supply sources | LA Aqueduct/E. Sierra = 18%* Metropolitan WD = 71% Groundwater = 10% Recycled wastewater = 1% | Croton Watershed = 10% Catskill watershed = 40% Delaware watershed = 50% |
| Distance from source to city | Owens Valley = 376 km (233 miles) Mono Lake = 544 km (338 miles) | Croton Reservoir = 201 km (125 miles) |
| Number of storage facilities | 114 (reservoirs and tanks) | 19 (reservoirs) |
| Water supplied/day | 1998.7 million liters (528 million gallons approx.) | 3792.9 million liters (1.2 billion gallons approx.) |
| Customers | 9 million | 4.1 million |
| Metered water rates[1] | $2.92 - 5.19/hundred feet$^3$ | $3.17/hundred feet$^3$ |

*In recent years, the LA Aqueduct from Owens Valley has supplied upwards of 35% of the city's water supply. However, mandated restoration of Mono and Owens Lakes has resulted in a reduction of supply of approximately half that annual delivery.
[1] New York charges a flat water rate while Los Angeles has a "tiered" or increasing block rate system wherein customers are charged a lower "base" if they stay within a designated conservation allotment. If they exceed that allotment (typically 79,287 liters or 2800 cubic feet/month), they are charged at the higher rate (Los Angeles Department of Water and Power (2009b).

Table 1. Water Supply Systems of Los Angeles and New York City

In the event, there is one major way their quest for regional dominion *differed*: Los Angeles sought external sources of supply mostly because its regional access to water was always precarious. After its first full century of settlement (c. 1880) the city suddenly aspired to grow, but found that its semi- arid region simply had few ground or surface water supplies available nearby. By contrast, New York was impelled toward the Croton Watershed in Westchester County, some 64.4 km (40 miles) to its north, by the poor quality and inadequate volume of its local supplies. A cholera epidemic in 1832, caused in part by degraded water quality and poor waste disposal, drove efforts to build a Croton Aqueduct. Declining well levels, which made fire fighting capacity inadequate, was also a factor (Koeppel, 2000: 6). Each city's respective quest for regional dominion reveals these intricate patterns.

### 4.1 Los Angeles – Early water development

From its founding in 1781, and for nearly a century afterwards, the Los Angeles River was the city's major water source. The first families who founded and settled the "pueblo" almost immediately set about constructing a brush "toma" or dam across the river, diverting water into a so-called "Zanja Madre," or mother ditch, which fed homes and irrigation canals into fields that, at first, were closely adjacent to the plaza - the civic center of the early settlement (Los Angeles Department of Water and Power (2010b).

This cooperative effort to develop and manage local water supplies was animated by two principal, and somewhat contradictory, goals. First, harvesting of the river was necessary to accommodate a population sufficient to allow the young pueblo to serve, within its nexus in hispanic colonial culture, as a trading center for neighboring ranchos and missions (Hundley, 2001). In effect, the young pueblo was a "service" community for a regional economy. Second, the pueblo needed to secure rights to water from regional competitors in order to protect its modestly growing population and its own agricultural and commercial activities - including trade with local native American tribes.

What made these goals somewhat contradictory was the fact that within a few short years after its founding, interdependence with the Gabrielino community, which served as a labor source for the city's own agricultural and commercial activities, threatened the authority of nearby missions and ranchos (Estrada, 2008). In effect, the Los Angeles River supported the small-scale agricultural and domestic needs of the pueblo, while the pueblo itself functioned as a commercial center for regional farming, ranching, and artisanal manufacturing.

In addition to developing an extensive system of locally-managed and maintained *zanjas*, to secure its local water rights, city officials sought and obtained ratification of a so-called "Pueblo" water right: an entitlement under traditional Spanish law to lay claim to all needed waters in the vicinity. This virtual ownership of water in the Los Angeles River was granted in perpetuity by King Carlos III of Spain in 1781 (Los Angeles Department of Water and Power, 2010b). However, various legal manevers were exercised during the 19th century - under Mexican and later U.S. rule - to ensure that this Pueblo water right was legally perfected. By the time the City of Los Angeles was incorporated in 1850 under the state's first constitution, following California's admission as the 35th state, the city of 1,600 was vested with all of the rights to the water of the river (Hundley, 2001).

Four years later (1854) the zanjas system became encapsulated into a city department which, within a few years, was leased to a private company and became the Los Angeles City Water Company. This company was purchased by the city in 1902 for some $2 million in order to facilitate municipal control of the system, and to help facilitate the financial arrangement needed to build an aqueduct. Privatization of the water system of Los Angeles was largely a reaction to the high costs of maintaining and repairing the *zanjas* which, by the late 19th Century, had grown to encompass not merely ditches and channels but an arrangement of water wheels to lift the water to gravity flow irrigation systems, along with some 300 miles of water mains, reservoirs, infiltration galleries and pumping facilities.

## 4.2 New York – Early water development

While unofficially founded as a Dutch trading post in 1624, and centered for decades on Manhattan Island, early New York - like early Los Angeles - relied on domestic water supplies obtainable from sources in the immediate vicinity. Initially, these consisted of shallow, privately-owned wells. Under Dutch rule – a period of about forty years, sanitation and water quality were decidedly poor: accumulations of human and animal waste were common, contaminated runoff into holding ponds was frequent, and there was no concerted effort to regulate harmful activities impinging on locally-adjacent well-users (Koeppel, 2001). Thus, New York's initial efforts to develop a water supply system were largely animated by three concerns: accommodating population growth, averting communicable disease, and achieving both objectives while saving money (Glaeser, 2011; 99).

Under English rule (1664 and after), improvements were only marginal. Foul, standing water was common, and outbreaks of epidemics stemming from poor water quality - including yellow fever and cholera – were not unknown (Koeppel, 2001). In 1677 the first general public-use well was dug near the fort at Bowling Green, while the first city reservoir was constructed on the east side of Broadway between Pearl and White Streets in 1776 - about the time the city's population grew to over 20,000 residents. Initially, water pumped from wells near the Collect Pond, east of the reservoir, was distributed through hollow logs laid along main thoroughfares in Manhattan (New York City, 2011).

As the city's population grew, pollution of wells became a serious problem, as did periodic supply shortages due to drought. These tribulations led to more concerted efforts to supplement local supplies through cisterns and springs in upper Manhattan (an area less developed at this time). Following the outbreak of a Yellow Fever epidemic in the last decade of the 18th Century (1798), New York (along with Philadelphia, one of the country's largest cities), sought to provide a safer, more secure, and disease free water supply.

In 1800 the Manhattan Company (forerunner of Chase Manhattan Bank) sank a well at Reade and Centre Streets, pumped water into a reservoir on Chambers Street and distributed it through wooden mains to a portion of the community. This venture became, in effect, the city's first quasi-public water utility and was a major enhancement to the earlier Collect Pond (Willensky and White, 1988: 18). In 1830, in an effort to enhance emergency supplies, the city built a tank for fire protection at 13th and Broadway which was replenished from a well. Water was distributed through 30.48 cm (12-inch) cast iron pipes.

As in Los Angeles, these advanced efforts to provide a safe and secure supply were largely privately funded and managed (Koeppel, 2000). Led, ironically, by two New York statesmen who soon became mortal enemies - Aaron Burr and Alexander Hamilton - the city's Common Council was persuaded to obtain state legislative endorsement of the Manhattan Company's charter. Burr and Hamilton had different motives in advocating the charter and the company: the former sought financial profit through transforming the "surplus" revenues of the firm to his own design, while the latter was swayed by the desire to un-burden city residents of a tax-supported public system (Glaeser, 2011: 99).

In any case, the result - as in Los Angeles - was to acquire sufficient revenue to enlarge and expand the water supply infrastructure of the city. Moreover, in both cities the motive of water security eventually led them to expand greater *public* control in order to construct massive aqueduct systems: a feat begun by New York in the 1840s and by Los Angeles in the early 1900s.

### 4.3 Public works and urban triumphalism – The aqueduct age in both cities

After weighing various alternatives for additional water supply Los Angeles and New York expropriated distant sources. The contentious history of Los Angeles' efforts to acquire the Owens Valley, in contrast to those of New York in its Croton and Catskill watersheds, has been well documented (Walton, 1992; Davis, 1993; Mulholland, 2002). In a later section we will discuss why reactions differed. At this juncture, the important point is that neither city pursued much consultation with regional decision-makers in undertaking these efforts.

After exploring various options for increasing supply to keep up with growing demands, New York officials sought to impound water from the Croton River, in today's Westchester

County, and to build an aqueduct to carry water from what became known as the "Old" Croton Reservoir to the City. In contrast to Los Angeles, the urgency of an aqueduct was not as readily apparent to many local residents, and initial political support was far from unanimous. According to one writer, a major fire in 1835 which consumed a sizeable portion of what is now lower Manhattan, convinced many wavering citizens of the need for an aqueduct (Koeppel, 2001). Moreover, even after the aqueduct was completed – in 1842 – not all city water users chose to connect themselves to the system, preferring to rely upon less reliable, but still cheaper, local supplies from wells and cisterns (Koeppel, 2000).

New York city's original aqueduct, known today as the *Old Croton Aqueduct*, had an initial capacity of about 90 million gallons per day and was placed into service in 1842. Distribution reservoirs were first located in Manhattan at 42nd Street (discontinued in 1890 - at the site where the present-day New York Public Library is located) and in Central Park, south of 86th Street (discontinued in 1925). Newer reservoirs were subsequently constructed to increase supply: Boyds Corner in 1873 and Middle Branch in 1878 (New York City, 2011).

In 1883, as the city's continued growth and commercialization taxed this supply source, a commission was formed to build a *second* aqueduct from the Croton watershed together with additional storage reservoirs. This conduit, known as the *New Croton Aqueduct*, was built between 1885-1893 and first placed in service in 1890, while still under construction. One of the biggest land use issues was the need to acquire land and right-of-way for the New Croton Dam and Aqueduct System – an effort begun in 1880 when seven thousand acres were acquired to harness the Croton River's three branches, while a twenty square mile area was needed by the city on which to build the New Croton Dam. Twenty-one dwellings and barns, one and a half dozen stores, churches, schools, grist mills, flour mills, saw mills, four towns, and over four hundred farms were condemned and taken over to build the dam – and some 1500 bodies were removed from six cemeteries and relocated along with their stones and fences. One local historical account states that "protests, lawsuits and some confusion preceded payment of claims" (Village of Croton, 2010).

At the same time, the present municipal system was consolidated from the various water systems in the communities now consisting of the Boroughs of Manhattan, the Bronx, Brooklyn, Queens and Staten Island. An important parallel with Los Angeles, here, is how water system consolidation became an important first step toward *municipal* annexation. For Los Angeles, completion of the first Owens Valley Aqueduct in 1913 leveraged the city's ability to force smaller communities coveting water (e.g., Hollywood) to accede to annexation as a condition for becoming connected to the distribution system.

A third phase development occurred after the turn of the century. In 1905, a Board of Water Supply established by the New York State Legislature cooperated with the city in developing the Catskill region as an additional water source – with the former planning and constructing facilities to impound Esopus Creek, and to deliver the water to the city via the Ashokan Reservoir and Catskill Aqueduct. This project was completed in 1915. It was subsequently turned over to the City's Department of Water Supply, Gas and Electricity for operation and maintenance. The remaining development of the Catskill System, involving the construction of the Schoharie Reservoir and Shandaken Tunnel, was completed in 1928.

A fourth and final effort to acquire water was the effort to allocate the Delaware River. In 1927 the Board of Water Supply submitted a plan to the state Board of Estimate and

Apportionment for the development of the upper portion of the Rondout watershed and tributaries of the Delaware within New York State. This project was approved in 1928. Work was subsequently delayed by an action brought by the State of New Jersey in the U.S. Supreme Court to enjoin the City and State of New York from using the waters of any Delaware River tributary (New York City, 2011). This case underscores the regional animosity brought about by the City's effort to seek water hegemony.

In May 1931 the Supreme Court upheld the City's right to augment its water supply from the Delaware's headwaters. However, a second Supreme Court ruling, in 1954, was required to adjudicate riparian allocation of the Delaware between New York, New Jersey, and Pennsylvania (Derthick, 1974: 48, 54). Construction of the Delaware System was begun in March 1937 and entered service in stages: the Delaware Aqueduct was completed in 1944, Rondout Reservoir in 1950, Neversink Reservoir in 1954, Pepacton Reservoir in 1955 and Cannonsville Reservoir in 1964. Figure 1 depicts the current New York water supply system.

Los Angeles took much longer, but followed a similar path in its efforts to build a major supply conduit from the Owens Valley. As the city's population rapidly grew after 1880, it became apparent that the Los Angeles River was simply was not large enough to support the city's transformation into a large metropolis. Its population doubled during the 1890s, from 50,000 to 100,000, and more than doubled again within five years (to over 250,000), all but depleting local groundwater. Moreover, the city's incorporated area doubled between 1890 and 1900 as many basin communities embraced annexation to ensure water supply.

Fred Eaton, one-time city engineer during the 1890s, mayor from 1899-1901, and superintendent of Los Angeles' municipal water system conceived of an Owens River aqueduct in the early 1900s (Davis, 1993: 5-9). Initial challenges proved to be fiscal, not logistical. The city - which had long sought to rationalize management and maintenance of the zanjas system – succeeded, under Eaton, in persuading voters to acquire public ownership of the vast, fragmented, and poorly maintained private network of water providers in 1902. Following consolidation of legal control over water in its immediate vicinity, the Owens Valley project was pursued.

After an unusually harsh drought in the summer of 1904, William Mulholland – a protégé of Eaton and now city engineer – asked his mentor to "show me this water supply" in the Owens Valley about which Eaton had often spoke. Following an intrepid journey both took through the region, which included a preliminary survey of an aqueduct route, events moved quickly. In September 1905, voters approved by a 10-1 margin a $1.5 million project to acquire right-of-way, and to build an aqueduct that would stretch from north of Independence some 376 km (234 miles) southeast to the San Fernando Valley – a recently incorporated area of the city.

At precisely the moment political forces in Los Angeles maneuvered to acquire Owens Valley water rights, the newly-formed U.S. Reclamation Service drafted a plan to irrigate the Owens Valley by constructing one or more dams in the vicinity of Long Valley. As a federal agency mandated to promote irrigation, the Service was inclined to support the people of the valley against those of a large city seeking to augment its water supply. However, the Reclamation Service's southwestern regional chief, Joseph P. Lippincott served (secretly) as a paid consultant to Los Angeles – abetting the city's plans, since Lippincott advocated for the city's interests in Washington, DC, not those of the Owens Valley. Lippincott also helped ensure that, while valley lands would be set aside for public purpose, no land rights would be secured: an action that abetted Eaton's efforts to set about buying up options on lands for aqueduct construction (Kahrl, 1982).

Within two years, two other efforts were completed in the city's favor: a successful campaign to obtain Congressional approval of the City's application to build the aqueduct

Fig. 1. New York City's Water Supply System

was effectuated in June 1906; while in 1907, Los Angeles voters approved a second bond measure authorizing $23 million for aqueduct construction. Construction began in 1908 and the project was completed in November 1913.

Like New York City, the Owens Valley was one phase in the city's water supply expansion. By the early 1920s, the Board of Public Service commissioners (the overseers of the Los Angeles Department of Water and Power or LADWP), became aware that the city would exceed the Owens Valley's supply by 1940 (thus, a second aqueduct was built in the Owens Valley all the way to Mono Lake - a project approved by voters in 1930 and completed in 1940).

A third phase was symbolized by the efforts of Mulholland to acquire water from the Colorado River. A four-year series of surveys began in 1923 to find an alignment that would bring the water of the Colorado River to Los Angeles. In 1925 the Department of Water and Power (LADWP) was established, and the voters of Los Angeles approved a $2 million bond issue to perform the engineering for the Colorado River Aqueduct. While the six-cooperating states of the basin sought a means to allocate the Colorado's flow - an effort that began with the 1922 Colorado River Compact and required Congressional passage of the Boulder Canyon Dam Act in 1928 - Los Angeles proactively sought to move events forward.

Needing allies in Washington, and help from neighboring Southern California cities who also coveted this water, in 1928 the city and LADWP got the state legislature to create the Metropolitan Water District of Southern California or MWD (Fogelman, 1993: 101-3; Erie, 2006). In 1931, voters approved a $220 million bond issue for construction, and work began on the ten-year 300 mile long project which now supplies 60% of Los Angeles, Orange, Ventura, San Bernardino, Riverside, and San Diego Counties' water. In the 1970s the regional cooperative also began importing water from Northern California via the State Water Project and the California Aqueduct. Figure 2 depicts Los Angeles' water system.

### 4.4 Post-aqueduct policies – Collaboration with external regions

Subsequent to completion of their respective aqueduct systems, both cities began to face a series of water-related environmental quality challenges which, unlike the efforts to initially acquire water, required unprecedented levels of regional collaboration to resolve. In Los Angeles' case, this collaboration emerged after a series of litigious actions resulting from adverse ecological and tribal-equity issues. In New York, they came about through harsh economic realities brought to the fore by a severe federal regulatory challenge.

As far back as 1913, the virtual draining of Owens Lake as a result of the opening of the first Los Angeles Aqueduct exposed the alkali lake bed to winds that lofted toxic dust clouds containing selenium, cadmium, arsenic and other elements throughout the region. Airborne particulates were often suspended for days during excessively dry periods – and have long posed a health hazard to local residents. They have even posed risks to communities further to the South. In the 1970s, the siphoning off of additional flows following completion of a second and larger aqueduct worsened the problem – igniting further protest.

These environmental impacts to Owens Lake - and to other, smaller watersheds within the Owens Basin (e.g., Lee Vining, Walker, and Parker Creeks) - dovetailed with concerns regarding water management in Los Angeles itself, beginning in the 1970s. Continuing drought and unrelenting population growth compelled the city to embrace a more adaptive approach to water management reliant on conservation, drought management, and a balance between augmenting supplies while providing incentives to lower demands: a method termed *integrated resource management.* This approach came to rely on non-structural, incentive-based, and education-driven methods to reduce water use and has been facilitated in part by concerns over climate change as well as the stresses and strains felt throughout its water importing regions (Los Angeles Department of Water and Power, 2010a).

Fig. 2. City of Los Angeles' Water Supply System

These issues came to a head in the 1990s through public protest, litigation, and federal intervention. In 1994, a settlement was reached between Los Angeles, Inyo and Mono Counties, and the U.S. EPA, and was enforced - in part - through a series of massive fines levied upon the LADWP. The settlement forced the agency to restore 62 miles of the lower Owens River, to "re-water" portions of Owens Lake and to allow the return of flows through Owens Gorge, and to restock bluegill, largemouth bass, fingerling trout, and other aquatic species.

Over time, native fauna are expected to return in significant numbers. In exchange, LADWP will receive 18,503 fewer cubic meters (15,000 fewer acre-feet) of Owens Valley water each year - reducing Los Angeles' reliance on the Owens Valley from some 35% of its total imported supply to approximately 18-20% (Linder, 2006; Hundley, 2001). As important as these changes in policy outcome may prove to be, of at least equal if not greater significance is the change in decision-making process by which they are being implemented. A Collaborative Aqueduct Modernization and Management Plan, or CAMMP, led by LADWP, the California Department of Fish and Game, and two environmental groups - California Trout and the Mono Lake Committee - has been undertaken to determine the means by which aqueduct operations can best be modified to facilitate changes in streamflowthat can satisfy environmental restoration needs on the one hand, while continuing to provide water to Los Angeles. Thus far, extensive data gathering, analysis, and drafting of prescriptions have been conducted, and the effort has entailed far more coperation among protagonists than in the past (McQuilkin, 2011).

While these environmental resotration activities involve consultation among intervenor groups, elected decisionmakers, and regulators, another collaborative effort has been conducted, off-and-on, regarding Native American water rights in the Owens Valley. Several Paiute Indian tribes lost their land and water rights in the region following white settlement in the mid-19th Century - and well-before the aqueduct was built. A partial restoration of water rights occurred in 1908 following a pivotal Supreme Court case – Winters vs. U.S. - which "explicitly affirmed water rights on Indian Reservations" by, in effect, setting aside correlative water rights on these reserved lands (Burton, 1991).

An Owens Valley Indian Water Commission – comprised of representatives of the Bishop, Big Pine, and Lone Pine Paiute tribes – are negotiating with LADWP to ensure they receive the water they are entitled to. While a final settlement has yet to be reached, when completed it will set relations regarding water use between Los Angeles and its surrounding region on another new footing (Owens Valley Indian Water Commission, 2009).

One of the notable benefits of New York's acquisition of much of the Catskill and Croton watersheds during the 19th Century was the opportunity to, in effect, ensure a virtually pristine source-water strategy. The storage reservoirs built by the city are surrounded by hardwood and evergreen forests that naturally filter water and retard erosion, thus averting sedimentation that would otherwise reduce drinking water quality. This asset also saves New York City billions of dollars in water treatment costs, according to the World Bank; has averted water-borne diseases; and, facilitates New York's distinction as the nation's largest city *without* a drinking water treatment plant (American Planning Association, 2011).

In the 1970s, water quality in these watersheds began to deteriorate as a result of contamination from sewage outfalls, leaky residential septic systems, agricultural runoff, and land cleared for residential development. The most significant issues that arose were: 1) sediment problems or turbidity within the Catskill Watershed, which can transport pathogens and interfere with the effectiveness of water filtration and disinfection; and, 2) excess nutrients, particularly phosphorus. The former can generate algae blooms that cause serious odor, taste and color issues, while excess phosphorus can cause eutrophic water conditions and increase carbon. Moreover, this water, mixed with chlorine, can result in the formation of "disinfection byproducts"suspected of being carcinogenic (New York State Department of Environment and Conservation, 2010b).

After years of study, environmental protection officials in New York City – and state officials representing the Department of Environmental Conservation – concluded that there were two feasible options to forestall threats of federal intervention, by EPA, to institute more strenuous remedial measures. The first was to build an artificial filtration plant, the city's first, at an estimated cost of between $8-10 billion, with an annual operating expense in the vicinity of some $360 million. The second option was to restore the Catskill/Croton watersheds through a combination of land purchases, compensation of existing private property owners for growth restrictions (e.g., conservation easements), and subsidies for septic system and other improvements. The city chose this much less-expensive option (at a total cost of approximately $200 million) – paid for through the sale of municipal bonds (NewYork State Department of Environment and Conservation, 2010b).

The second option - now known as the *New York City Watershed Protection Plan,* has been effective in complying with federal drinking water standards and delaying the need for a filtration plant. It is based on explicit, legally binding agreements – a Filtration Avoidance Determination (FAD) agreement, and a Memorandum of Agreement (MOA), concluded in January 1997 between several federal, state, New York county and city agencies, as well as various educational and non-profit organizations and watershed coalitions to provide regulatory oversight, perform environmental monitoring, protect water quality, educate the public, communicate about issues pertaining to pollution and watershed stewardship, and provide funding and other assistance to watershed communities (Westchester County Department of Planning, 2009: 2-26).

This partnership acknowledges the common interest of both public and private entities - in the city and within the two watersheds - in abating pollution through working together, especially given the limited power of any single entity to abate non-point pollution. Unlike the Los Angeles case, where collaboration on environmental quality issues initially emanated from an adversarial clash of interests, this partnership came about more amicably, while its composition has been similarly diverse. Members include New York City agencies, upstate communities in the twin watersheds, the U.S. EPA and other federal agencies, the New York State Department of Environmental Conservation (DEC) other state agencies, and various environmental groups.

One explanation for this comparatively amicable partnership is political realism: most watershed communities would have been adversely affected had New York City been forced to build a drinking water filtration system. This is so for two reasons: 1) the plant would have been paid for by all water users (and, in all likelihood, by regional taxpayers); and, 2) the state - if not the City itself as eminent domain tenant - would have been forced to impose more onerous land-use controls over the watershed if a partnership had not been formed. In effect, the indirect threat of having to pay for a water filtration plant was exactly the incentive needed to collaborate. Moreover, the choice of a multi-party partnership best suited the goals of all protagonists. It offered a viable, effective solution at manageable cost and through largely voluntary action (Croton Watershed Clean Water Coalition, 2009). However, given continued growth in rural areas throughout the region, and continued problems with turbidity, it has been necessary to revisit this plan.

In 2004, the city began construction of a $2 billion underground filtration plan in Van Cortlandt Park, Bronx designed to filter water from the Croton system, which is scheduled to be completed in 2012. It has also continued to acquire sensitive lands in the Catskills/ Delaware watersheds to further buffer their reservoirs from contamination, and thus, to remain in compliance with the state/EPA approved FAD agreement (New York City Department of Environmental Protection, 2010).

In sum, for both Los Angeles and New York City, local collaboration was abetted to some degree by federal and state government action. For the former, EPA intervention forced Los Angeles to rectify the condition of Owens Lake (and thus, indirectly, also improve the condition of other valley watersheds affected by adverse flows). Ironically, violation of the Clean Air Act (not the Clean Water Act) forced the city to work with state agencies, local valley officials and intervenor groups. For New York City, it was the *threat* of EPA (and state regulatory) intervention under the Safe Drinking Water Act (the Croton and Catskills are, after all, potable water sources) which compelled the city and its neighbors to collaborate to avert further sewage and non-point runoff contamination of the region's reservoirs.

## 5. Conclusion

Two fundamental questions are prompted by our discussion of Los Angeles' and New York City's diversion of water from their surrounding regions. The first is: why the absence of overt political conflict in the latter case as compared with the former? The second (as earlier noted) is: what can other megacities learn from these cities' experiences?

Taking the first of these questions – the attenuation of conflict in New York, and its intensity in Los Angeles, it is important to parse the question somewhat. An often assumed difference in the two cases is socio-economic: the Croton and Catskill watersheds are closer to New York City than the Owens Valley is to Los Angeles, and far better integrated into the former's economy. In the present-era, for example, evidence of the strong integration of the Croton Watershed's economy with that of New York City's five boroughs is offered by commuter traffic patterns - some 17,000 Croton Watershed workers commute from New York City daily - nearly 40% of the region's workforce, while some 18,000 workers living in the watershed commute to the city daily (about 35% of the workforce - see Westchester County Department of Planning, 2009: 2-27, 8).

However, this explanation is a bit trickier than might at first appear. New York and Los Angeles share profound socio-economic contrasts with their importing watersheds, which remain highly rural in character. While this is obvious with regards to the Owens Valley - a rural region initially dependent on farming and ranching before Los Angeles diverted its water - it is just as true for the Croton and Catskill watersheds. When initially settled, the upper Croton watershed, for example, was a remote and economically self-reliant region. Its residents developed separate and distinct ways of life initially dependent on dairy and crop farming (Westchester County Department of Planning, 2009: 2-26). Only in the late 19th Century, after completion of the aqueduct system, did the region's economy become more closely integrated with that of New York City.

A better explanation for the seeming absence of inter-regional conflict in the Croton and Catskill watersheds is the fact that New York's efforts to develop the water resources of these basins were, by comparison with those of Los Angeles in the Owens Valley, far more transparent and politically above-board. There is no evidence that the former sought to buy up watershed lands in secret, or to secure both surface and groundwater rights exclusively for its own use (and with federal government help). By comparison, the well-documented resistance to Los Angeles' activities in the Owens Valley, evidenced in part by the militancy of opposition, including acts of sabotage against the aqueduct during the 1920s, and tacit complicity in these acts displayed by many valley residents, dramatize the deep resentments generated by Los Angeles' actions. Many Owens Valley residents believed they had become a virtual colony of Los Angeles (Walton, 1992: chapter 5).

Their animosity was strengthened by what they believed was national-level collusion in the city's actions. President Theodore Roosevelt personally interceded in the Owens Valley case, persuaded that the future growth of Los Angeles was more important than the interests of Valley settlers. He not only ordered the eastward extension of the Sierra National Forest to discourage additional homesteading, thus ensuring protection of the aqueduct's right-of-way, but he further stated that the interests of Los Angeles exemplified ". . . the greatest benefit of the greatest number and for the best building up of this section of the country" (Los Angeles Department of Water and Power, 2010b).

Given all this, one must remain cautious about putting too fine a point on these differences. Opposition to New York City's efforts in the Croton Watershed, while infrequently reported, nevertheless existed. As early as 1837, some Westchester County residents lamented the implications of a Croton Aqueduct on their welfare. As one writer stated: "If the rivers of Westchester County are to be taken from it, how is it to rise in arts, manufacturing, and farming" (Quoted in Koeppel, 2001: 8)? Clearly, some residents acknowledged the long-term economic implications of diverting water.

There are two other reasons to avoid drawing too radical a contrast between New York and Los Angeles with regards to inter-basin conflicts. First, both cities have experienced intense *interstate* water conflicts, in both cases entailing Supreme Court litigation. And eventual water apportionment. Conflict between California and Arizona, spurred mostly by Los Angeles' utilization of the Colorado River as a major source of water after 1940, led to the important case of *Arizona v. California* (1964) by which the court reduced the amount of Colorado River water available to California, and further ruled that lower basin states (e.g., Arizona) were entitled to reasonable uses of tributary flows (U.S. Department of the Interior, 2008). Similarly, conflict between New York, Delaware and Pennsylvania led to two U.S. Supreme Court decisions allocating water among protagonists. Initially, the court upheld New York City's right, as an upstream riparian, to use a portion of the Delaware watershed. In a later case, the Court acknowledged the rights of all three states to an equitable apportionment of the Delaware River (Derthick, 1974). Environmental concerns under the Endangered Species Act have likewise prompted federal courts to reduce water deliveries from the Sacramento-San Joaquin Delta in recent years (Erie, 2006).

A second reason for caution is that both cities have experienced intense political conflict over the respective roles of private, market-driven water development efforts on the one hand and advocates for public control on the other. As noted in section 4, while both cities' preoccupation with water security led them to seek expanded public control of their local water systems to permit construction of massive aqueduct systems, originally, things began quite differently. In their early civic histories, both Los Angeles and New York viewed private water provision as the most desirable way to achieve water security. In fact, private provision was the norm throughout much of the 19th century. Incorporated in 1799, New York's Manhattan Company was inefficient and scandal-ridden. Yet, until 1834, it conspired with water cart owners to block the New York legislature's creation of a board of water commissioners, which ultimately bought out the company and built the Old Croton Aqueduct (Erie, 2006: 174).

Recall that Los Angeles, in 1902, acquired its private water company in part to amass the finances to build an Owens Valley Aqueduct. Even after acquiring its water company, however, Los Angeles never succeeded in eliminating the sway of private capital over water-supply. As is widely known, a syndicate of land investors sought to enrich themselves through the Los Angeles Aqueduct project by purchasing lands in the San Fernando Valley. Contrary to widespread belief, William Mulholland – the project's principal engineer - did not share this syndicate's avaricious motives. He sought to free the city from dependence upon erratic water sources in order to permit orderly growth. While he only conveyed knowledge of plans to build an Owens Valley Aqueduct to the Board of Water Commissioners and a few local officials, he did so simply to avert a stampede of speculators into the valley that would cause land prices to skyrocket (Mulholland, 2002).

## 5.1 Lessons for other megacities

So, what can other megacities learn from the experiences of New York and Los Angeles in regards to collaboration on regional issues and impacts of water development? The basic answer to this question brings us back to where we started this chapter - the challenge of water stress. As we have seen, Los Angeles and New York historically experienced stress, took various actions to address it which impacted their hinterlands, and continue to reckon with it through efforts to conserve water, improve infrastructure, and plan for climate change. While neither city has "solved" the problem of stress, their efforts to manage it harbor lessons for other megacities.

Since the 1970s, Los Angeles' conservation efforts have principally revolved around metering, conservation pricing, low-flow water appliance mandates, and efforts to compensate low-income groups for the costs of installing the latter. Water use has been considerably reduced - average water demands in period 2004 - 2010 are comparable to those of 1980, even though some 1.1 million additional people now live in Los Angeles (Los Angeles Department of Water and Power, 2010a: 8).

In 1988, New York City began metering to induce conservation and to ensure that larger volume water users pay their fair share. By the 1990s, water use declined some 28 percent as compared to 1979 (Shultz, 2007). Like Los Angeles, New York has also invested nearly $400 million in a 6.0 liter (1.6-gallon) per-flush toilet rebate program, which reduced water demand and wastewater flow by 342.96 million liters (90.6 million gallons) per day, a seven-percent reduction. One effect of this rebate program, aside from saving some $600 million, is delaying by about 20 years the need for water supply and wastewater-treatment expansion (U.S. EPA, 2010).

From the standpoint of regional collaboration, these experiences hold important lessons for other megacities in one important respect: conservation efforts lessen impacts on outlying communities - including the same communities from whence water supply originates. For Los Angeles, the more water that is conserved, the easier it becomes to reduce reliance upon both Owens Valley imports and those from other regions. For New York, similarly, the less water used, the less likely it is that stored water supplies will be depleted - thereby stretching available water andmaking less urgent the completion of various infrastructure improvements to deliver water to the city. Both cities are pursuing additional "active" conservation measures - with Los Angeles emphasizing stormwater capture and wastewater reuse and New York focusing on drought management and distribution system leak detection (Los Angeles Department of Water and Power, 2010a; New York City Department of Environmental Protection, 1998). While New York will continue to rely on incremental improvements to achieve conservation goals - more metering and the like - it, too, is likely to experience the same economic pressures as Los Angeles. It is likely that other megacities will look to both cities for assessments of these innovations' effectiveness.

As for infrastructure, issues related to stress may be far more problematical. Both cities suffer from aging and deteriorating water distribution systems. New York City is rebuilding its aqueduct system - and is currently engaged in construction of "Tunnel n. 3", an upgrade of the Croton aqueduct system, which loses millions of gallons annually. New components are also being added to its Delaware Aqueduct - all at a cost of some $2 billion. Los Angeles is rebuilding - piece-by-piece - its oldest distribution network components. However, the city faces a unique megacity challenge - continuing to deliver water in the event a major seismic eventruptures the Colorado and/or State Water Project Aqueducts. This is a major preoccupation for the Metropolitan Water District (MWD) which is the primary importing water agency for the region. While Los Angeles aspires to reduce reliance on MWD, during dry years it cannot do so. Moreover, it has made numerous investments in MWD projects under the assumption that it will continue to be a beneficiary of its supply (Los Angeles Department of Water and Power, 2010a).

Finally, as regards climate change, both cities are devoting enormous efforts in embracing climate issues in water resource planning. In New York City's case, sea-level rise threatens water infrastructure, especially for water treatment (Beller-Simms, et. al., 2008: 104-5). For Los Angeles, climate change threatens the robustness of already precarious imports - the aforementioned Metropolitan Water District, for example is already concerned that climate change will complicate its ability to engage in water trading schemes with rural, agricultural water users in the future (Erie, 2006).

In conclusion, it is not far-fetched to suggest that the massive water diversion projects Los Angeles and New York have pursued have had a symbolic as well as practical significance. For Los Angeles, the Owens Valley Aqueduct, Colorado River and State Water Project Aqueducts, and Port of Los Angeles all became symbols of the city's rise to eminence, and its ability to surmount the difficulties of being located in an insular region not readily blessed by a natural port or source of abundant freshwater. Similarly, for New York City, the Croton Aqueduct - the city's oldest imported water project - became part of a tradition of "grand civic projects" that, in the 19th Century, included the Erie Canal, Brooklyn Bridge, and IRT subway - all of which made the city the greatest metropolis in North America (Hood, 1993: 92). A final lesson here is that all these projects were not just civic activities, but publically-funded ones financed through bond markets, reminding us that neither the private nor public sectors alone can solve urban water problems - an important reminder in a political climate increasingly ambivalent about "privatizing" water supply.

## 6. Nomenclature – Key terms

| | |
|---|---|
| CAMMP | Collaborative Aqueduct Modernization and Management Plan |
| DEC | Department of Environmental Conservation (New York State) |
| FAD | Filtration Avoidance Determination Agreement |
| IRT | Interborough Rapid Transit (New York City's original subway) |
| LADWP | Los Angeles Department of Water and Power |
| MOU | Memorandum of Understanding |
| MWD | Metropolitan Water District (of Southern California) |
| US EPA or EPA | United States Environmental Protection Agency |

## 7. References

Adekalu, K.O., et. al. (2002). "Water sources and demand in SW Nigeria: implications for water development planners and scientists," *Technovation* 22: 799-805.

Alcamo, J., et. al., (2003). "Development and testing of the WaterGAP 2 global model of water use and availability," *Hydrological Sciences–Journal–des Sciences Hydrologiques* 48 (3) June: 317-337.

American Museum of Natural History. (2011). *The New York Water Story.* http://www.amnh.org/education/resources/rfl/web/nycwater/AMNH_Water.php

American Planning Association (2011). *How Cities Use Parks to Improve Public Health.* City Parks Forum Briefing Paper #7. http://www.planning.org/cityparks/briefingpapers/physicalactivity.htm

Beller-Simms, Nancy, et. al., (2008). *U.S. Climate Change Science Program Synthesis and Assessment Product 5.3 -- Decision-Support Experiments and Evaluations using Seasonal to Interannual Forecasts and Observational Data: A Focus on Water Resources.* National Oceanic and Atmospheric Administration, November.

Burton, Lloyd (1991). *American Indian Water Rights and the Limits of Law.* Lawrence, KS: University Press of Kansas.

Croton Watershed Clean Water Coalition (2009). *Updated 2009 Croton Watershed Management Plan.* New York: CWCWC. http://www.newyorkwater.org/pdf/managementPlan/MPlanNOV309.pdf

Davis, Margaret Leslie (1993). *Rivers in the Desert: William Mulholland and the Inventing of Los Angeles.* New York: Harper Collins.

Derthick, Martha (1974). *Between State and Nation - Regional Organizations of the United States.* Washington, DC: Brookings Institution.

Downs, T., M., et. al. (2000). "Sustainability of least cost policies for meeting Mexico City's future water demand," *Water Resources Research* 36.8: 2321-2339.

Earth Science Educational Resource Center. (2011). *New York City Water Supply.* http://www.eserc.stonybrook.edu/cen514/info/nyc/watersupply.html

Erie, Steven P. (2006). *Beyond Chinatown: The Metropitan Water District, Growth, and the Environment in Southern California.* Palo Alto: Stanford University Press.

Estrada, William David (2008). *The Los Angeles Plaza: Sacred and Contested Space.* Austin: University of Texas Press.

Feldman, David L. (2009). "Preventing the repetition: Or, what Los Angeles' experience in water management can teach Atlanta about urban water disputes," *Water Resources Research* 45, W04422, doi:10.1029/2008WR007605.

Fogelman, Robert M. (1993). *The Fragmented Metropolis : Los Angeles, 1850-1930.* Berkeley: University of California Press.

Furumai, H. (2008). "Rainwater and reclaimed wastewater for sustainable urban water use" *Physics and Chemistry of the Earth* 33: 340-346.

Gandy, M. (2008). "Landscapes of Disaster: water, modernity, and urban fragmentation in Mumbai," *Environment and Planning* 40: 108-130.

Glaeser, Edward (2011). *Triumph of the City : How Our Greatest Invention Makes us Richer, Smarter, Greener, Healthier, and Happier.* New York : Penguin Press.

Gumprecht, B. (2001). *The Los Angeles River: Its Life, Death, and Possible Rebirth.* Baltimore: Johns Hopkins University Press.

Hood, Clifton. (1993). *722 Miles – the Building of the Subways and how they transformed New York.* Baltimore: Johns Hopkins University Press.

Hundley, Norris Jr. (2001). *The Great Thirst - Californians and Water: A History.* Berkeley: University of California Press.

Kahrl, William M. (1982). *Water and Power: The Conflict over Los Angeles' Water Supply in the Owens Valley.* Berkeley: University of California Press.

Koeppel, Gerard T. (2000). *Water for Gotham: A History* (2000). Princeton, NJ: Princeton University Press.

Koeppel, Gerard T. (2001). *The Water Supply of New York City* (PDF manuscript, dated October 6th).

Linder, Michael (2006). "Water Wars! The Battle for the Owens Valley" *A KNX Exclusive Investigative Report, September 21*
http://www.knx1070.com/pages/86152.php?videoEpisodeId=42

Los Angeles Department of Water and Power (2009). *Mandatory Water Conservation - Fact Sheet,* May 4.

Los Angeles Department of Water and Power (2010a). *Urban Water Management Plan.* www.ladwp.com

Los Angeles Department of Water and Power (2010b). *The Story of the Los Angeles Aqueduct.* http://wsoweb.ladwp.com/Aqueduct/historyoflaa/

McQuilkin, Geoffrey (2011). « Stream Restoration Discussions picking up pace - implementation requires answering many, many questions, » *Mono Lake Newsletter.* Summer: 5

Mulholland, Catherine (2002). *William Mulholland and the Rise of Los Angeles.* Berkeley: University of California Press.

New York City (2011). *History of New York City Water Supply.*
http://www.nyc.gov/html/dep/html/drinking_water/history.shtml

New York City Department of Environmental Protection (2010). *New York City 2010 Drinking Water Supply and Quality Report.* Flushing, NY: NY DEP.
http://www.nyc.gov/html/dep/pdf/wsstate10.pdf

New York City Department of Environmental Protection (1998). *Drought Management Plan and Rules,* December.

New York State Department of Environmental Conservation (2010a). *New York City Watershed Program.* http://www.dec.ny.gov/land58597.html

New York State Department of Environmental Conservation (2010b). *Facts about the New York City Watershed* http://www.dec.ny.gov/lands/58524.html

Owens Valley Indian Water Commission, (2009). *Protecting, Preserving, Our Water Rights.* http://www.oviwc.org/thecrusade.html

Satterthwaite, David (2000). "Will Most people live in Cities?" *British Medical Journal* 321 (7269) November 4: 1143-1145.
http://www.pubmedcentral.nih.gov/articlerender.fcgi?artid+1118907

Shultz, Harold (2007). *Some Facts on the New York City Water and Sewer Supply System.* New York: Citizens' Housing and Planning Council.

Tortajada, C. & E. Casteian. (2003). "Water Management for a Megacity: Mexico City Metropolitan Area," *Ambio* 32.2: 124-129

UNPF (United Nations Population Fund). (2007). *State of World Population 2007: Unleashing the Potential of Urban Growth.* New York: United Nations Population Fund.

(UN) *United Nations World Water Development Report 3,* chapter 2 (2009). *Chapter 2: Demographic, Economic, and Social Drivers.* Paris: World Water Assessment Program, UNESCO Division of Water Sciences, UNESCO Publishing.

U.S. Department of the Interior, Bureau of Reclamation (2008). *The Law of the River.* Lower Colorado Region, Bureau of Reclamation.
http://www.usbr.gov/lc/region/g1000/lawofrvr.html

(U. S. EPA) Environmental Protection Agency (2010). *Water Sense – New York Water Fact Sheet.* EPA 832-F-10-104. June www.epa.gov/watersense

VanderBrug, Brian (2009). "In the Owens Valley, resentment again flows with the water," Los Angeles Times, May 16: B-1.

Village of Croton (2010). *History of the New Croton Dam.* http://village.croton-on-hudson.ny.us/Public_Documents/CrotonHudsonNY_WebDocs/HistoricalSociety crotondam?textPage=1

Walton, John (1992). *Western Times and Water Wars: State, Culture, and Rebellion in California.* Berkeley: University of California Press.

Westchester County Department of Planning (2009). *The Croton plan for Westchester – the Comprehensive Croton Watershed Water Quality Protection Plan.* September.
www.westchestergov.com/crotonplan

Willensky, Elliot & White, Norval. (1988). *AIA Guide to New York City, 3rd edition.* New York: Harcourt, Brace, Jovanovich.

World Water Council (2005). *Water Crisis* www.worldwatercouncil.org/index.php?id=25
(WMO) World Meteorological Organization (1997) *Comprehensive Assessment of the Freshwater Resources of the World.* Stockholm, Sweden: WMO and Stockholm Environment Institute.

(WRI) World Resources Institute in collaboration with the United Nations Development Programme, United Nations Environment Programme, and the World Bank (1998) *World Resources 1998-99: A Guide to the Global Environment: Environmental Change and Human Health*. New York, New York: Oxford University Press.

Yusuf, K.A. (2007). "Evaluation of Groundwater Quality Characteristics in Lagos City," *Journal of Applied Sciences* 7.13: 1780-1784.

Zérah, M.H.. (2008). "Splintering urbanism in Mumbai: Contrasting trends in a multilayered society," *Geoforum* 39 (2008): 1922-1932.

# Analysis of the Current German Benchmarking Approach and Its Extension with Efficiency Analysis Techniques

Mark Oelmann[1] and Christian Growitsch[2]

*[1]University of Applied Science Ruhr West in Muelheim an der Ruhr;*
*[2]Institute of Energy Economics at the University of Cologne*
*Germany*

## 1. Introduction

The analysis of the current German Benchmarking approach and an extension with efficiency analysis techniques seems to be a very specific topic at first sight. However, at second sight it turns out that this question is relevant for many European countries. The reason is that many countries share a similar structure of their water sector which then implies similar challenges.

A first similarity in most European countries is that local governments are the responsible bodies for providing water services. They can decide if they want to perform the service themselves or if they contract it out to private companies. Figure 1 shows that they predominantly transfer the task to publicly owned companies. It is worth noting that Figure 1 displays the percentage of *population served* by either a public, a private or a mixed operator. If it would show the number of companies then the percentage of private companies in relation to all would diminish drastically. The same holds true if the term "privatization" would be specified more clearly. In Germany, for example, it ought to be distinguished if a company is only formally privatized, which means that the shareholders remain public, or if a company is really materially privatised.

A second observation for the various European countries is that the water sector is very fragmented.[1] EUREAU (2009, p. 94) is counting 600,000 jobs for more than 70,000 water services operators. On average a water service provider would employ less than 10 people. The structure is thus mainly publicly organized and rather fragmented.

At the same time, tariffs are now supposed to cover all costs.[2] In addition, immense investments will be needed in many European countries to fulfill the various European Directives.[3] Both developments will lead to significantly higher water prices in the future. It

---

[1] The Netherlands, England/Wales or Scotland are examples of countries where rather large companies prevail.

[2] This cost recovery principle is introduced in Art. 9 of the Water Framework Directive 2000/60/EC.

[3] A good description of the various important European directives is available under http://ec.europa.eu/environment/water/water-dangersub/76_464.htmDirectives.

can be expected that the public in many European countries will hold the companies accountable to show that they are performing efficiently.

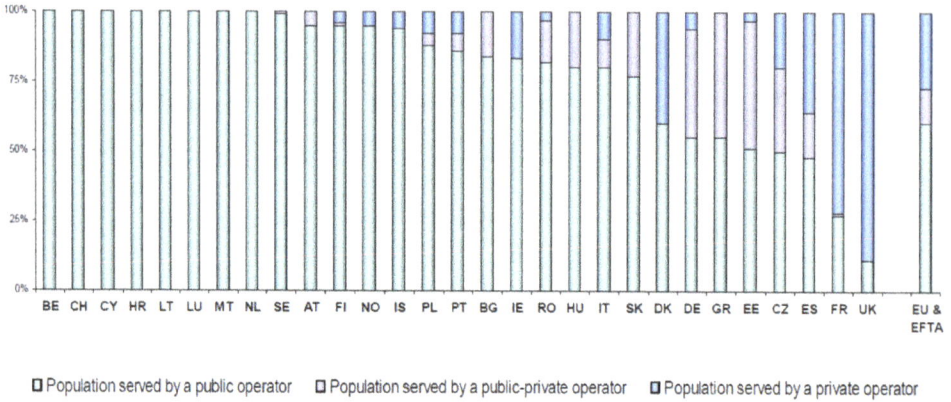

□ Population served by a public operator      □ Population served by a public-private operator      □ Population served by a private operator

Fig. 1. Ownership structure of European water service providers 2008 (EUREAU, 2009, p. 93)

The provision of water supply is a natural monopoly which implies that companies are not sanctioned if they are inefficient. A water utility regulator and the economic regulation of water supply and wastewater companies is thus an important issue which is discussed internationally.[4] The problem, however, is that in countries with a very fragmented structure of the industry a complex economic regulatory framework, like in England & Wales or Scotland, is not applicable for the majority of providers. At the same time, privatisation is not always worth considering: a precondition for a successful privatisation is that public authorities have the knowledge and the data to supervise the private service provider. Otherwise, a public monopoly is only transferred into a private one (Newberry, 2003, p.4). Many European countries are thus considering a third approach: benchmarking. Such a benchmarking system compares companies with one another according to certain indicators. It generally serves two purposes: First, it shall be a measure to increase transparency in the sector. Displaying reports which are publicly available are supposed to enhance accountability of companies. Second, performance benchmarking systems evolve which analyze certain processes within the company in more detail and give, therefore, insight to companies where they could enhance their efficiency.[5]

The main question for this paper is to analyze the potential of benchmarking. We will use the German experiences with the current approach and will answer the question if the current system can be enhanced by applying efficiency analysis techniques. Can we expect that such an enhanced benchmarking will imply that a regulator is redundant?

The remainder of this article is structured as follows. In the following, second, section we will briefly present the concept of benchmarking as well as the current use of benchmarking in Germany. Its deficiencies imply the need, in the third section, to portray alternative

---

[4] International regulatory approaches for water and wastewater services are portrayed in Marques (2010).

[5] For a short portray of European benchmarking approaches see Marques/De Witte (2007).

techniques like Data Envelopment Analysis (DEA) and Stochastic Frontier Analysis (SFA) as means to better analyze data. In the fourth section we introduce the employed database. Section five displays the best models to explain cost differences for small, middle and large water service providers in the distribution of water in Germany – something which has never been done for the German water supply sector. In section six we practically describe what kind of information a company, which is participating in such an enhanced benchmarking approach, can expect. The paper ends with a brief conclusion and outlook.

## 2. Current benchmarking in the German water supply sector

Benchmarking can be defined as "the process whereby a company compares and improves its performance by learning from the best in a selected group" (BDEW, 2010, p. 4). 36 of such, so called *process benchmarking* projects are carried out in the German water and wastewater sector (ATT et al., 2011, p. 94ff.). Parts of the value chain are analyzed in detail mainly between a limited number of companies. Up to 20 companies are participating in the various projects (ATT et al., 2011, p. 94ff.). The concept is displayed in the following figure.

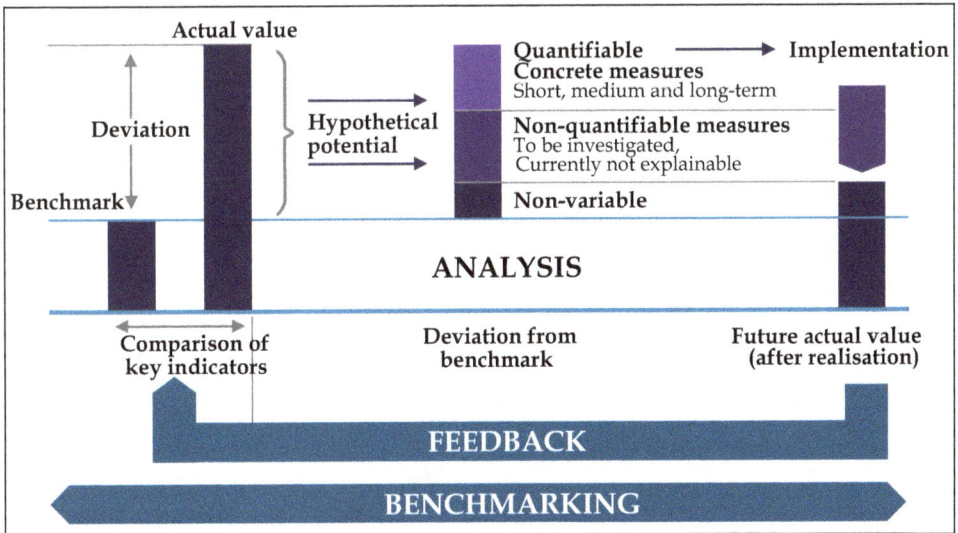

Fig. 2. Concept of Benchmarking (BDEW, 2010, p.4)

The process starts out with a comparison of key indicators. For each single company the deviation between its actual value and the benchmark is determined. The different factors which may explain the difference are then intensely discussed between the specialists of the companies for the particular process. Quantifiable measures which are then implemented shall diminish the gap between own value and benchmark. The relative efficiency of the company within this particular process increases. Process benchmarking is, therefore, characterized by a continuous process to learn from the best.

The 36 water and wastewater programmes have approximately 12 participants, on average. Very often the same companies take part in several projects covering different processes. For the 16 water supply projects 100 different companies might participate. Compared to more than 6,000 German operators, this number is quite negligible.

The question, therefore, arose of how to activate more companies to participate. Particularly, since the Federal Government and the Bundestag have submitted its so called "modernization strategy", *metric benchmarking* projects increased in number.[6] The "modernization strategy" – approved by the German Parliament on the 22nd of March 2002 - acknowledged the benchmarking concept and asked the German water associations to continue implementing them in the various *Bundesländer*. Benchmarking projects in the water supply sector are now performed in each of them (see Figure 3). Public reports are available for 12

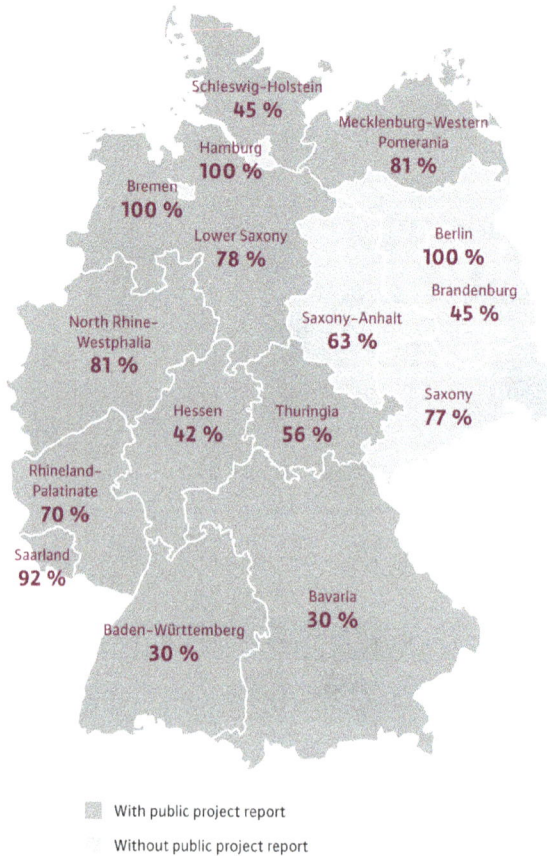

Fig. 3. Metric benchmarking in German Bundesländern (BDEW, 2010, p.9)

---

[6] A metric benchmarking system is not going into such detail as a process benchmarking system does. It merely compares companies by employing key performance indicators. The link to the German "modernization strategy", however, does not imply that metric benchmarking, as such, is a new invention. The *Betriebsvergleich kommunaler Versorgungsunternehmen* (Benchmarking of public water supply utilities), run by the German Water Association VKU, was first installed about 50 years ago.

out of the 16 *Bundesländer* (ATT et al., 2011, p. 92f.). Based on drinking water quantities 30 % (Baden-Württemberg and Bavaria) up to 100 % (city states of Berlin, Hamburg and Bremen) are covered by the projects. Based on the number of companies it is far less.

However, these kinds of metric benchmarking projects are also activating a number of additional water service providers which are not participating in the very intense performance benchmarking projects. Its main intention is to give the companies a first insight of how good they actually seem to perform. Similar to the performance benchmarking projects, discussions between the relative good and bad companies are intended to take place. The problem, however, is to distinguish a good and a bad company. The current approach shows that, where ones costs in a certain part of the value chain are solely compared with those of others without actually taking into account differences in basic conditions, benchmarking is not as efficient as it could be. Due to very different conditions for companies to deliver water services, costs can be very different between companies for good reason. A company with rather unfavorable conditions and higher costs might be more efficient than another one with more favorable conditions and lower costs. As a result, current metric benchmarking projects seem not to fullfill the high expectations. In nearly all of the metric benchmarking projects the number of participating companies remains either constant, over time, or diminishes (ATT et al., 2011, p. 90ff.).

Current metric benchmarking approaches should, therefore, employ techniques which are able to assess costs, taking into account the relevant environmental conditions in which the company actually operates. The Data Envelopment Analysis (DEA) and the Stochastic Frontier Analysis (SFA) are the scientifically established tools, which are giving a good indication about the relative efficiency of a company. Water suppliers which are performing badly according to both DEA and SFA – given their particular, not influenceable environmental conditions – ought to have potentials to improve efficiency. Such an enhanced benchmarking can thus improve the information a company may receive from participating in a benchmarking project.

It is worth noting that such an enhanced metric benchmarking project is better displaying the relative performance of a company. It is, however, not giving advice on how a company might increase its efficiency. In order to determine the correct measures a company might participate in a process benchmarking project, install certain working groups within its company or employ consulting companies. A metric benchmarking project is thus very often a necessity for a company to deal with its own performance relative to others. After detecting certain inefficiencies, the company should encounter incentives to install programs which help in improving their performance. Time series data of a company's performance should thus be collected.

All European countries which are employing metric benchmarking systems will, therefore, sooner or later face the necessity to decide which kind of information they want to display publically and whether companies should be obliged to participate. The Netherlands, for example, made it compulsory to take part in such programs whereas Germany is very reluctant to do so.

There are also, however, other means to give incentives to companies to participate in enhanced metric benchmarking systems. Those German water suppliers, which are setting prices, are currently under the supervision of cartel offices. Currently, these regulatory institutions are investigating those companies which have high prices per m³. This is particularly ridiculous because, due to very different conditions, a company with high water prices might be much more efficient than a company with low ones. An incentive for companies to participate in metric benchmarking projects could, therefore, be to either start investigations in companies which are not participating in metric benchmarking projects at all or which seem to be relatively inefficient at first sight. For other European countries it might be worth considering attaching the granting for subsidies to a successful participation in benchmarking projects.

## 3. Brief introduction into efficiency analysis techniques

Scientific efficiency and productivity analysis can be differentiated into parametric and non-parametric methods (Coelli et al., 2005). Parametric approaches, like Ordinary Least Squares (OLS) or Stochastic Frontier Analysis (SFA), estimate cost or production functions and an (in-) efficiency value per observation. Therefore, one has to specify a functional form (like log-linear, Cobb Douglas or Translog). This, indeed, leads to implicit assumptions about the underlying production technology (Jamasb and Pollitt, 2003), for instance, about factor substitution etc. A major advantage of parametric methods is that they allow for statistical inference and their robustness against outliers and statistical noise (Coelli et al, 2003). Non-parametric techniques like the Data Envelopment Analysis (DEA) rather calculate than estimate multi-input/multi-output productivities. The major advantage of Data Envelopment Analysis is its flexibility, i.e. that the analyst does not have to specify a functional form (Coelli et al, 2003). This section briefly discusses the different methods of productivity analysis.[7]

The statistical method of Ordinary Least Squares (OLS) is a parametric method estimating the explanatory power of so called exogenous variables (regressors) on an endogenous variable (regressand). The parameters are estimated by minimizing the squared deviances of modeled to actual values (sum of squared residuals). A widespread application of this relatively easy method is the linear regression analysis. The central problem of the linear regression model is, however, that the deviation of one firm's value to the regression line is declared to result from relative efficiencies, which does not always have to be the case.

But, even if the linear regression analysis provides substantially better information to a firm than the average cost approach used up until now, further improvement in efficiency evaluation is in order. For "operational distribution costs", as well as for "total costs" and the other most important costs along the value chain "operational costs production and treatment", "administrative costs" and "capital costs", two additional analyses should be employed to make the linear regression results more robust when analyzed in detail.

Stochastic Frontier Analysis (SFA) is another parametric method to determine the efficiency frontier and an advancement of the OLS method in some ways. It requires assumptions about the functional form of the relationship between costs and output values.[8] Essentially, the actual costs of one firm are compared to the minimum (efficient) costs of another firm.

---

[7] For a detailed description, see Coelli et al. (2005).
[8] Different models are used nationally and internationally in benchmarking grid connected infrastructure services. Next to Cobb Douglas and translog specifications, mostly log-linear and standardized functions, using only one input variable obtained by division, are used.

Here, in contrast to the linear regression model, the deviation from the optimum need not be resulting purely from inefficiencies, but also from so called "White Noise". Hence, interpreting these deviations purely as efficiency potentials may be misleading and should be avoided.

The aim of the Data Envelopment Analysis (DEA) is also to measure the efficiencies of respective firms relative to a threshold firm. The productivity of single entities is compared to an efficiency frontier, which is derived from a linear connection between efficient firms (so called "peers"). The DEA is a non-parametric method so that the efficiency frontier is not estimated empirically but calculated by a linear optimization program.

In other grid-reliant sectors (like electricity, gas, telecommunications and even water supply in other countries) the DEA and SFA methods are well established, while the linear regression model does not provide robust and consistent results.

## 4. Data set

We use the dataset of Rödl & Partner, the biggest consultancy which conducts metric benchmarking for German water supply utilities. The original data set comprised 612 observations from the years 2000 to 2007. Each of these observations contained 179 firm specific units of information. First, all observations from different years of the same company were eliminated, keeping the most current one.[9] Second, all observations from before 2006 were deleted in order to minimize the problems of inflating cost data from older years to the base year of 2007. Third, all companies without any distribution network, or with mainly bulk water supply, were removed from the dataset. Fourth, all observations where crosschecks revealed inconsistencies were deleted.[10] 196 observations remained.

2007 served as the base year. Using the producer-price index "Water and Water Services" from the German Federal Statistical Office, the data were made comparable by restating 2006 data in terms of 2007 prices. To reach a maximum of comparability we then deducted the concession levy from the operational distribution costs.[11]

The sample is as close in line with the overall structure of the German water supply sector as possible. However, Figure 4 shows that the distribution, according to the size of the companies between our sample and the overall situation, differs. 30.2 % of approximately 6,400 water supply utilities (ATT et al., 2008, p. 12) in the German water sector supply more than 500,000 m³ annually. In our sample this percentage of companies, which supply more than 500,000 m³ annually, is nearly 80 %. In the whole German water supply sector 92.6 % of water output is supplied by companies with an annual water delivery of more than 500,000 m³. The figure for our sample is nearly 99 %. This implies that our sample contains relatively bigger companies than the overall German average.

---

[9] Panel data might be interesting in the future to follow the efficiency development of a single company over time.
[10] Rödl & Partner have been very cautious to crosscheck, in particular, all cost data. No inconsistencies were found. Over time however, the set of data slightly changed. Particular older observations with lacking structural variables were, therefore, removed from the data set.
[11] For our calculations in the production/treatment segment we deduct the water abstraction charges. DEA and SFA for total costs imply that concession levy, water abstraction charges and compensatory payments for agriculture have to be subtracted.

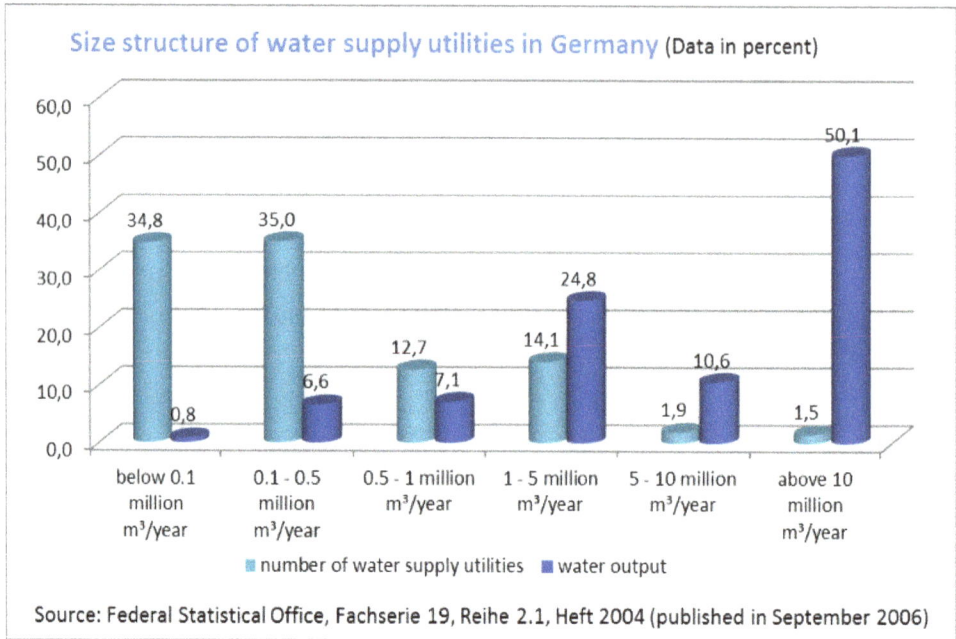

Fig. 4. Size structure of water supply utilities in Germany (ATT et al., 2009, p. 14)

## 5. Methodological approach

To achieve robust modelling results, we follow a three step approach. First, we cluster the observations with regard to their size. By that, we implicitly assume that small companies have a different production technology than larger ones. Secondly, we perform some theoretical and empirical analyses on the potential variables to develop reasonable input-output combinations for our latter modelling. This is then performed in the third subsection on section 5.

### 5.1 Clustering

Rödl & Partner, the consultancy which performs the benchmarking for several German *Bundesländer* and which provided the data for calculating the efficiencies, has been clustering all participating companies according to the annual accounted water. In workshops with water supply companies they agreed to form three groups. The first cluster comprises 38 companies with a water delivery of 500,000 m³ annually, the second one comprises 97 companies with water delivery between 500,000 m³ and 2,500,000 m³ and for the last one, all remaining companies with annual water delivery up to 50,000,000 m³ (61 companies).

Such a differentiation, according to the size of companies, is extremely important. Our models will later reveal that the production functions of the three different groups vary. Thus, a data set should always contain enough observations in order to be able to form groups.

## 5.2 Variables and cost driver analysis

We assume that the objective of a water service provider is not solely to produce drinking water. The objective rather is to provide the option for clients to use as much drinking water as they wish at any time. This implies that the set-up of the network with transportation and distribution pipes, tanks, pumps, valves and service areas should also be considered as outputs, which at least in the short-run cannot be influenced by the company.

Correlation analysis provides an initial determination of the statistical relationship between costs on the one hand and potential explanatory variables reflecting the specific frameworks faced by each water supplier on the other hand. Such correlation analyses are the basis for estimating costs as a function of multiple drivers, i.e., regressors, as they help specifying the efficiency-analyzing models later on. In this step it is made sure that the exogenous variables, like outputs and cost drivers, explain the endogenous variable, costs, sufficiently.[12]

Analyses revealed, particularly for the bigger companies, that the five variables of group one in Table 1 are highly correlated, both with operational distribution costs as well as with one another. Both the technical common understanding and the analysis of the empirical literature stress the explanatory power of these variables.[13] It thus makes sense to always have at least one of these variables in the DEA- or SFA-functions in a cost or production function model. Variables of group two to four were tested for additional explanatory value.

Walter et al. (2010, p. 228) refer to a number of studies which display the significance of "water losses" as an explanatory variable. For countries like Brazil, Spain or Peru this might certainly be of importance due to high variations in the quality of the network. For a country like Germany however, where the level of water losses is only about 6.5 % (ATT et al., 2011, p. 56) on average,[14] water losses cannot serve as a good proxy for the quality of the network or associated operational costs, respectively.

The two variables "downturn of demand since 1992" and "downturn of demand since 1998" are surely interesting for explaining the development of prices. Many companies, which face a significant decrease of demand due to various reasons, need to increase prices if they lack the appropriate tariff models. Too often only a minor share of the total fixed costs is actually covered by earnings, which are independent from actual demand. However, for a cost benchmarking – particularly the operational distributional costs - these variables are insignificant.

Whereas all variables of the fourth group were not taken into account any longer, the variables of the third group were tested in DEA- and SFA-functions, where a certain combination of variables made sense from a technical water perspective. Particularly, the client structure ("Household supply relative to accounted water") is quite often used to explain differences in both operational distribution costs as well as total costs. We, however,

---

[12] Tests for heteroskedasticity (Breusch-Pagan/Cook-Weisberg test) and multicollinearity (Variable Inflation Factor, VIF) have been applied to fulfill general conditions of multivariate regression analysis and specifically Ordinary Least Squares conditions.
[13] Besides the literature discussed in Walter et al. (2010) also see Lin (2005), Picazo-Tadeo et al. (2009) and Coelli & Walding (2006). All of them, however, only apply either DEA or SFA. Due to rather bad data quality they were also not able to analyze other than total costs.
[14] For more detailed data on German water losses see IGES (2010, p.30).

encountered that this criterion did not have any significant explanatory power. The reason for it might be that the companies within our three groups are actually quite homogenous, which again stresses our hypothesis that companies need to be analyzed according to groups. Our findings might have been different if we would have followed the same path as other researchers which have not had such a detailed database, both quality and quantity wise, and therefore were not able to cluster their observations.

| Variable | Unit |
|---|---|
| **Group one:** | |
| Number of household connections | No. |
| Accounted water | € |
| Transportation and distribution pipes | Km |
| Distribution pipes | Km |
| Inhabitants | No. |
| | |
| **Group two:** | |
| Tanks | No. |
| Tank capacity | $m^3$ |
| Valves | No. |
| Service areas | No. |
| Height difference | m |
| Distribution and transportation pipes to accounted water (excl. re-distribution) | $m/m^3$ |
| Distribution pipes to accounted water (excl. re-distribution) | $m/m^3$ |
| Distribution and transportation pipes per household connection | m |
| Distribution pipes per household connection | m |
| | |
| **Group three:** | |
| Supply to re-distributors | $m^3$ |
| Household supply relative to accounted water (excluding re-distribution) | % |
| Pipe damages | No. |
| Peak supply relative to supply of the day | % |
| Energy consumption per transported and distributed $m^3$ of water | $kWh/m^3$ |
| | |
| **Group four:** | |
| Area | $km^2$ |
| Inhabitants per $m^3$ (area) | No. |
| Water losses | $m^3$ |
| Downturn in demand since 1992 | % |
| Downturn in demand since 1998 | % |
| Area | $km^2$ |
| Supply (adjusted for re-distribution) per tank | $m^3$ |
| Household connections per tank | No. |

Table 1. Variables for explaining operational distribution costs

## 5.3 Methods

Based on the definition of relevant cost drivers we apply a parametric and a non-parametric benchmarking approach, namely SFA and DEA (compare Section 3). Because DEA is sensitive towards extreme values, an outlier analysis is applied in addition. Therein, firms that are most efficient in many of the observations are iteratively taken out of the sample and, hence, the efficiency analysis. The process stops when the average value of efficiency of all transmission system operators, including the potential outliers, is statistically indifferent (at 95% confidence) to the average value of efficiency excluding the potential outliers. A t-test (according to Satterthwaite) is used to compare the expected values. Identified outliers are removed from the sample.

Multiple specifications of SFA models are estimated to compare specifications given by similar correlation coefficients in earlier phases of the analysis. To conclude on an improved goodness of fit of one specification against the other, Akaike's and Schwarz's information criteria are used as well as a comparison of the log-likehood values. Given insignificant parameters, a likelihood ratio test is performed. Also, we test different functional forms: Cobb-Douglas, Translog, and log-linear models.

## 5.4 Results

The best model for the largest companies is displayed in the following table.

| Variable | Coefficient | Standard Deviation |
|---|---|---|
| ln (Distribution pipes to accounted water (excl. re-distribution)) | 0,861*** | 0,133 |
| ln (Distribution pipes per household connection) | 1,27*** | 0,229 |
| ln (Distribution pipes) | 1,377*** | 0,119 |

Table 2. SFA-Model Large Companies (2.5-50 Mill. m³ per year) for operational distribution costs

The results of the DEA- and SFA-analysis have shown that a combination of the variables Distribution pipes, Distribution pipes per household connection and Distribution pipes to accounted water (excl. re-distribution) suits particularly well for an efficiency evaluation of operational distribution costs in the group of the largest water suppliers (2.5 mio. m³ up to 50 mill. m³ annual supply): All three indicators are significant with a minimum confidence level of 99%. Besides, the combination of those three variables explains about 70% of operational distribution costs in this group of firms ($R^2 = 0,706$). The English water regulation authority OFWAT, in comparison, uses models sometimes with less than 30% explanatory power. Last, but not least, the sign on the coefficients for Distribution pipes and Distribution pipes per household connection is positive, as expected. Increasing absolute, as well as relative grid length, independently increases costs. Only the regressor Distribution pipes to accounted water (excl. re-distribution) is expected to have a negative influence on costs as it increases with population per km². Because of simultaneous modeling of Distribution pipes per household and Distribution pipes to accounted water (excl. re-distribution) the result can additionally be interpreted in the way that costs of a one unit increase in grid length overcompensate the grid density advantages.

The best models for the two other groups of companies are as follows:

| Variable | Coefficient | Standard Deviation |
|---|---|---|
| ln (Distribution pipes per household connection) | 0,6709** | 0,283 |
| ln (Distribution pipes) | 2,019*** | 0,188 |

Table 3. SFA-Model Medium Companies (0.5-2.5 mill. m³ per year) for operational distribution costs

| Variable | Coefficient | Standard Deviation |
|---|---|---|
| ln (Number of household connections) | 1,390*** | 0,048 |
| ln (Distribution pipes to accounted water (excl. re-distribution)) | 0,509*** | 0,186 |

Table 4. SFA-Model Small Companies (< 0.5 mill. m³ per year) for operational distribution costs

The DEA and SFA results for the largest group of companies are plotted in the figure below to verify consistency of the efficiency analyses. Because the efficiency measures are not always comparable due to the different methods used, the rank correlation of the results are determined (according to Spearman) and plotted. A value of, for example, 0,78, means that the ranks of a firm resulting from DEA and SFA analyses correlate with 78%.[15]

## 6. Additional value for companies

The problem of current metric benchmarking has been to distinguish between a good and a bad company. So far a water supplier with low costs per m³ is supposed to be a good company. Such a company might however face very favourable conditions which would actually be the reason for this low relative costs compared to other operators. Applying the concept "Learning from the Best" is thus misleading. As mentioned earlier the same holds true for the classical OLS. If a company is, however, good according to both the DEA as well as the SFA (in Figure 5: the company marked by the bottom arrow) this operator would quite surely be an interesting candidate for a discussion with those laggards which should have rather high efficiency potentials (in Figure 5 the company marked by the upper arrow).

Besides, in better identifying a good and a bad company, this approach also helps to quantify an efficiency potential. This potential can of course be displayed in various ways. It is possible to list the DEA-, the SFA- or for example the average of DEA- and SFA-result. The table below could thus be read as follows: Whereas no. 79 is the most efficient company and thus encounters only minor options to decrease operational distribution costs, no. 136 seems to face major inefficiencies. Approximately 2.3 million € per year could be saved according to these first calculations.

---

[15] The rank correlation results for the medium companies are 75% and for the small ones 41%. To put these values in context with rank correlations of other sectors and other countries see Sumiscid AB (2007, p. 34). As seen there, first rank correlations from German energy models are between 70 and 75%. This value is said to be very high compared to other models used internationally, like for example in Sweden with rank correlations of 40%.

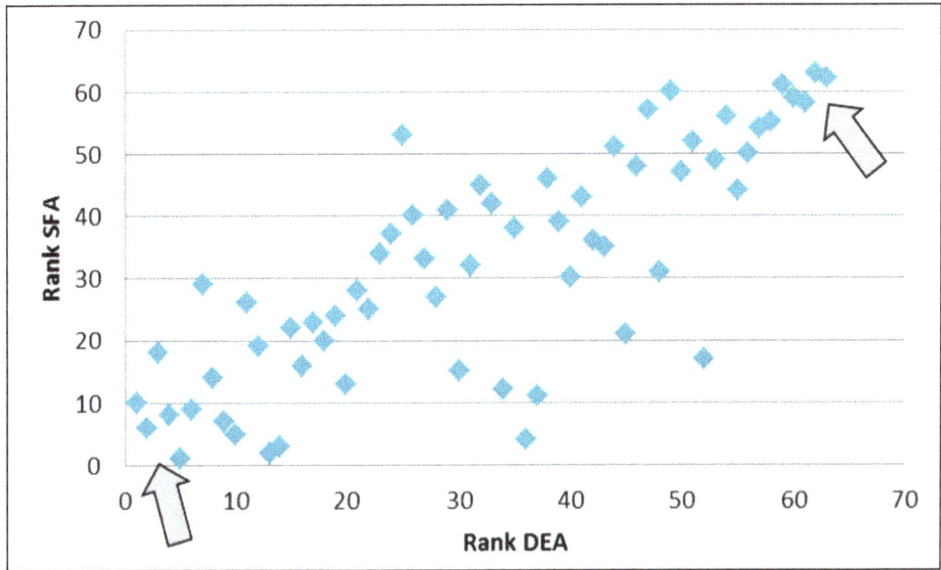

Fig. 5. Rank comparison DEA and SFA, Cluster "Large companies" (Bottom [Upper] arrow: Relatively efficient [inefficient] companies according to both DEA and SFA]

| # Company | Actual Costs | Expected Costs | Efficiency Potential in % |
|-----------|--------------|----------------|---------------------------|
| 79 | 2.884.076 | 2.721.218 | 5,98% |
| 115 | 4.929.511 | 3.608.453 | 36,61% |
| 119 | 1.833.783 | 1.670.595 | 9,77% |
| 136 | 6.551.907 | 4.243.461 | 54,40% |
| 155 | 2.261.552 | 1.574.094 | 43,67% |
| 226 | 2.169.827 | 1.953.611 | 11,07% |

Table 5. Efficiency analysis techniques and implications for individual efficiency potential

We would certainly always suggest to not only analyze the results of the "operational distribution costs". It is worth doing the same calculations for "total costs" and the remaining sub-categories "operational production costs", "capital costs" and "administration costs". In such a way potential trade-offs between, for example, operational and capital costs can be observed and interpreted. In addition, an analysis would also need to take into account different quality provision between companies which would need to be backed by willingness-to-pay-studies.

## 7. Summary and outlook

Benchmarking is an already well established concept in the German water supply sector. It is performed in all of the German *Bundesländer*. However, if we compare the number of companies which take part in such a metric benchmarking with the total number of water suppliers the percentage will be less than 2 %. One reason for the companies to not participate might be the quality of the feedback they receive. Currently, the structural variables which a company faces are not taken sufficiently into account. Many reasons, besides inefficiency, can explain why one company encounters higher costs than another water supplier.

Efficiency analysis techniques, which imply not only a comparison of companies according to key performance indicators but a measurement of performance, are tools which better analyze the existing benchmarking data. They take into account differing structural conditions of the companies and, therefore, give a more valuable feedback to a company in which parts of the value chain they have potentials to increase their efficiency. Using the data of Rödl & Partner, the most prominent metric benchmarking consultant in the German water supply industry, we showed for the first time that these methods can be easily applied. Our results show that the rank correlations between DEA- and SFA-results are much higher than in other network sectors and other countries. They also display that companies should be clustered. A small and a very large company seem to have different production functions and can hardly be compared with one another. It, therefore, makes sense to cluster the companies according to groups. Overall, we may conclude that the enhancement of the current metric benchmarking systems by performance measurement is displaying if a company seems to have potentials to increase its efficiency in the operational distribution costs. It may also give a quantitative indication of the extent of inefficiency.

The analysis of the operational distribution cost, which has been performed here, certainly is only the first step. The feedback for the companies will further increase if a performance measurement is offered for all parts of the water supply value chain and the total cost.[16] An introduction of performance measurement into the current benchmarking is, therefore, a big leap forward. We may hope that such an improved benchmarking system is giving incentives for more German companies to participate voluntarily.

Otherwise, the question certainly arises of how to proceed. Due to the situation, that companies in a natural monopoly sector are particularly accountable to the public, the German water sector faces similar questions than those ones in other European countries which are employing metric benchmarking systems. Sooner or later the countries face the necessity to decide which kind of information they want to display publically and whether companies should be obliged to participate in benchmarking. The Netherlands, for example, made it compulsory to take part in such programs. Every three years they are publishing reports which also display the performance development of a water supply company both in relation to all other companies as well as over time. By publishing these data, companies did not only detect efficiency potentials but also faced the public pressure to use measures to actually improve their performance. Between 1997 and 2005 the average efficiency of a Dutch water supply company increased by 23 % (Dijkgraaf et al., 2006, p. 8). Such incentives in benchmarking systems to actually increase performance are essential that benchmarking could be regarded as an alternative to an introduction of an economic water utility regulator.[17]

---

[16] First, preliminary results for the total cost calculations reveal that the rank correlations between DEA and SFA-results are even higher than for the operational distribution costs.

[17] Cross-country comparisons of benchmarking systems in the drinking water sector in the Netherlands, England and Wales, Australia, Portugal and Belgium recently revealed that the average efficiency correlates with the incentives of such benchmarking systems (De Witte/Marques, 2010).

## 8. References

ATT et al. (2011). Branchenbild der deutschen Wasserwirtschaft 2011, Berlin, Germany, available from:
http://www.bdew.de/internet.nsf/id/40873B16E2024175C125785A00350058/$file/110321_Branchenbild_dt_WaWi_2011_Langfassung_Internetdatei.pdf

ATT et al. (2008). Profile of the German Water Sector 2008, Berlin, Germany, available from:
http://www.bdew.de/internet.nsf/id/DE_Profile_of_the_German_Water_Industry/$file/Profile_German_Water_Industry_2008.pdf

BDEW (The German Association of Energy and Water Industries) (2010). Benchmarking: 'Learning from the best' - Comparison of performance indicators in the German water industry, Berlin, Germany, available from:
http://www.bdew.de/internet.nsf/id/DE_Benchmarking_Learning_from_the_best_Comparison_of_performance_indicators_in_the_German_water_ind/$file/100301_BDEW_Benchmarking_Broschuere_WA_in_ENGLISCH.pdf

Brenck, A., Grenz, M. & T. Beckers (2010). Auswertung und Begutachtung aller öffentlichen Projektberichte Benchmarking (Trinkwasser), IGES Report fort the BDEW (The German Association of Energy and Water Industries), available from :
http://www.bdew.de/internet.nsf/id/DE_IGES_TU-Berlin-Kurzstudie_Auswertung_und_Begutachtung_aller_oeffentlichen_Projektberichte_Benchm/$file/101019_IGES_TU_Berlin_Kurzstudie_Benchmarking_final.pdf.

Coelli, T. & Walding, S. (2006). Performance Measurement in the Australian Water Supply Industry: A Preliminary Analysis, in: Coelli, T. & Lawrence, D. (eds.): Performance Measurement and Regulation of Network Utilities, Edward Elgar Publishing, Cheltenham, pp. 29-61.

Coelli, T. J., Estache, A., Perelman, S. & Trujillo, L. (2003): A Primer on Efficiency Measurement for Utilities and Transport Regulators, Washington D. C.

Coelli, T. J., Prasada Rao, D. S., O'Donnell, C. J. & Battese, G. E. (2005): An Introduction to Efficiency and Productivity Analysis, 2nd edition, New York.

De Witte, K. & Marques, R. (2010). Designing performance incentives, an international benchmark study in the water sector, CEJOR (2010) 18, pp. 189–220.

Dijkgraaf, E., van der Geest, S.A. & Varkevisser, M. (2006). The efficiency gains of benchmarking

Dutch water companies. Working Paper Erasmus University Rotterdam, Draft Version, available from:
http://www.tilburguniversity.edu/research/institutes-and-research-groups/tilec/events/seminars/dijkgraaf.pdf .

EUREAU (European Federation of National Associations of Water & Wastewater Services) (2009). EUREAU Statistics Overview on Water and Wastewater in Europe 2008 - Country Profiles and European Statistics, Brussels, Belgium, available from:
http://eureau.org/sites/eureau.org/files/news/EUREAU_statistics-edition_2009.pdf.

Jamasb, T & Pollitt, M. (2003): International Benchmarking and Yardstick Regulation: An Application to European Electricity Utilities,: Energy Policy, Vol. 31, S. 1609-1622.

Lin, C. (2005). Service Quality and Prospects for Benchmarking: Evidence from the Peru Water Sector, Utilities Policy, 13(3), pp. 230-239.

Marques, R.C. (2010). Regulation of Water and Wastewater Services – An International Comparison, IWA Publishing, ISBN 978-184-3393-41-2, London, UK.

Marques, R.C. & De Witte, K. (2007). Towards a benchmarking paradigm in the European public water and sewerage services, available from: https://www.econ.kuleuven.be/ew/academic/econover/Papers/Benchmarking %20paradigm%2023-03.pdf

Newberry, D. M. (2003). Privatising Network Industries, CESifo Working Paper No. 1132, February 2003.

Picazo-Tadeo, A.J., Sáez-Fernández, F.J. & González-Gómez, F. (2009). The Role of Environmental Factors in Water Utilities' Technical Efficiency. Empirical Evidence from Spanish Companies, Applied Economics, 41, pp. 615 628.

Sumiscid AB (2007). Development of benchmarking models for German electricity and gas distribution, Project Gerner/AS 6 – Final Report, 1.1.2007.

Walter, M., Cullmann, A., von Hirschhausen, C., Wand, R. & Zschille, M. (2010). Quo Vadis Efficiency Analysis of Water Distribution? A Comparative Literature Review. *Utilities Policy*, Vol.17, No. 3-4, pp. 225-232.

# The Willingness to Pay of Industrial Water Users for Reclaimed Water in Taiwan

Yawen Chiueh[1], Hsiao-Hua Chen[2] and Chung- Feng Ding[3]

*[1]Department of Environmental and Cultural Resources,
National Hsinchu University of Education,
[2]Environment and Development Foundation, Hsinchu,
[3]Water Problems Enterprise Information & Incubation Center,
Tainan Hydraulics Laboratory, Water Resources Agency,
Ministry of Economic Affairs/ National Cheng Kung University,
Taiwan*

## 1. Introduction

Water resource in Taiwan is mainly delivered through diversion from river, regulation of reservoir and groundwater extraction. Diversion from river is dependent on hydrological conditions on yearly bases. Reservoir regulation is unpredictable and subject to actual conditions of the watershed, and moreover, building a reservoir involves huge funding and takes a long time in addition to frequent confrontation of questioning and protests of environmental groups. As for groundwater extraction, land subsidence resulting from excessive groundwater extraction leads to national land conservation issues. Considering natural restrictions and environmental policies, wastewater reclamation has become an important topic when the government seeks for new water sources. Aiming to resolve the supply of water resource, the government is progressively promoting the wastewater reclamation industry and exploiting the corresponding market. With provision of statutes that favor the development of wastewater reclamation industry as well as financial setup that supports relevant technologies, the government expects to counsel the private sector to participate in establishing a wastewater reclamation industry that conforms to the trend of green ecology.

Taiwan has had a wastewater reclamation industry of the produce/use model instead of the produce/supply model. The produce/use mode is the situation that the supplier and user of the reclaimed water are the same. The reduce/supply mode is the situation that the supplier and user of the reclaimed water are different. The main reason is that the government has encouraged the industry to achieve "process recycle ratio" and "total plant recycle ratio" for more than 10 years, and there have been operators in Taiwan engaged in wastewater reclamation plants of "produce/use" model. However, wastewater reclamation plants of the "produce/supply" model still remain in the Model Plant stage and require successive fostering. Even in advanced countries such as EU, USA or Japan, there had been a number of barriers hindering the promotion of wastewater (sewage) reclamation; one of them is the dominant insufficiency of user confidence over quality and safety of the reclaimed water. In

Taiwan, except adopting the produce/use model - in which factories who promote water saving within the industrial park reclaim their own wastewater for reuse, the environmental assessment requires that wastewater or sewerage within a building to be reclaimed by the building, or a wastewater/sewage treatment plant reclaims a portion of its own effluent for in-plant miscellaneous use - no other industrial development or application reference has been seen. Furthermore, affected by the lacking of experience and the rather low water price, the willingness to use reclaimed wastewater has been fairly low.

The Hydraulic Planning and Experimental Institute made a preliminarily research in 2009 and found that potential candidates for the use of reclaimed water include: 1. Water for secondary livelihood use: using effluent of urban wastewater/sewage treatment plant for watering nearby golf courses, to enhance flexibility of the local supply of water resources. 2. Water for agricultural use: treating effluent from the urban wastewater/sewage treatment plant to meet the standard of "water quality for irrigation" and using the reclaimed water for agricultural irrigation in areas having a water shortage. 3. Water for conservation: using water reclaimed from urban wastewater/sewage treatment plants for groundwater recharge, for artificial recharge of disaster prevention purposes, for agricultural use in substitution for the groundwater which would have originally been extracted, so as to alleviate groundwater extraction. And 4. Water for industrial use. Organizations that may increase water consumption in the future include: Hsinchu Science Park Yilan Base, Taoyuan Aviation City, Taipower Letzer Industrial Park Power Plant, Taipei Harbor Power Plant, Expansion Project of Dragon Steel Corporation, Middle Taiwan Science Park Taichung Base and Houli Base, Taichung Harbor Proprietary Areas (including power plant, petrochemical and industrial areas), Hsinchu Science Park Phase IV Tongluo Base, Yunlin Offshore Fundamental Industrial Park, Taiwan Petroleum Corporation Third Naphtha Cracker Renovation Project, Tainan County Great Hsinyin Industrial Park Development Project, Southern Taiwan Science Park Phase II, Development Project of Southern Taiwan Science Park LCD TV District (Tree Valley Park), Tinnan Industrial Park, and China Steel Corporation. The study carries out questionnaire interview with industrial water users to comprehend their willingness towards paying for the reclaimed water as well as their methods to use the same.

## 2. Method

### 2.1 The theoretical model

The study use Contingent Valuation Method (CVM) to analyze the "willingness to pay (WTP)" of future industrial water users for using reclaimed water. The CVM method surveys user valuation over non-existing transactions of goods or services in the market by way of a direct questionnaire designed basing on hypothetic conditions in the market, therefore is a valuation method over non-market resources. The major feature of the CVM model is the forward-looking (ex ante) decision which evaluates a future event in advance. The price and quality level of the reclaimed water supply mechanism are presumed by the study without actually putting into operation; they are preliminary assumptions of the future supply mechanism of the reclaimed water which may be applicable to the existing factories of the industrial and science parks, for further understanding the WTP level of users regarding water price and the quality level.

The major difference of the CVM approach contrasting with a direct valuation approach is that CVM is specific in combining the survey practice with theories. Popular use of the CVM approach began in 1970's when the Forest Act of UK and the presidential directive #12291 were promulgated; a number of researches were seen conducted with the CVM approach over economic benefits of natural resources. During the Exxon tanker oil spoil incident in 1989, the federal court of USA ordered compensation to be paid by Exxon appraised with the CVM approach, enhancing the authenticity of the same. By 1993, since the US government extensively used CVM to make public policies that concern natural resources, the National Oceanic and Atmospheric Administration (NOAA) therefore promulgated guidelines on the use of the CVM approach, regulating the use of Contingent Valuation Surveys. Research papers show that CVM is applicable to offering a rational valuation over public goods or environmental goods (Smith, 1993). Hutchinson, et al.(1995) also pointed out that as long as the questionnaire is duly designed, CVM is a highly credible means for price evaluation.

The CVM appraisal can be conducted with a Random Utility Model (Hanemann, 1984) or an Expenditure Function (Cameron, 1988). However, Cameron(1988) uses Censored dichotomous choice model to directly estimate the parameter of the Expenditure Function, it directly and easily obtaining the WTP of the public over environmental goods. The microeconomic theory also demonstrates that the indirect utility function has a dyadic relationship with the Expenditure Function; therefore it can also represent the utility preference of the consumer. In order to prevent excessive biases and to make adequate use of all the data acquired from the questionnaire survey, the study employs a close-ended dichotomous choice method design for the questionnaire, and uses the Expenditure Function model (proposed by Cameron (1988) and Cameron & James (1987)) for calculating the WTP function of the reclaimed water.

This study use the questionnaire survey in determining the price level that the factories are willing to pay for the reclaimed water, and to provide incentives for the factories becoming willing to use the reclaimed water, a hypothetic market must be conceptually established for the factories, to create a bidding function based on individual social and economical characters and the level of bidding prices. The main method is to estimate the acceptable price by way of Cameron's expenditure function model based on the WTP of the questionnaire and the percentage of factories that are willing to pay or willing to accept. Follow the defined of resource value by Freeman (1993). We set up the empirical model of reclaimed water as follows:

$$Y(Q0,Q1,U0,S) = E(Q1,U0,S) - E(Q0,U0,S) \quad\quad (1)$$

$Y(Q0,Q1,U0, S)$ is the bidding function of the factories for the reclaimed water;
$E(Q0,U0,S)$ and $E(Q1,U0,S)$ are the Expenditure Function.
In the formula,
$Q0$ is the situation that the factory do not get reclaimed water;
$Q1$ is the situation that the factory got reclaimed water;
$U0$ is the utility function of the factory;
$S$ is the price vector of market goods and individual social and economic characteristics vectors.

If the price suggested by the CVM questionnaire is $T$,

$$Y(Q0,Q1,U0, S) \geq T \quad\quad (2)$$

the probability for the interviewee to check this bid can be expressed by formula (3):

$$Pr=Pr[Y^*(Q0,Q1,U0,S)-T>u] \qquad (3)$$

Where $Y^*$ is observable component, u is observable random component, as shown in Formula (4):

$$Y(Q0,Q1,U0,S)=Y^*(Q0,Q1,U0,S)+u \qquad (4)$$

The Bidding Function can be estimated based on the probit model by Cameron & James(1987) as shown below:

Ii=1 if Yi >Ti
= 0 otherwise

$$\begin{aligned} Pr(Ii=1) &= Pr(Yi>Ti) = Pr(ui>TI-Xi'B) \\ &= Pr(ui/\sigma > (TI-Xi'B)/\sigma) \qquad (5) \\ &= 1-\phi((Ti-Xi'B/\sigma) \end{aligned}$$

where Xi'B is exclaiming variable, $\phi$ is accumulated probability of intensity function, then the interviewee's bidding valuation can be shown as formula (6) :

$$Yi = Xi'B+ui \qquad (6)$$

Yet standard binary probit model shall be

Ii=1 if Yi>0
=0 otherwise

$$\begin{aligned} Pr(Ii=1) &= Pr(Yi>0i) = Pr(ui>-wi'\delta) \\ &= Pr(zi>-wi'\delta/v) \\ &= 1-\phi(-wi'\delta/v) \end{aligned}$$

at this time,

$$Yi = wi'\delta + ui$$

using the following transformation

$$-(Ti, Xi') \begin{bmatrix} -1/\sigma \\ B/\sigma \end{bmatrix} = -wi'\delta$$

$$\delta^* = (\alpha,\gamma) = (-1/\sigma, B/\sigma)$$

we obtain

$$B = -\gamma/\alpha$$
$$\sigma = -1/\alpha \qquad (7)$$
$$Yi^* = Xi'B$$

Where Yi* is the price of reclaimed water estimated by the supplier under the standard binary probit model; this can be used for the calculation of a reasonable price for the reclaimed water.

Assuming $u$ to be the logistic distribution, the empirical result can be calculated based on the logistic model by Cameron(1988).

$$P(Y)=[1+e-[Yi-Ti]]-1$$

Similar to the probit model appraoch, we can obtain

$$Yi^* = Xi'B \tag{8}$$

Where Yi* is the price of reclaimed water estimated by the supplier under the logistic model; this can be used for the calculation of a reasonable price for the reclaimed water.

## 2.2 Questionnaire design

To the demand end, quality and price of the reclaimed water are the major concern. We detail as follows: Water reclaimed from effluent of large scale wastewater treatment plant by reverse osmosis: capable of reaching quality standard of Taiwan Water Works. For the selection of questionnaire valuation method, the study employs the most easy-to-operate and time saving "Single-bounded dichotomous choice elicitation method" (Boyle & Bishop, 1988) to carry out interviews based on NOAA suggestions.

The scenario of this study is as follow: We assumption that "the government guarantees that quality of reclaimed water conforms with city water specifications, no interruption of supply 365 days a year with assured quality and loss indemnification on supply interruption," and that "dedicated pipeline to be installed for reclaimed water delivery, plus with free-of-charge pipe connection," and that "50% deduction on wastewater treatment charge if total consumption of reclaimed water exceeds 40% of total industrial water consumption of the company". Than we ask the manager or boss of the factory " Are you willing to pay for the reclaimed water for the "T" price we suggested on the questionnaire[1] under the assumption scenario?"

The value of reclaimed water depends on its water quality. The quality of "city water" is just the basic requirement of the customer when comparing with more expensive and better quality of "soft water, 1μS/cm". Besides, the assumptions of the following are not yet done but they are the requests of the factories. So we set up the approximate realistic assumption of the scenario.

The selection of the price we suggested on the questionnaire, i.e. "T" in formula(2) in each questionnaire of the scenario is determined based on the current city water price in Taiwan and the costs for reclaiming the wastewater. Furthermore, one or several extreme and median values are set to meet theoretical requirements of the Contingent Valuation Method, the scenario having 12 kinds of "T" prices as shown in Table 1. In another word, the study employs 12 different questionnaires, $Q_A$ through $Q_L$, with different assignment of the "T" prices for each type of questionnaires scenario.

---

[1] we give different "T" price in different type of questionnaires which shows on Table1.

| No. of Questionnaire | Scenario |
|---|---|
| A | 3 |
| B | 5 |
| C | 7 |
| D | 9 |
| E | 11 |
| F | 12 |
| G | 13 |
| H | 15 |
| I | 16 |
| J | 18 |
| K | 20 |
| L | 24 |

Table 1. The price "T" we suggested on the questionnaire (in NT\$/ton)

## 2.3 Sampling design

The study takes industrial and science parks having a higher water consumption as survey objects, including factories in Hsinchu Industrial Park, Chungli Industrial Park, Taichung Industrial Park, Linyuan Industrial Park, Hsinchu Science Park, Central Taiwan Science Park and Tainan Technology Industrial Park. The above industrial and science parks comprise the sampling zone of the study. For industrial parks, factories having a water consumption exceeding 200CMD are selected as survey objects, a total of 347 factories are included. Then, we call factories one by one to verify their water consumption and exclude those having a low water demand or those lacking willingness to participate our survey. After verification, a total of 205 factories are included in the survey. The survey schedule covers September and October of 2009.

| Areas \ Count | Count | Selected as survey objects | Send after verify their water consumption and exclude those having a low water demand or those lacking willingness to participate our survey | Return |
|---|---|---|---|---|
| Hsinchu Industrial Park | 20 | 12 | | 5 |
| Chungli Industrial Park | 13 | 8 | | 2 |
| Taichung Industrial Park | 1 | 1 | | 0 |
| Tainan Technology Industrial Park | 6 | 3 | | 2 |
| Linyuan Industrial Park | 8 | 4 | | 4 |
| Hsinchu Science Park | 245 | 145 | | 26 |
| Central Taiwan Science Park | 54 | 32 | | 8 |
| Total | 347 | 205 | | 47 |
| Return Ratio | | 22.93% | | |

Table 2. Statistics on questionnaire count of sampling zone factories

Questionnaire via fax or mail is adopted for the survey. Every returned questionnaire is checked for completion; in case of miss or obvious mistake of key items, a telephone re-check will be made against the particular factory. For factories that fail to return the questionnaire, urging telephone calls will be made. The study sends out a total of 205 questionnaires and receives 47 returns of which 2 are null; the return ratio is 22.93%. Table 3 shows questionnaire distribution of the sampling zone. For further understanding of the scenario, Table 3 is compiled to statistically manifest the WTP selected for the scenario, and the percentage of factories that are willing to use the reclaimed water at different WTPs corresponding to respective questionnaires of the study.

| Scenario | Questionnaire No | A | B | C | D | E | F | G | H | I | J | K | L |
|---|---|---|---|---|---|---|---|---|---|---|---|---|---|
| | T price | 3 | 5 | 7 | 9 | 11 | 12 | 13 | 15 | 16 | 18 | 20 | 24 |
| Scenario | Return Count | 3 | 7 | 3 | 3 | 4 | 4 | 1 | 6 | 4 | 6 | 2 | 2 |
| | Ratio of Willingness | 67 | 71 | 33 | 67 | 75 | 50 | 0 | 33 | 50 | 33 | 0 | 0 |
| Total number of questionnaire | | 29 | 29 | 29 | 29 | 29 | 29 | 29 | 29 | 29 | 29 | 29 | 28 |

Note: Data of null questionnaires are not included in this table. Unit of T: NT$/ton. Ratio of Willingness is in %.

Table 3. Return ratio of difference type of questionnaires

## 3. Results

For empirical analysis of factories' Willingness to Pay (WTP), the study employs the valuation method developed by Cameron & James (1987) and Cameron (1988) using the software called LIMDEP. In order to avoid influences from extreme sample values, Logit Model is used to establish the valuation formula of WTP (Willingness to Pay); Approximation of Newton's method and maximum likelihood estimation (MLE) are used to evaluate the WTP; accuracy of the prediction exceeds 75% although the number of sample is small. Table 4 shows results of WTP:

| | Scenario |
|---|---|
| WTP of factories (NT$/ton) | 13.97 |
| Accuracy of the prediction model | 82.22% |

Note: The estimation models are under the 0.5% significant level

Table 4. Results of WTP of factories

Under the assumption that "the government guarantees that quality of reclaimed water conforms with city water specifications, no interruption of supply 365 days a year with assured quality and loss indemnification on supply interruption," and that "dedicated pipeline to be installed for reclaimed water delivery, plus with free-of-charge pipe connection," and that "50% deduction on wastewater treatment charge if total consumption of reclaimed water exceeds 40% of total industrial water consumption of the company", factories are willing to purchase the reclaimed water at an average price of 0.48$/ton (13.97NT/ ton), reclaimed water demand is 131,000 Cubic Meter per Day (CMD[2]), 22.8% of factories in the sampling zone are willing to use reclaimed water, and a ratio of 47% of the

---

[2] Cubic Meter per Day (CMD) is the flow rate of water

returned effective samples. Average consumption of reclaimed water per factory is 291.55CMD, 48.86% to the total consumption of industrial water. Table 5 shows statistics of potential usage and consumption of reclaimed water by factories.

| Application of reclaimed water | Average ratio of factories willing to accept | Average potential maximum consumption of reclaimed water per factory (CMD) |
|---|---|---|
| Process Water | 24% | 160 |
| Boiler Feed Water | 20% | 50 |
| Cooling Water | 76% | 65.75 |
| Washing Water | 58% | 10.6 |
| Firefighting Water | 51% | 5.2 |
| Total consumption of reclaimed water | | 291.55 |

Note: Average potential maximum consumption of reclaimed water per factory = Total potential consumption of reclaimed water / number of factories that are willing to accept

Table 5. Statistics of potential reclaimed water application and consumption by factories

Table 6 shows values of model parameters of the Scenario. In which MAA indicates surveyed "T" price (N.T.D./ton); MAW indicates the product of ratio of maximum reclaimed water to total industrial water acceptable to the factory multiplied by the total consumption of industrial water in 2008(CMD); MAN indicates the amount of washing water the factory is willing to use(CMD). If the MAA coefficient is negative and MAW and MAN coefficients are positive, the theoretical expectation is deemed met.

| Variable explain | Variable Name | Coeff. | Std.Err. | t-ratio | P-value |
|---|---|---|---|---|---|
| Intercept | ONE | 0.170238 | 1.04835 | 0.162387 | 0.871001 |
| The surveyed "T" price (N.T.D./ton) | MAA | -0.20879 | 0.099374 | -2.10108 | 0.035634 |
| The amount of washing water the factory is willing to use (CMD) | MAN | 2.73873 | 1.01492 | 2.69846 | 0.006966 |
| The product of ratio of maximum reclaimed water to total industrial water acceptable to the factory multiplied by the total consumption of industrial water in 2008 (CMD) | MAW | 0.005193 | 0.002371 | 2.19045 | 0.028491 |

Table 6. Value of model parameters of Scenario

## 4. Discussion and conclusions

The study assumed the scenario "the government guarantees that quality of reclaimed water conforms with city water specifications, no interruption of supply 365 days a year with assured quality and loss indemnification on supply interruption," and that "dedicated pipeline to be installed for reclaimed water delivery, plus with free-of-charge pipe connection," and that "50% deduction on wastewater treatment charge if total consumption of reclaimed water exceeds 40% of total industrial water consumption of the company", factories are willing to purchase the reclaimed water at an average price of 0.48$/ton. The Scenario mach the policy that Taiwan government wants to promote to use of reclaimed water for new water source other than diversion from river, reservoir water, building new reservoir or groundwater extraction.

The study results show the WTP under the Scenario exceed the existing price of city water indicate that the assumption of senior as about are the works if which need to be done in the future. We presume the reason behind this are that: 1) the factories that are willing to assist the survey had suffered from water shortage in the past operation and therefore are willing to procure the reclaimed water at a cost higher than the city water under the assumed scenario. 2) It appear to be under the changing climate, the Industrial water users are more concern about the stable water source. 3) We speculated that the Scenario of "50% deduction on wastewater treatment charge" have a great incentives to use reclaimed water.

## 5. Lessons learned

The result of this study implies that the appropriate water management policy design could really encourage the use of reclaimed water. In other word, appropriate water management policy design could change the structure of water use. Well water management policy or incentives mechanism, such of deduction on wastewater treatment charge, could bring about good water conservation patterns. Furthermore, the willing to pay for the reclaimed water price is higher than the city water also show that the wastewater reclamation industry have good future prospects. If there has appropriate water management policy, the reclaimed water could be good water source other than diversion from river, reservoir water, building new reservoir or groundwater extraction.

## 6. Acknowledgments

The article was extracted from detailed project "The Research on the Strategic Development of Specialists Training of the Wastewater Reclamation and Reuse Industry" that sponsored by the Water Resources Agency, Ministry of Economic Affairs, Taiwan (project code: Moeawra0980052) and "A game theory approach to evaluation the irrigation water transfer" that sponsored by the National Science Council, Taiwan (project code: 100-2221-E-134-001). The authors would like to thank the anonymous reviewers and all the participants of this project for their efforts.

## 7. References

Asano, T., Burton, F., Leverenz, H., Tsuchihashi, R. and Tchobanoglous, G. (2007). "Water Reuse: Issues, Technology and Applications", McGraw Hill, New York, 1570pp.

Cameron , T. A. and M. D. James, 1987. "Estimating Willingness to Pay from Survey Data: An Alternative Pre-Test-Market Evaluation Procedure." Journal of Marketing Research. ,24:389-95.

Cameron, T. A., 1988. "A New Paradigm for Valuing Non-market Goods Using Referendum Data: Maximum Likelihood Estimation by Censored Logistic Regression." Journal of Environmental Economics and Management, 15:355-79.

Choon Nam et al. (2002). "Singapore Water Reclamation Study- Expert Panel Review and Findings", June 2002.
http://www.pub.gov.sg/NEWater_files/download/view.pdf

Department of Health Services, DHS (2001). "California Health Laws Related to Recycled Water", State of California, Department of Health Services, Sacramento CA, June 2001 Edition.

Department of Water Resources of Georgia (2007). http://www.mde.state.md.us/assets/document/WatRuse07-7%20Gelot.pdf

El Paso Water Utilities (2007). http://www.epwu.org/wastewater/fred_hervey_reclamation.html

Freeman A..Myrick(1993). The Measurement of Environmental and Resource Values. Resources for the Future, Washington,, D. C.

Hanemann, Michael W., 1991."Willingness To Pay and Willingness To Accept: How Much Can They Differ?" The American Economic Review 81 (3 ), June :635-647

Hanemann, W. M., 1984." Welfare Evaluations in Contingent Valuation Experiments with Discrete Responses." American Journal of Agricultural Economics. 66:332-341.

Idelovitch E., Icekson T.N., Avraham O., and Michail M. (2003). "The long-term performance of Soil Aquifer Treatment (SAT) for effluent reuse", Water Science Technology Water Supply, Vol. 3, No. 4, pp. 239-246.

Juanico, M. (2008). "Israel as a case study", Chap. 27, in Jimenez and Asano (Ed.) Water reuse, an international survey of current practice, issues and needs, pp483-502.

Khan and Roser (2007). "Risk assessment and health effect studies of indirect potable reuse schemes", Center for Water and Waste Technology, School of Civil and Environmental Engineering, University of New South Wales, Australia.

Law, I. B. (2003). "Advanced Reuse- From Windhoek to Singapore and beyond", Water Vol.30(5), pp31-36

McEwen, B. (1998). "Indirect potable reuse of reclaimed water", in Asano T. (Ed.) Wastewater Reclamation and Reuse. Chapter 27. Water Quality Management Library, Vol.10, Technomic Publishing, Lancaster, PA.

Mekorot, Israel National Water Co. (2003). "Dan Region Project – Groundwater Recharge with Municipal Water", Brochure of Introduction.

Mitchell, R. C. and R. T. Carson, 1989. Using Surveys to Value Public Goods: The Contingent Valuation Method. Washington, D. C. : Resources for the Future.

Morrison, Gwendolyn C. 1998., "Understanding the disparity between WTP and WTA: endowment effect, substitutability, or imprecise preferences?" Economics Letters 59:189-194.

Sheikh, B., Cooper, C. and Israel, K. (1999). "Hygienic evaluation of reclaimed water used to irrigate food crops – A case study", Water Science and Technology, Vol.40 (4-5), pp261-7.

Smith, V. K., 1993. "Nonmarket Valuation of Environmental Resources: An Interpretive Appraisal" Land Economics, 69(1):1-26.

State of California (1978). "Wastewater Reclamation Criteria", Title 22, Division 4, California Code of Regulations, State of California, Department of Health Services. Sanitary Engineering Section. Berkeley, California.

State of Forida (2004). "2003 Reuse Inventory", Department of Environmental Protection, Division of Water Resources Management, Tallahassee, Florida.

Swayne, M., Boone, G., Bauer, D. and Lee, J. (1980). "Wastewater in receiving waters at water supply abstraction points", EPA-60012-80-044, US Environmental Protection Agency, Washington, D.C.

USEPA (2004). "Guidelines for Water Reuse", EPA 625/R-04/108, Washington, D.C.

Water Pollution Control Federation (1989). "Water Reuse (2nd Edition)", Manual of Practice SM-3, Water Pollution Control Federation, Alexandria, Virginia.

# Permissions

The contributors of this book come from diverse backgrounds, making this book a truly international effort. This book will bring forth new frontiers with its revolutionizing research information and detailed analysis of the nascent developments around the world.

We would like to thank Dr. Uli Uhlig, for lending his expertise to make the book truly unique. He has played a crucial role in the development of this book. Without his invaluable contribution this book wouldn't have been possible. He has made vital efforts to compile up to date information on the varied aspects of this subject to make this book a valuable addition to the collection of many professionals and students.

This book was conceptualized with the vision of imparting up-to-date information and advanced data in this field. To ensure the same, a matchless editorial board was set up. Every individual on the board went through rigorous rounds of assessment to prove their worth. After which they invested a large part of their time researching and compiling the most relevant data for our readers. Conferences and sessions were held from time to time between the editorial board and the contributing authors to present the data in the most comprehensible form. The editorial team has worked tirelessly to provide valuable and valid information to help people across the globe.

Every chapter published in this book has been scrutinized by our experts. Their significance has been extensively debated. The topics covered herein carry significant findings which will fuel the growth of the discipline. They may even be implemented as practical applications or may be referred to as a beginning point for another development. Chapters in this book were first published by InTech; hereby published with permission under the Creative Commons Attribution License or equivalent.

The editorial board has been involved in producing this book since its inception. They have spent rigorous hours researching and exploring the diverse topics which have resulted in the successful publishing of this book. They have passed on their knowledge of decades through this book. To expedite this challenging task, the publisher supported the team at every step. A small team of assistant editors was also appointed to further simplify the editing procedure and attain best results for the readers.

Our editorial team has been hand-picked from every corner of the world. Their multi-ethnicity adds dynamic inputs to the discussions which result in innovative outcomes. These outcomes are then further discussed with the researchers and contributors who give their valuable feedback and opinion regarding the same. The feedback is then collaborated with the researches and they are edited in a comprehensive manner to aid the understanding of the subject.

Apart from the editorial board, the designing team has also invested a significant amount of their time in understanding the subject and creating the most relevant covers. They scrutinized every image to scout for the most suitable representation of the subject and create an appropriate cover for the book.

The publishing team has been involved in this book since its early stages. They were actively engaged in every process, be it collecting the data, connecting with the contributors or procuring relevant information. The team has been an ardent support to the editorial, designing and production team. Their endless efforts to recruit the best for this project, has resulted in the accomplishment of this book. They are a veteran in the field of academics and their pool of knowledge is as vast as their experience in printing. Their expertise and guidance has proved useful at every step. Their uncompromising quality standards have made this book an exceptional effort. Their encouragement from time to time has been an inspiration for everyone.

The publisher and the editorial board hope that this book will prove to be a valuable piece of knowledge for researchers, students, practitioners and scholars across the globe.

# List of Contributors

**Ralph A. Wurbs**
Texas A&M University, United States

**José Pinho, José Vieira, Rui Pinho and José Araújo**
Department of Civil Engineering, University of Minho, Portugal

**Dejan Komatina**
International Sava River Basin Commission, Croatia

**Enrique Troyo-Diéguez, Arturo Cruz-Falcón, Alejandra Nieto-Garibay, Bernardo Murillo-Amador and Alfredo Ortega-Rubio**
Programa de Agricultura en Zonas Áridas, Centro de Investigaciones Biológicas del Noroeste, S.C. (CIBNOR, S.C.). La Paz, Baja California Sur, Mexico

**Ignacio Orona-Castillo and José Luis García-Hernández**
Facultad de Agricultura y Zootecnia (FAZ), Universidad Juárez del Estado de Durango, Gómez Palacio, Durango, México

**Nikki Funke and Inga Jacobs**
Council for Scientific and Industrial Research (CSIR), South Africa

**M.S.M. Amin, M.K. Rowshon and W. Aimrun**
Smart Farming Technology Centre of Excellence, Department of Biological and Agricultural Engineering, Faculty of Engineering, Universiti Putra Malaysia (UPM), 43400 UPM Serdang, Selangor DE, Malaysia

**Vikas Rai and A. M. Sedeki**
Department of Mathematics, Faculty of Science, Jazan University, Jazan, KSA, India

**Rana D. Parshad**
Applied Mathematics and Computational Science, King Abdullah University of Science and Technology, Thuwal 23955 – 6900, KSA, India

**R. K. Upadhyay and Suman Bhowmick**
Department of Applied Mathematics, Indian School of Mines, Dhanbad, Jharkhand, India

**Muthukrishnavellaisamy Kumarasamy**
School of Civil Engineering Surveying & Construction, University of KwaZulu-Natal, Durban, South Africa

**Sandrine Simon**
Open University, United Kingdom

**Okeyo Joseph Obosi**
Department of Political Science & Public Administration, University of Nairobi, Kenya

**Antonio A. R. Ioris**
University of Aberdeen, United Kingdom

**Rania A. Abdel Khaleq**
Ministry of Water and Irrigation, Jordan

**David L. Feldman**
University of California, Irvine, California, USA

**Mark Oelmann**
University of Applied Science Ruhr West in Muelheim an der Ruhr, Germany

**Christian Growitsch**
Institute of Energy Economics at the University of Cologne, Germany

**Yawen Chiueh**
Department of Environmental and Cultural Resources, National Hsinchu University of Education, Taiwan

**Hsiao-Hua Chen**
Environment and Development Foundation, Hsinchu, Taiwan

**Chung- Feng Ding**
Water Problems Enterprise Information & Incubation Center, Tainan Hydraulics Laboratory, Water Resources Agency, Ministry of Economic Affairs/National Cheng Kung University, Taiwan